Smart Control of Ship Propulsion System

Smart Control of Ship Propulsion System

Editors

Marco Altosole
Maria Acanfora
Flavio Balsamo
Bowen Xing

MDPI

Basel • Beijing • Wuhan • Barcelona • Belgrade • Novi Sad • Cluj • Manchester

Editors

Marco Altosole
Department of
Industrial Engineering
University of Naples
Federico II
Naples, Italy

Maria Acanfora
Department of
Industrial Engineering
University of Naples
Federico II
Naples, Italy

Flavio Balsamo
Department of
Industrial Engineering
University of Naples
Federico II
Naples, Italy

Bowen Xing
College of Engineering
Science and Technology
Shanghai Ocean University
Shanghai, China

Editorial Office
MDPI
St. Alban-Anlage 66
4052 Basel, Switzerland

This is a reprint of articles from the Special Issue published online in the open access journal *Journal of Marine Science and Engineering* (ISSN 2077-1312) (available at: https://www.mdpi.com/journal/jmse/special_issues/smart_ship_propulsion).

For citation purposes, cite each article independently as indicated on the article page online and as indicated below:

Lastname, A.A.; Lastname, B.B. Article Title. *Journal Name* **Year**, *Volume Number*, Page Range.

ISBN 978-3-0365-9356-2 (Hbk)
ISBN 978-3-0365-9357-9 (PDF)
doi.org/10.3390/books978-3-0365-9357-9

Contents

Journal of *Marine Science and Engineering*

Article

Research on Multi-Energy Integrated Ship Energy Management System Based on Hierarchical Control Collaborative Optimization Strategy

Yuanjie Ren [1], Lanyong Zhang [1,*], Peng Shi [2,3] and Ziqi Zhang [1]

1 College of Intelligent Science and Engineering, Harbin Engineering University, Harbin 150001, China
2 School of Information Science and Technology, Fujian University of Technology, Fuzhou 350100, China
3 National Research Base of Intelligent Manufacturing Service, Chongqing Technology and Business University, Chongqing 400067, China
* Correspondence: zhanglanyong@hrbeu.edu.cn

Abstract: The propulsion systems of hybrid electric ship output and load demand have substantial volatility and uncertainty, so a hierarchical collaborative control energy management scheme of the ship propulsion system is proposed in this paper. In a layer of control scheme, the traditional perturbation algorithm is improved. Increasing the oscillation detection mechanism and establishing the dynamic disturbance step length realizes the real-time stability of maximum power point tracking control. In the second-layer control scheme, the power sensitivity factor and voltage and current double closed-loop controller is introduced. By designing a two-layer coordinated control strategy based on the dynamic droop coefficient, the problem of voltage and frequency deviation caused by load switching is solved. In the third-layer control scheme, due to the need of the optimal scheduling function, the multi-objective particle swarm optimization algorithm was improved through three aspects: introducing the mutation factor, improving the speed formula, and re-initializing the strategy. Compared with other algorithms, this algorithm proves its validity in day-ahead optimal scheduling strategy. The superiority of the hierarchical collaborative optimization control schemes proposed was verified, in which power loss was reduced by 39.3%, the overall tracking time was prolonged by 15.4%, and the environmental cost of the diesel generator was reduced by 8.4%. The control strategy solves the problems of the steady-state oscillation stage and deviation from the tracking direction, which can effectively suppress voltage and frequency fluctuations.

Keywords: multi-energy integrated ship; energy management strategies; hierarchical control; dynamic droop control; improved PSO algorithm

Citation: Ren, Y.; Zhang, L.; Shi, P.; Zhang, Z. Research on Multi-Energy Integrated Ship Energy Management System Based on Hierarchical Control Collaborative Optimization Strategy. *J. Mar. Sci. Eng.* **2022**, *10*, 1556. https://doi.org/10.3390/jmse10101556

Academic Editor: Mohamed Benbouzid

Received: 20 September 2022
Accepted: 18 October 2022
Published: 20 October 2022

1. Introduction

Ship electric propulsion systems have grown significantly over the last century, but at present advanced new energy ship propulsion technologies require lower pollutant emissions from ships [1]. Clean energy (fuel cells, photovoltaic power generation, wind power generation), advanced control technology, and power energy management technology are being introduced into ship power systems [2–4]. As these technologies develop and change, future marine power systems are expected to include generator sets and other alternative power sources with different characteristics. Therefore, electric propulsion systems will become multi-energy complementary, and new hybrid-energy ship propulsion systems will be built to meet the new demands [5–7].

Mixing the two types of energy storage systems, Fang, S., et al., proposed a two-step multi-objective optimization method for optimizing the management of all-electric ships, striving to minimize the total operating cost [8]. In order to optimize the operating cost of diesel generators and energy storage systems, Anvari, M., et al., extended the principles of optimal planning and economic dispatch problems to shipboard systems in order to realize the coordinated power supply of diesel generators and energy storage systems [9]. Mistress, G., et al., focused on the parameter identification method of the onboard battery storage system electrical model, which solves the energy storage system degradation problem [10]. Lee, K., et al. realized green ship and low-carbon operation by showing the experimental results of a prototype green ship and achieved the demanding output on the load side with the minimum cost and low volatility [11]. In order to minimize fuel consumption and verify the feasibility of its scheme through simulation, Accetta et al. proposed an energy management system for the electrical system of yachts [12], but most of the existing research focuses on optimizing energy storage. Energy systems and controllers ignore the overall scheduling of power system energy. In addition, previous work has rarely noted that for the hybrid propulsion systems of photovoltaic power generation systems, wind turbines, and diesel generators. It is necessary to carry out the maximum power point tracking control of wind and photovoltaic power generation systems and the system connection of distributed power sources.

Hybrid electric ships are generally designed to improve energy utilization: Brizuela-Mendoza, J.A., et al. [13] analyzed in detail the efficiency of a ship's DC hybrid power system, constructed an optimization strategy aiming at improving energy efficiency, and provided the optimal on-load operating point of the generator set. Zhu, L., et al. [14] proposed an energy management strategy based on fuzzy logic for a hybrid ship that uses fuel cells, batteries, and supercapacitors as energy sources, aiming to optimize the power distribution among generating units and optimize the performance and fuel economy of the hybrid system.

Fully electric ships in the military field, equipped with sophisticated electronic instruments, require high power quality and fault handling capabilities: Feng, X., et al. [15] proposed a multi-agent system coordination controller for the Marine MVAC power system, which can balance load and power generation in real-time while meeting system operation constraints while considering load priority and reducing the influence of pulse load on power quality by coordinating pulse load and propulsion load. Seenumani, G., et al. [16] proposed a hierarchical control method to meet the real-time requirements, aiming at the power failure problem of all-electric ships for military applications. The top-level controller realized the suboptimal power decomposition of batteries and online power supply to meet the system power demand, while the local controller regulated the power command of each independent power supply. Nelson, M., et al. [17] proposed the use of graph theory to provide complete autonomous control for the power system when the power system fails or the ship task changes.

There are also some energy management strategies for specific needs: Mensah, E., et al. [18] built a model for the design of energy management systems for simulation and including system reconfiguration and load shedding functions, allowing the simulation and analysis of power systems with discrete events. Hou, J., et al. [19] evaluated the interaction of multiple power sources in the ship electric propulsion system with a hybrid energy storage system on the basis of model analysis and revealed the important role of system-level energy management strategy. Kanellos, F. D., et al. [20] proposed an optimization method of load-side management and power generation planning, in which the load-side management is realized by adjusting the power consumption of the propulsion motor and the dynamic programming algorithm is used to solve the optimization problem according to the constraints of ship operation and environment, so as to ensure the operation mode with minimum cost. López, A. R., et al. [21] proposed the adoption of load-shedding technology based on expert system rules in order to control the voltage level of the ship power grid, and the application was verified in a ship power grid. Paran, S., et al. [22] proposed the energy management of ship DC power system based on model predictive control to ensure reasonable load-sharing among generators while maintaining a stable DC bus voltage.

There have been several identified optimization methods for hybrid schemes, such as the use of artificial intelligence, as the appropriate way to enhance the optimization process.

Hatata, A. Y., et al. [23] used clonal selection techniques to optimize hybrid solar and wind power schemes with batteries to utilize their outputs with minimal cost and small volatility. Hadidian, M. J., et al. [24] used an intelligent pollination algorithm to optimize the power output of a hybrid system (solar/wind/hydrogen) so as to minimize the total operating cost. Sanajaoba Singh, S., et al. [25] analyzed and studied the effectiveness of cuckoo search using a hybrid solar/wind/battery scheme design problem in remote areas of India. De la Cruz, J., et al. [26] used a well-known heuristic algorithm based on simulated annealing to optimize independent wind power and photovoltaic systems. In another paper, the influence of using predicted load information on the performance of small independent hybrid power was studied [27]. Maleki, A., et al. [28] optimized the size of the battery pack, the area of the PV system, and the fuel consumption of the diesel generator in the hybrid system to minimize the cycle operation cost of the system and used a heuristic algorithm based on tabu search. These hybrid methods provide a new way for microgrid optimization and show certain advantages and application prospects.

In the master–slave control, one of the distributed power supply (or energy storage unit) controllers is selected as the master controller, and the others are used as the slave controllers. In this mode, the distributed power supply of the main control unit needs to track the change of charge, which requires its power output to be controlled within a certain range, and it can also respond quickly to charge fluctuation. In peer-to-peer control, all distributed power controllers are in the same position [29]. Therefore, when the power required by the load changes, the distributed power supply can make the microgrid return to a stable state by properly adjusting its voltage and frequency, but its control results have certain deviations.

According to the existing research literature, the strategies of paralleling multiple distributed power sources are generally divided into master–slave control and peer-to-peer control. The master–slave control method is simple and easy to implement, mainly relying on high-speed communication units, but the communication cost is high and the system redundancy is poor. The peer-to-peer control in the form of point-to-point, such as droop control, does not require high-bandwidth communication and is easy to implement using the "plug and play" of micro-sources, which meets the needs of the distributed system. Therefore, droop control can be used as an ideal load voltage and frequency control method for the ship propulsion system. There is a conflict between current sharing and voltage regulation due to the virtual impedance introduced by droop control. Additionally, there are problems with the steady-state oscillation phase and deviation from tracking direction. In order to solve the above contradictions, a hierarchical control collaborative optimization strategy for propulsion systems with multiple distributed power sources is proposed.

2. Typical System Structure and Mathematical Modeling of Hybrid Electric Ship

2.1. Typical System Structure

The typical system structure of a ship propulsion system consists of the wind turbine, photovoltaic power generation system, diesel generator set, energy storage system, load, electric energy conversion device, and energy management system. A typical system structure diagram of a hybrid ship is shown in Figure 1. In the following sections, the working principle and mathematical model of the leading equipment in the system will be presented.

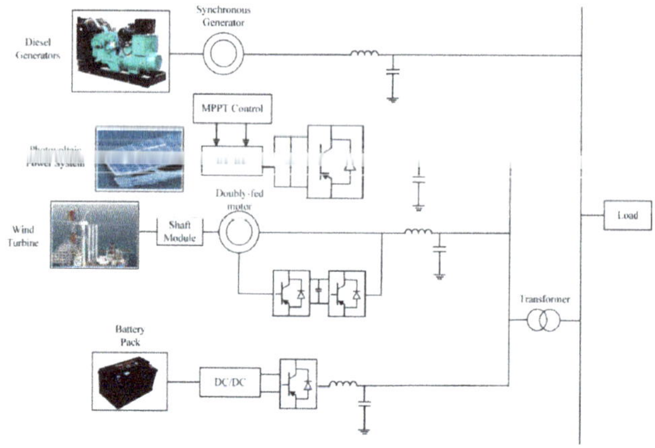

Figure 1. Typical system structure diagram.

2.2. *Distributed Power Mathematical Model*

2.2.1. Diesel Engine and Speed-Governing System Model

The diesel engine converts the released chemical energy into thermal energy by burning diesel oil and then converts the thermal energy into the mechanical energy of the rotating shaft to provide the motive power for the synchronous generator, forming a closed-loop speed to ensure the stability of the generator speed [30]. The block diagram of the system structure is shown in Figure 2.

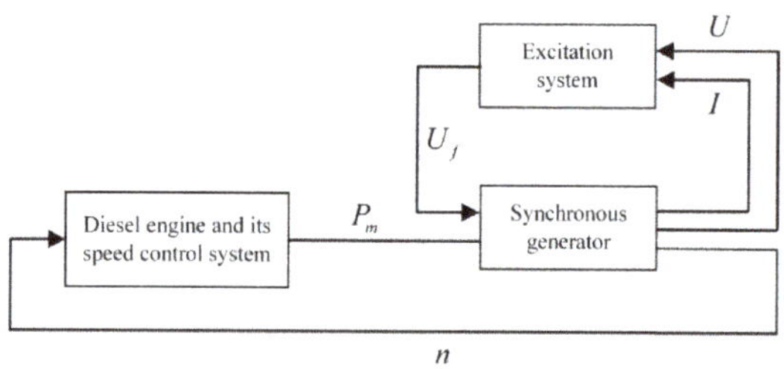

Figure 2. Schematic diagram of the diesel engine system.

According to the external characteristic curve of the diesel engine (torque–speed characteristic curve), when the diesel engine is in a standard and operation state, the resistance torque is balanced with the output torque of the diesel engine. The load power driven by the diesel engine changes, the fuel injection quantity of the diesel engine will change, and the output torque will also change until a new equilibrium state is reached.

According to D'Alembert's principle, the equation of motion of the unit is:

$$J\frac{d\omega_g}{dt} + Kp\omega_g = M_d - M_c \tag{1}$$

where K represents the damping coefficient, which is related to the damping winding; ω_g is the angular velocity of the diesel engine; p is the number of pole pairs of the synchronous generator; M_d and M_c represent, respectively, the output torque and resistance torque of the diesel engine.

When the load carried by the diesel engine decreases, the rotor speed increases. Therefore, in order to keep the speed change of the diesel engine within a reasonable range, a speed control link should be installed on the diesel generator to ensure that the speed of the diesel engine does not fluctuate considerably. The diesel engine speed adjustment system is shown in Figure 3.

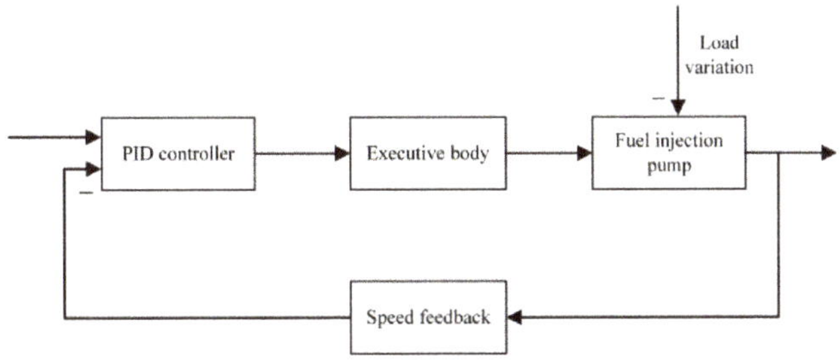

Figure 3. Block diagram of diesel engine speed regulation system.

When the load increases, the speed of the diesel engine will decrease. Then by comparing with the given speed, the speed feedback unit obtains a speed difference signal and acts on the speed control unit, transmits the control amount to the actuator, and finally injects the oil pump increases the amount of fuel injected, so that the speed of the diesel engine rises to the rated speed. The model of the diesel engine speed control system in Power Systems Computer Aided Design (PSCAD software) is shown in Figure 4.

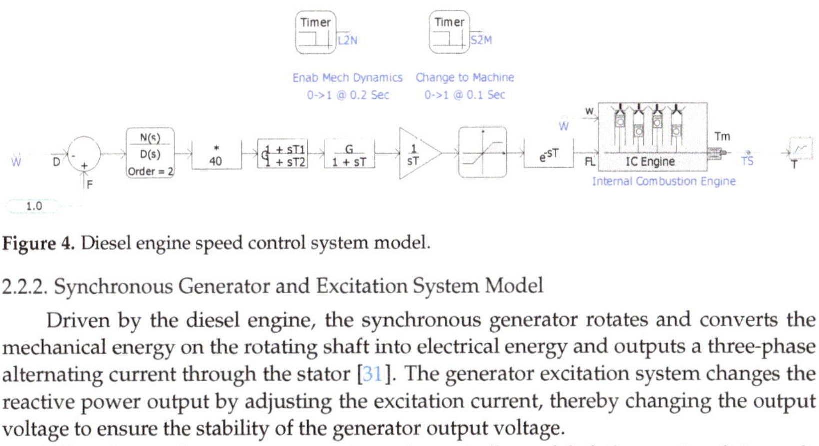

Figure 4. Diesel engine speed control system model.

2.2.2. Synchronous Generator and Excitation System Model

Driven by the diesel engine, the synchronous generator rotates and converts the mechanical energy on the rotating shaft into electrical energy and outputs a three-phase alternating current through the stator [31]. The generator excitation system changes the reactive power output by adjusting the excitation current, thereby changing the output voltage to ensure the stability of the generator output voltage.

When the synchronous generator is mathematically modeled, the rotating dq0 coordinate system is adopted through Park transformation, which can simplify the model and reduce the amount of calculation. The transformation from the abc coordinate system to the dq0 coordinate system is realized through the Park transformation, which can be written as a matrix:

$$\begin{pmatrix} i_d \\ i_q \\ i_0 \end{pmatrix} = \frac{2}{3}\begin{pmatrix} \cos\alpha & \cos(\alpha-\frac{2}{3}\pi) & \cos(\alpha+\frac{2}{3}\pi) \\ \sin\alpha & \sin(\alpha-\frac{2}{3}\pi) & \sin(\alpha+\frac{2}{3}\pi) \\ \frac{1}{2} & \frac{1}{2} & \frac{1}{2} \end{pmatrix}\begin{pmatrix} i_a \\ i_b \\ i_c \end{pmatrix} \tag{2}$$

The potential equation of the dq0 system after Park transformation is:

$$\begin{cases} u_d = -\dot{\varphi}_d - \omega\varphi_q - ri_d \\ u_q = -\dot{\varphi}_q - \omega\varphi_d - ri_q \\ u_0 = -\dot{\varphi}_0 - ri_0 \end{cases} \tag{3}$$

The flux linkage equation of the dq0 system after Parker transformation is:

$$\begin{pmatrix} \varphi_a \\ \varphi_b \\ \varphi_c \\ \varphi_f \\ \varphi_D \\ \varphi_Q \end{pmatrix} = \begin{pmatrix} L_d & 0 & 0 & m_{af} & m_{aD} & 0 \\ 0 & L_q & 0 & 0 & 0 & m_{aQ} \\ 0 & 0 & L_0 & 0 & 0 & 0 \\ \frac{3}{2}m_{fa} & 0 & 0 & L_f & L_{fD} & 0 \\ \frac{3}{2}m_{Da} & 0 & 0 & L_{Df} & L_D & 0 \\ 0 & \frac{3}{2}m_{Qu} & 0 & 0 & 0 & L_Q \end{pmatrix} \begin{pmatrix} i_d \\ i_q \\ i_0 \\ i_f \\ i_D \\ i_Q \end{pmatrix} \tag{4}$$

The synchronous generator excitation system model adopts the AC1A excitation model recommended by IEEE preset in PSCAD, which further stabilizes the output voltage of the synchronous generator by adjusting the magnitude of the excitation current. The model of the synchronous generator and excitation system in PSCAD is shown in Figure 5.

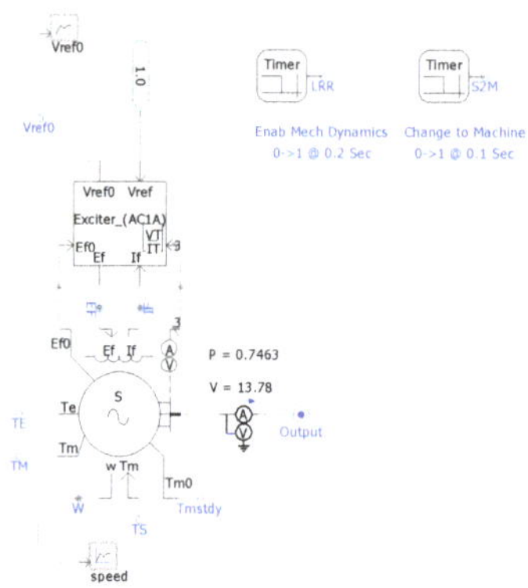

Figure 5. Simulation diagram of synchronous generator and excitation system.

2.2.3. Photovoltaic Power Generation System Model

The photovoltaic cell is the core device of photovoltaic power generation. Its primary function is to convert the light energy of solar radiation into electrical energy through the photovoltaic effect. Since the single-diode model is relatively simple and can be applied to higher irradiance conditions, it can better simulate the actual loss and charge diffusion effect inside the photovoltaic cell, so this paper chooses the single-diode model, and its equivalent circuit is shown in Figure 6.

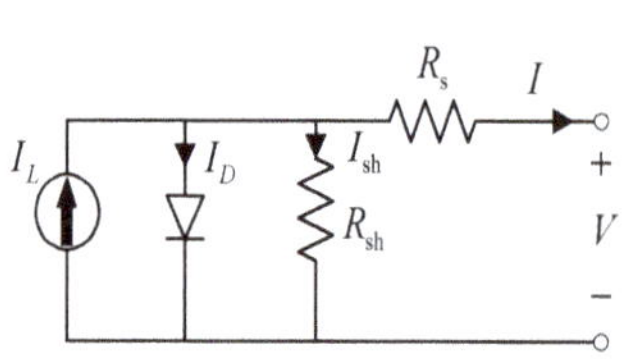

Figure 6. Equivalent circuit diagram of single diode model.

where I_L is the constant current source that produces the photogenerated current and I_D represents the current passing through the diode.

A mathematical model of photovoltaic cells is established under standard irradiance $R_{ref} = 1\ \text{kW/m}^2$ and ambient temperature conditions $T_{ref} = 25\ °\text{C}$. Through the above brief analysis of the circuit diagram, the output current characteristics of the photovoltaic cell can be obtained, and the following formula can express its mathematical function:

$$I = I_{SC}\left(1 - A\left(\frac{U - D}{e^{BU_{OC}}} - 1\right)\right) + C \tag{5}$$

$$A = \left(1 - \frac{I_m}{I_{SC}}\right)e^{-\frac{U_m}{BU_{oc}}} \tag{6}$$

$$B = \frac{\frac{U_m}{U_{oc}} - 1}{\ln\left(1 - \frac{I_m}{I_{SC}}\right)} \tag{7}$$

$$C = \frac{\alpha R}{R_{ref}}E + \left(\frac{R}{R_{ref}} - 1\right)I_{SC} \tag{8}$$

$$D = -\beta E - R_s C \tag{9}$$

$$E = T_m - T_{ref} \tag{10}$$

where α represents the temperature coefficient, I_m and U_m represent the current and voltage corresponding to the maximum power output, U represents the voltage to the photovoltaic array, U_{oc} represents the open circuit voltage, and T_m represents the temperature of the photovoltaic array, and the formula is calculated as follows:

$$T_m = T_a + t \cdot R \tag{11}$$

where T_a is the ambient temperature, t is the temperature of the photovoltaic array, and R is the actual solar irradiance. Therefore, the following formula can specifically describe the output power of the photovoltaic array:

$$P = IU = \left[I_{SC}\left(1 - A\left(\frac{U - D}{e^{BU_{OC}}} - 1\right)\right) + C\right]U \tag{12}$$

The *I–V* characteristic curve and the *P–V* characteristic curve under ambient temperature and irradiance can be obtained by fitting them with the above formula, as shown in Figure 7.

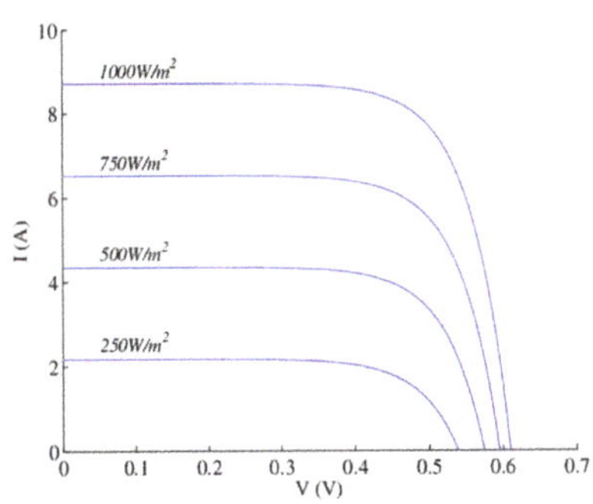

(**a**)

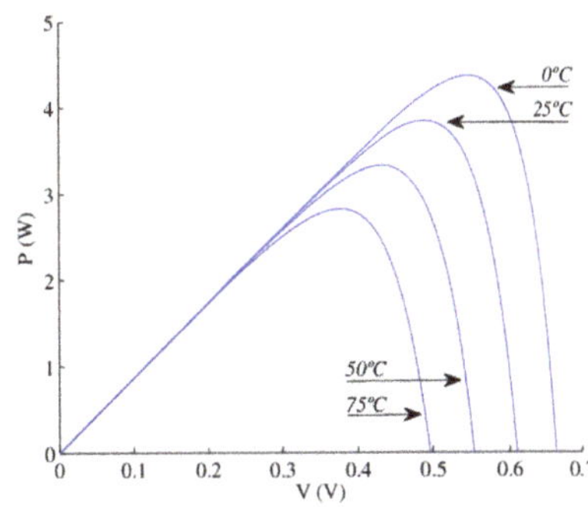

(**b**)

Figure 7. Simulation curves of *I–V* (**a**) and *P–V* (**b**) for PV power generation system.

The model of the photovoltaic power generation system in PSCAD is shown in Figure 8:

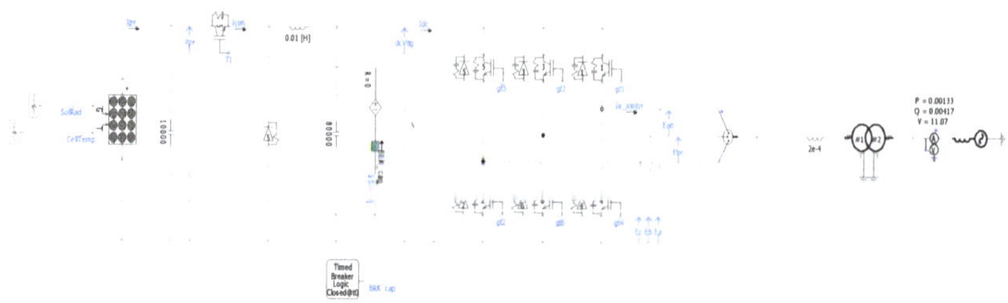

Figure 8. Photovoltaic power generation system simulation diagram.

2.2.4. Doubly-Fed Wind Power Generation System Model

The doubly-fed wind power generation system comprises wind turbines, shafting modules, doubly-fed wind turbines, and back-to-back double PWM converters. The external wind drives the wind turbine and converts the wind energy into mechanical energy. After the wind turbine rotor passes through the shafting module, the rotational speed is converted to a rotational speed suitable for the doubly-fed generator. The conversion of mechanical energy to electrical energy is realized in the doubly-fed generator [32,33]. The doubly-fed generator generally uses a three-phase wound asynchronous motor. The power system and the stator winding are directly connected, and the rotor winding is connected to the power system through the back-to-back double PWM converter. The schematic diagram of the doubly-fed wind power generation system is shown in Figure 9.

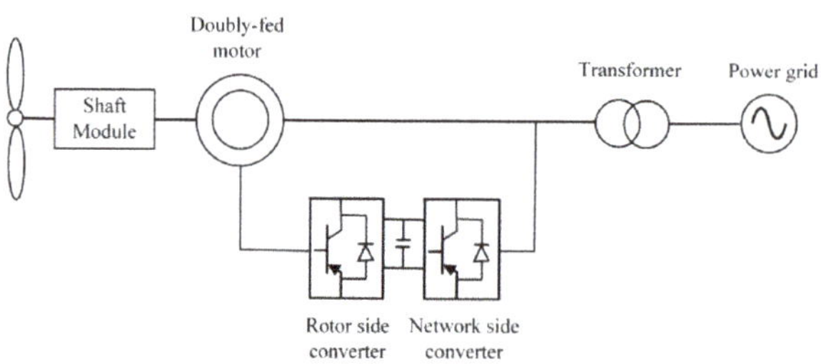

Figure 9. Schematic diagram of the doubly-fed wind power generation system.

According to Bates's theory, the calculation formula of the wind energy utilization coefficient $C_p(\beta,\lambda)$ is as follows:

$$\begin{cases} C_p(\beta,\lambda) = 0.5176\left(\frac{116}{\lambda^*} - 0.4\beta - 5\right)e^{-\frac{21}{\lambda^*}} + 0.0068\lambda \\ \frac{1}{\lambda^*} = \frac{1}{\lambda+0.08\beta} - \frac{0.035}{\beta^3+1} \end{cases} \tag{13}$$

where β is the pitch angle of the blade and λ is the tip speed ratio, which can be expressed as:

$$\lambda = \frac{\omega_T R}{V_0} \tag{14}$$

where ω_T is the rotational speed, R is the radius of the rotor, and V_0 is the wind speed.

The maximum power point tracking control of wind turbines means that under different wind speed conditions, C_P can always be kept at the maximum value by controlling the speed of wind turbines. The mechanical power P_m captured by the wind turbine can be expressed as:

$$P_m = \frac{1}{2}\rho S {V_0}^3 C_p(\beta,\lambda) \tag{15}$$

where ρ is the air density and S sweeps the area of the rotor. The model of the doubly-fed wind power generation system in PSCAD is shown in Figure 10.

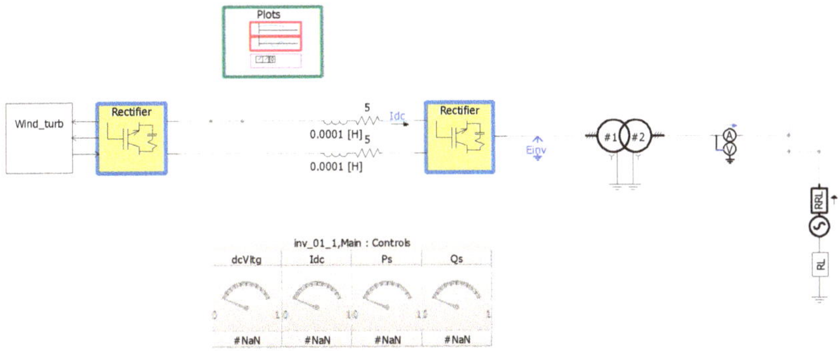

Figure 10. Model of a doubly-fed wind power generation system.

2.3. Mathematical Model of Energy Storage System

At present, the lithium-ion battery is the most widely used power battery, so a lithium-ion battery is selected as the energy storage system in the ship propulsion system in the research process [34,35]. The equivalent modeling of the lithium battery and the functions of its voltage and capacity in the charging and discharging process are as follows.

When discharging:

$$V_{BT} = E_0 - KQ_t\frac{Q}{Q-Q_t} - KI^*\frac{Q}{Q-Q_t} - IR + AIe^{-BQ_t} \tag{16}$$

When charging:

$$V_{BT} = E_0 - KQ_t\frac{Q}{Q-Q_t} - KI^*\frac{Q}{Q_t-0.1Q} + IR + AIe^{-BQ_t} \tag{17}$$

and

$$Q_t = Q(t-1) - Q(t) = I \cdot \Delta t \cdot \varepsilon \tag{18}$$

$$SOC(t) = \frac{Q(t)}{Q} \tag{19}$$

where V_{BT} is the lithium battery voltage (V); E_0 is the constant voltage source voltage (V); K is the polarization resistance proportional constant; Q is the battery capacity (Ah); Q_t is the lithium battery charge/discharge capacity (Ah); I^* is the Filtered current (A); R is the internal resistance of lithium battery (Ω); A is the amplitude coefficient of the exponential region; B is the time the inverse proportion of the exponential region; ε is the charge-discharge efficiency; and $SOC(t)$ is the state of charge of the lithium battery. The simulation model of the energy storage system in PSCAD is shown in Figure 11.

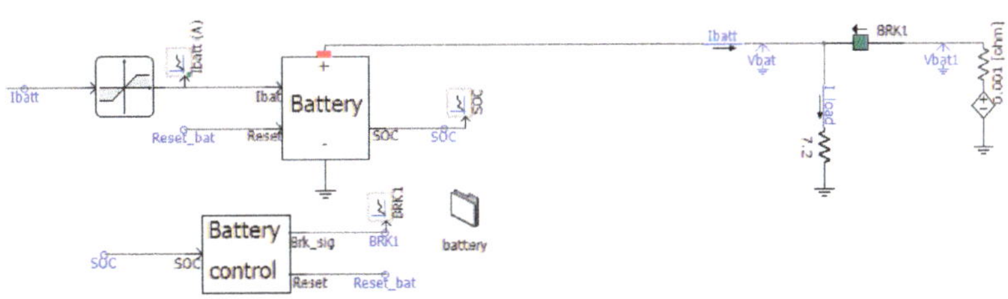

Figure 11. Simulation model of energy storage system.

3. Hierarchical Control Scheme and Optimization Algorithm Design for Hybrid Ships

3.1. Hierarchical Control Scheme

As the core of the ship propulsion system, the energy management system is not only responsible for the management of the generation and scheduling of the entire ship's electrical energy but also needs to manage the operation status of the propulsion system. To ensure that the ship propulsion system can provide continuous, stable, and economic power support during operation [36–39]. Based on the above functional requirements, this paper designs the overall scheme of the layered control of the ship propulsion system, as shown in Figure 12.

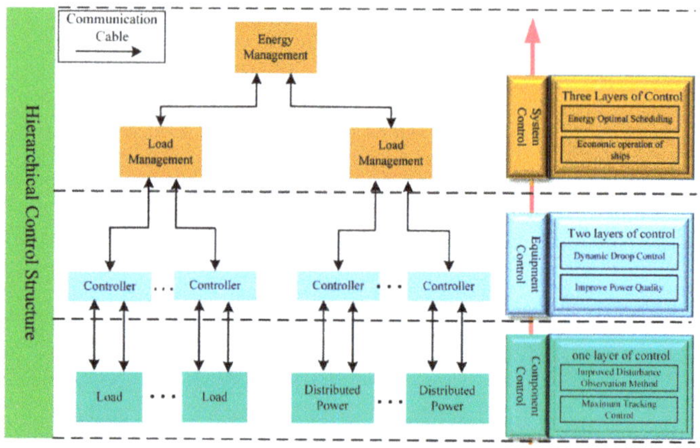

Figure 12. Design diagram of the hierarchical control scheme.

This hierarchical control structure defines three primary control levels, component-level control, device-level control, and system-level control, and communication lines connect them. Among them, the first-level control mainly focuses on controlling a single unit (load, distributed power supply), and the second-level and third-level controls, respectively, focus on coordinated and stable control and the improvement of power quality. The adjustment speed needs to be faster and more responsive to meet the fast dynamic characteristics of the system.

3.2. Maximum Power Point Tracking Control Strategy for Ship Propulsion Systems

The traditional disturbance observation method has two main problems: steady-state oscillation and deviation from the tracking trajectory [40]. If the system wants to achieve the fast and stable tracking of the maximum power point, it needs to achieve the following: (1) MPPT must be able to automatically locate the maximum power point; (2) the tracking must be dynamic, and the operating point must be continuously adjusted according to the measured irradiance and temperature conditions.

Based on the above two viewpoints, this paper designs a P&O algorithm based on dynamic perturbation step size, which improves the tracking efficiency of the P&O algorithm by reducing steady-state oscillation and adding perturbation steps [41,42]. The improved algorithm adds two features: (1) The algorithm includes a built-in oscillation detection mechanism to ensure the consistent detection of oscillations and changes the size of the disturbance accordingly to achieve adaptive performance. (2) The algorithm establishes a dynamic disturbance step long to ensure that the working point not deviate from its tracking trajectory.

The initial operating voltage V_{MPP} is set to 65% V_{OC}, and V_{OC} is the open-circuit voltage of the photovoltaic array. ΔP and ΔV represent the amount of slope change. The value is the sign of the two quantities multiplied and normalized, as shown in Table 1.

Table 1. Slope value table.

ΔP Symbol	ΔV Symbol	Slope Value
+	+	+1
+	−	−1
−	+	−1
−	−	+1

The oscillation detection mechanism detects oscillations by recording three consecutive slope values. When the sign of the disturbance voltage is the same as the sign of the power variation, the value of the slope is +1. Otherwise, it is −1. From this, the following oscillation state detection formula can be obtained:

$$\left|\sum slope\right| = \begin{cases} 3, & \text{no converge to steady state oscillation} \\ 1, & \text{Convergence to Steady-State Oscillation} \end{cases} \quad (20)$$

Dynamic perturbation step adjustment strategy: The initial perturbation size is set to 2% of V_{OC}, and when the operating point converges to near the maximum power point, the perturbation size is reduced by 0.5% of V_{OC} in each iteration step. The perturbation step size is continuously reduced until it reaches 0.5% of V_{OC}. This tiny oscillation allows the algorithm to consume almost zero power while simultaneously making it immune to the circuit's radiation and noise.

Under normal conditions, the maximum irradiance variation can reach up to 0.027 $KW/m^2/s$. To enhance the algorithm robustness, this paper choose T_1 and T_2 to be 0.001 s and 0.05 s, and any change in irradiance between 1.0 $W/m^2/s$ and 50 $W/m^2/s$ is considered a gradual change. When the perturbation step size changes, if a value greater than 50 $W/m^2/s$ is detected, it is considered a rapid step change and the perturbation step size is increased to 2%. The algorithm flowchart is shown in Figure 13.

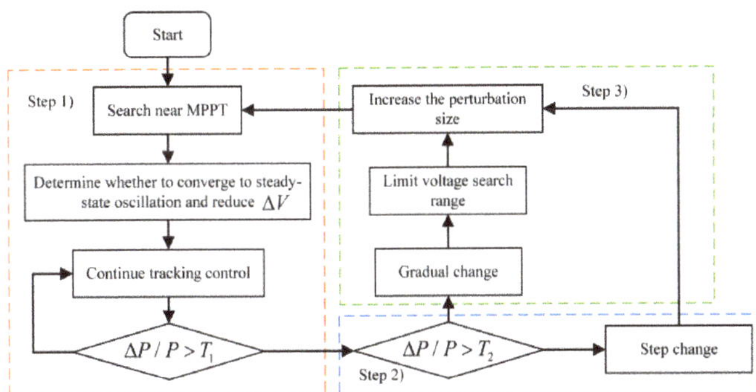

Figure 13. Flowchart of the improved P&O algorithm.

The algorithm adopts dynamic perturbation step size to reduce the oscillation and introduces boundary conditions to prevent it from deviating from the maximum power point. The performance of the P&O algorithm is improved by eliminating steady-state oscillation and preventing the algorithm from deviating from the maximum power point trajectory.

3.3. Design of Two-Layer Coordinated Control Strategy Based on Dynamic Droop Controller

A dynamic droop controller is designed to aim at the problem of voltage and frequency deviation caused by one-layer control and switching loads. This control strategy can quickly adjust the fixed droop coefficient, realize the adaptive change of the droop coefficient, and introduce power at the same time. The sensitivity factor improves the system dynamic performance. A voltage and current double closed-loop controller is designed to further improve immunity and power quality of the inverter and reduce the influence of random disturbances on the voltage and frequency of the ship propulsion system [43–45]. The schematic diagram of the droop controller is shown in Figure 14.

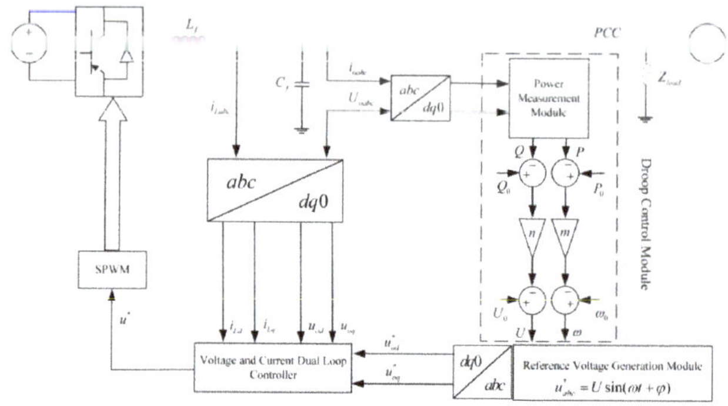

Figure 14. Block diagram of the droop controller.

where i_{oabc} and U_{oabc} represent the three-phase current and voltage output by the inverter; P and Q are the actual active and reactive power of the inverter; P_0 and Q_0 are the reference active and reactive power of the control system; m, n is the droop coefficients of active and reactive conditions; ω and U represent the given actual voltage frequency and voltage; ω_0 and U_0 represent the given reference frequency and voltage; and u^* is the SPWM modulation signal.

Droop control plays an essential role in the entire inverter control system. When the propulsion system inverter adopts traditional droop control, the droop coefficient is fixed. Fluctuations are likely to occur in the propulsion system at the moment of load switching on the system-connected side, resulting in the distributed circulating current occurring between the power supplies. This paper adopts a real-time dynamic change control method of the droop coefficient, which can effectively suppress the circulation phenomenon. Through this method, the droop coefficient and power are combined to form a new droop coefficient term with dynamic adjustment ability, suppressing the circulation phenomenon and enhancing frequency and voltage stability. Designing the power sensitivity factor can reduce the transient impact caused by power fluctuations during system load switching. The formula of this dynamic droop control strategy is:

$$\begin{cases} f = f_0 - \left(k_{f1} + k_{f2}P\right)P - \eta_f\sqrt{\left|\frac{P}{P_0} - 1\right|} \\ U = U_0 - (k_{U1} + k_{U2}Q)Q - \eta_U\sqrt{\left|\frac{Q}{Q_0} - 1\right|} \end{cases} \tag{21}$$

where k_{f1} and k_{f2} represent the dynamic droop coefficient during active power regulation, k_{U1} and k_{U2} represent the dynamic droop factor during reactive power regulation, and η_f and η_U are sensitivity factors. Using the minimum and maximum values of voltage and frequency to solve the dynamic droop coefficient formula is as follows:

$$\begin{cases} k_{f1} = \frac{f_{\max} - f_{\min}}{P_{\max}}, k_{f2} = \frac{0.6k_{f1}}{P_{\max}} \\ k_{U1} = \frac{v_{d\_\max} - v_{d\_\min}}{Q_{\max}}, k_{U1} = \frac{0.6k_{U1}}{Q_{\max}} \end{cases} \tag{22}$$

where $f_{\max}$ and $f_{\min}$ are the maximum and minimum frequencies due to the instantaneous change of the load with values of 50.5 Hz and 49.5 Hz; $P_{\max}$ and $Q_{\max}$ are the maximum active power and reactive power values of the system overload, which are 5 kW and

2 kW; and v_{d_max} and v_{d_min} are the maximum and minimum values of the d-axis voltage component, which are 150 V and 145 V.

3.4. Research on Day-Ahead Optimal Dispatching Strategy of Ship Power System

In this part, based on the third-layer control of the ship power system, the relevant research of the optimization scheduling method is carried out. In order to give full play to the advantages of renewable energy, a multi-objective optimization model of the power system under the cooperative work of multiple power generation systems is established.

3.4.1. Mathematical Model of Day-Ahead Optimal Scheduling Problem in Power System

The day-ahead optimal scheduling of the power system is to predict the output power of the wind and photovoltaic power generation system, which, based on the wind and solar data of the next day, formulate the load power usage curve according to the load usage and schedule the renewable energy reasonably.

The following formula can describe the mathematical model of the day-ahead optimal scheduling problem. These include the operating costs F_L, F_B, F_R of diesel power generation systems, energy storage systems, and renewable power generation systems, and the overall benefit F_L of the ship load.

$$y(P,u) = F_D(P_D,u_D) + F_B(P_B,u_B) + F_R(P_{pv},P_{wt}) - F_L(u_L) \tag{23}$$

Diesel power generation system:

Without considering the influence of external factors, such as vibration and friction, it is assumed that the diesel generator operate in a stable state. Therefore, its fuel cost can be described by the following mathematical formula:

$$f_D = p_f \sum_{t=1}^{T} v_f(t) \tag{24}$$

where p_f is the fuel price and v_f is the total fuel consumption. On the other hand, in the wind–solar diesel-storage ship power system, only diesel generators emit pollutants, and the following mathematical formula can describe its environmental cost:

$$g_D = \sum_{j=1}^{n} E\varepsilon_j V_{F,j} \tag{25}$$

where n = 4 represents four air pollution sources, SO_2, CO_2, CO, and NO_X; E is the total output power of diesel generators per day (kWh); ε_j is the emission coefficient of pollution source (g/kWh); $V_{F,j}$ represents the penalty amount for the pollution source (CNY/kWh), and its specific parameters are shown in Appendices A and B.

Energy storage system:

For the ship power system, the charging and discharging function of the energy storage system is essential, so the cost of the energy storage unit must be calculated. The following mathematical formula can describe its running cost:

$$F_B = \sum_{t=1}^{T} \sum_{j=1}^{n_B} u_{B,j}(t) C_B \left| P_{B,j}(t) \right| \tag{26}$$

where n_B = 4 is the number of lithium-ion battery packs; $u_{B,j}(t)$ is 1 or 0 to indicate two states of operation and shutdown; C_B is the operating cost coefficient (CNY/kW); and $P_{B,j}(t)$ is the lithium-ion battery pack output power.

Renewable energy power generation system:

The core power generation sector in the ship power system is the renewable energy power generation system, and its operating cost can be described by the following mathematical formula:

$$F_R(P_{pv}, P_{wt}) = \sum_{t=1}^{T}\left(\sum_{m=1}^{n_{wt}}[f_{wt}P_{wt,m}(t)] + \sum_{n=1}^{n_{pv}}[f_{pv}P_{pv,n}(t)]\right) \tag{27}$$

where f_{wt} and f_{pv} are the operating cost coefficients of wind and photovoltaic power generation systems; n_{wt} and n_{pv} are the numbers of wind and photovoltaic power generation units; and $P_{wt,m}(t)$ and $P_{pv,n}(t)$ are the output power of wind and photovoltaic power generation systems at the moment.

Overall benefit of load:

The overall benefit of the load refers to the sum of the benefits obtained from the sale of electricity through marine loads, and its mathematical description is as follows:

$$F_L = \sum_{t=1}^{T}\sum_{k=1}^{n_L} f_{L,sale}P_{L,k}(t) \tag{28}$$

where F_L is the income value; n_L is the number of loads; $f_{L,sale}$ is the electricity price; and $P_{L,k}(t)$ is the power of the load.

Restrictions:

The constraints are set according to the power generation characteristics of the distributed power generation. In order to obtain more accurate optimization results, the system must meet the following constraints.

Power balancing constraints:

$$P_D(t) + P_B(t) + P_{wt}(t) + P_{pv}(t) = P_L(t) \tag{29}$$

Energy storage power constraints:

$$P_c \leq P_B(t) \leq P_d \tag{30}$$

Energy storage electric constraints:

$$Soc_{B,\min} \leq Soc_B(t) \leq Soc_{B,\max} \tag{31}$$

Diesel generator operating power constraints:

$$\alpha P_{D,r} \leq P_D(t) \leq P_{D,r} \tag{32}$$

Renewable energy operating constraints:

$$0 \leq P_{wt}(t) \leq P_{wt}^{p}(t) \tag{33}$$

$$0 \leq P_{pv}(t) \leq P_{pv}^{p}(t) \tag{34}$$

3.4.2. Research on Day-Ahead Optimization Scheduling Method Based on Improved PSO

In this section, a particle swarm optimization algorithm based on mutation particles is designed to solve the multi-objective function problem [46–48].

Introduce mutation factor: improve the particle swarm optimization algorithm in each generation to evolve a population, which can be expressed as $X_{i,G}$, where $i = 1, 2, 3 \ldots, NP$ and NP is the population size. After initialization, padding is randomly generated within the search domain. The algorithm mixes the set information of some top-ranked vectors with the new mutation particles generated by mutation and generates the following mutation vectors:

$$X_{ipso\_best,G} = \sum_{k=1}^{m} w_k \cdot X_{k,G} \tag{35}$$

where m represents the number of top-ranked vectors and w_k represents the weight of the k selected vectors $X_{k,G}$. The weights for each selected vector are calculated as follows:

$$w_k = \frac{m-k+1}{1+2+\ldots+m}, k = 1,2,\ldots m \tag{36}$$

Improvement of the speed formula: In order to improve the search performance of the algorithm, the method of improving the speed formula in sections is adopted in the algorithm. To make better use of the mutated particles in IPSO, the velocity update formula should be changed to:

$$v_i(t+1) = \omega \cdot v_i(t) + \lambda c_1 r_1 (p_i(t) - x_i(t)) + c_2 r_2 (p_g(t) - x_i(t)) \tag{37}$$

where λ is the adjustment function:

$$\lambda = \begin{cases} 2\eta \cdot \frac{g}{g_{\max}}, & 0 < g \leq \frac{g_{\max}}{2} \\ \eta \cdot \left(2 - 2 \cdot \frac{g}{g_{\max}}\right), & \frac{g_{\max}}{2} < g \leq g_{\max} \end{cases} \tag{38}$$

where η is a random coefficient between 0 and 1, g is the number of iterations, and $g_{\max}$ is the maximum number of iterations.

Re-initialization strategy: Since the current IPSO suffers from premature and stagnant problems, there will be premature convergence when the population falls into a local optimum. A re-initialization strategy is designed to deal with premature convergence and stagnation. Both of the above cases can be detected by the following formula:

$$stg = \begin{cases} 0 & f(U_i) < f(X_i) \forall i \in \{1,2,\ldots,NP\} \\ stg & f(U_i) \geq f(X_i) \exists i \in \{1,2,\ldots,NP\} \end{cases} \tag{39}$$

where $f(U_i)$ is the test vector function, $f(X_i)$ is the target vector function, and stg is an indicator used to monitor whether the current population reaches a stagnation state in a particular generation.

4. Simulation Verification and Analysis

4.1. Simulation and Verification of MPPT Control Strategy Based on Improved Disturbance Observation Method

In order to verify the feasibility and effectiveness of the algorithm, a mathematical model of the related system is built in this paper, in which the maximum output power of the wind and photovoltaic power generation systems are 200 kW and 120 kW and the wind speed drive simulates the input wind speed. The MPPT module is used to achieve the maximum wind mechanical power tracking.

1. Simulation of photovoltaic power generation system:

The irradiance change curve shown in Figure 15a is used as the test input signal, and the simulation duration is set to 400 s to verify the effectiveness of the control strategy in dealing with the gradual and sudden change of irradiance.

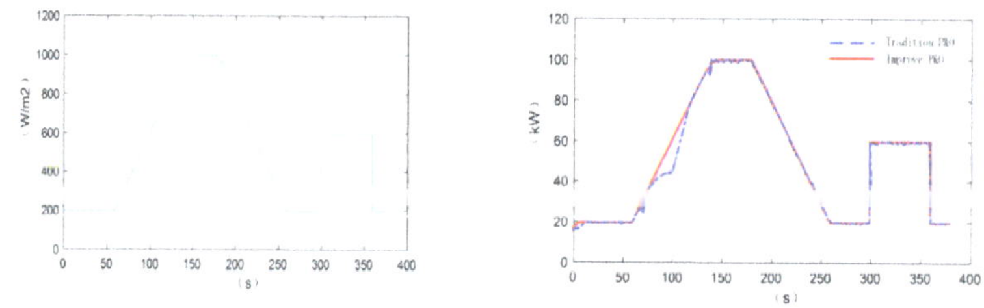

(a) (b)

Figure 15. Photovoltaic power generation system simulation diagram. (**a**) Irradiance variation diagram; (**b**) output power curve diagram.

The two curves in the figure are the output power curves of the traditional and improved P&O algorithms, and the voltage disturbance is fixed at 2% of V_{OC}. From the overall simulation results, the improved algorithm only consumes 6.3 s in approaching the maximum power point at the beginning, the oscillation amplitude is relatively small, and the trajectory can be tracked well.

The tracking effects of the two algorithms are shown in Table 2. It can be seen more intuitively that the tracking time of the improved P&O algorithm is increased by 15.4% compared with the traditional P&O algorithm, and the power loss is reduced by 39.3%. Therefore, the improved P&O algorithm can effectively reduce power loss to achieve the full tracking of the maximum power point, which will improve the efficiency of renewable energy generation.

Table 2. Comparison of algorithm effects.

Algorithm	Track Time (s)	Maximum Power Loss (kW)
traditional P&O	341.1	15.46
improved P&O	393.7	4.74

2. Wind Power System Simulation

Based on the improved P&O algorithm to verify the performance of the system in dealing with sudden changes in wind speed and in steady state, two main working conditions are set to detect in the simulation verification; step signal detection is working condition one and steady output signal is working condition two. The simulation time is set to 4 s, and the initial wind speed is set to 8 m/s. Entering the step signal detection part at 1 s, the wind speed suddenly changes from 8 m/s to 12 m/s, and then enters the steady-state output part of working condition two.

At the beginning of the simulation, the wind turbine is in the state of starting. Since the wind speed has not stabilized, the power will fluctuate in a small range, but it is still within the controllable range. At 1 s, the system enters the step signal detection stage of working condition one, the wind speed suddenly increases from 8 m/s to 12 m/s, the output power fluctuates slightly, and the output power remains stable at 180 kW. Then enter the steady state stage of working condition two, and the output power curve does not fluctuate.

Figure 16b is the simulation diagram of the wind energy utilization coefficient curve. The transient fluctuation occurs at the time of 1 s. After the dynamic adjustment of the algorithm, it quickly recovers and stabilizes, and finally stabilizes at a maximum value of about 0.48.

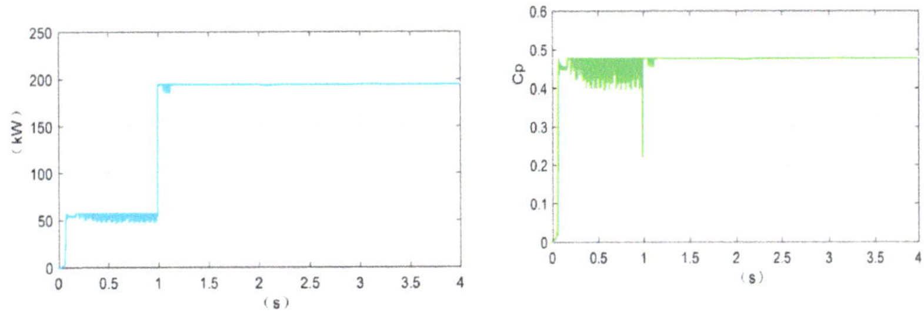

(a) (b)

Figure 16. Simulation diagram of wind power generation system. (**a**) Wind power system output power; (**b**) C_p curve.

4.2. Simulation Verification of Control Strategy Based on the Dynamic Droop Controller

In the system-connected mode of distributed power generation, the process of adding and cutting off the load of the propulsion system is simulated. The simulation time is set to 4 s, and the initial value of the load power is set to 10 kW. The load power suddenly changes from 10 kW to 70 kW in 2 s and then drops back to 10 kW in 3 s, and the system-connected side voltage, current waveform, and frequency fluctuation are detected.

It can be seen from the figure that the output voltage and current are standard sinusoidal waveforms. The system enters a stable state at about 0.2 s, the current amplitude increases to about 16.5 A after a sudden load is applied at 2 s, and the current waveform stabilizes around 16.5 A in 2~3 s; Figure 17b shows the voltage curve of the system-connected side, which is maintained at around 300 V after the system is stable, and the voltage waveform not change significantly after the load suddenly changes; Figure 17c displays the frequency change curve when the system is running, where the system frequency can be maintained at around 50 Hz and the load switching process can be realized smoothly and quickly.

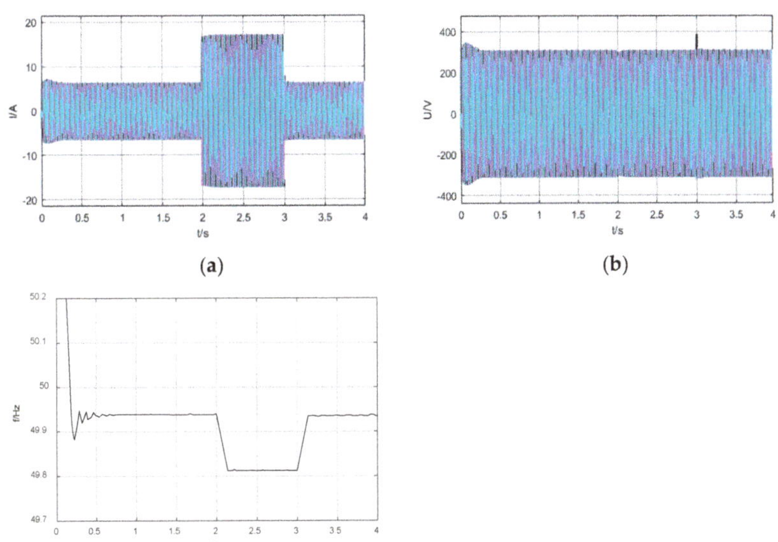

(c)

Figure 17. Simulation result of control system. (**a**) Current waveform; (**b**) Voltage waveform; (**c**) System frequency.

It can be seen from the simulation results that the designed two-layer coordinated control strategy of the propulsion system, which is based on the dynamic droop coefficient, has a significant effect on eliminating the voltage and frequency deviation. The system has the characteristics of rapid dynamic response and strong stability.

4.3. Simulation Verification of Day-Ahead Optimal Scheduling Control Strategy

Through the wind and solar forecast data and the electricity consumption of the shipload, the forecast curve of the day-ahead load power and the output power of the renewable energy is given, as shown in Figure 18.

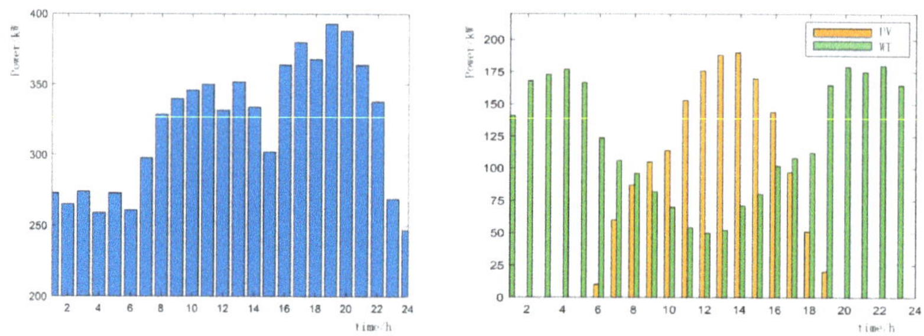

(a) (b)

Figure 18. Load day-ahead power forecast (**a**) and renewable energy day-ahead power forecast (**b**).

The day-ahead load power and renewable energy output power are input into system, and the day-ahead optimal dispatch curve and the energy storage system SOC curve can be obtained through the control calculation of each layer, as shown in Figure 19. It can be seen that the diesel generator set is basically maintained near the minimum operating power, which effectively reduces the pollutants produced by diesel generator sets.

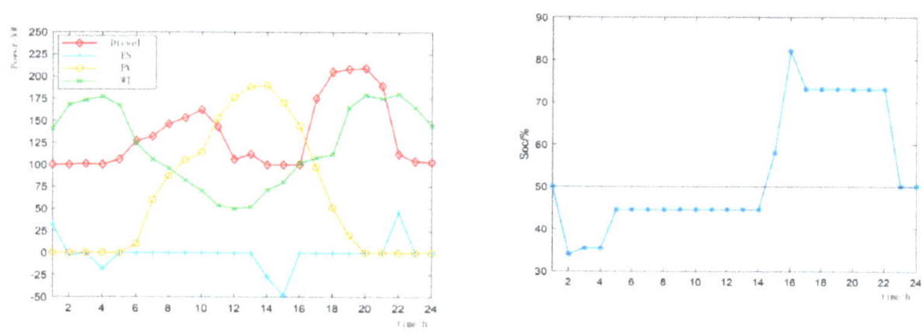

(a) (b)

Figure 19. Power curve of each power generation (**a**) and SOC curve of energy storage system (**b**).

During the period from 0:00 to 5:00, since the light intensity is zero, renewable energy is supplied by the wind power generation part, and through the charge and discharge control of the energy storage system, the output power of the diesel unit is maintained at the minimum operating power of about 100 kW. From 6:00 to 11:00, the first peak period of electricity consumption is ushered in. The load power increases, and the output power of the diesel generator needs to be started to make up for the remaining power required by the load. From 12:00 to 16:00, since the light intensity reaches the highest value, the total output power of renewable energy reaches the maximum. From 17:00 to 21:00, the second peak period of electricity consumption is ushered in. The load power of the ship power system increases. Due to the reduction of the light intensity to zero, the total output power of renewable energy begins to decrease. After 21:00, as the power required by the ship power system load decreases, SOC returns to around 50% and the SOC value remains within the optimal range.

In order to verify the economy of the IPSO algorithm proposed in the optimal scheduling problem in this paper, the day-ahead optimal scheduling strategy of traditional logic is used as a comparison with the strategy proposed in this paper. In the fixed logic day ahead optimal scheduling strategy, the energy storage unit only charges and discharges according to the predetermined fixed time period (selects the charging and discharging periods according to the wind and solar load day ahead prediction results to ensure the life of energy storage) and uses the diesel generator set as the main power source for distributed discharge to track the ship load data and the total charging and discharging power of the energy storage system. The economic cost is shown in Table 3.

Table 3. Economic cost.

Day-Ahead Planned Cost	Improvement Strategy	Fixed Strategy *
Diesel generator fuel cost	7663.2	8371.2
Environmental cost of diesel generator	751.51	820.93
Operating cost of diesel generator	28.74	28.74
Cost of energy storage system	280	400
Renewable energy cost	99.19	99.19
Load profit	1215	1215
Total cost	10,037.64	10,935.06

* Calculated from basic data.

The above economic operation cost formula is used to calculate the economic costs of the two methods. Since renewable energy always maintains the maximum power output and no load is in the state of power loss, the renewable energy operation cost and load income of the two strategies are consistent. Through calculation, it can be concluded that compared with the fixed logic strategy, the diesel generator fuel cost of the proposed strategy is reduced by 8.4%, the energy storage cost is reduced by 30%, and the total cost is saved by 8.2%. It not only saves energy but also reduces the pollution to the environment, which proves the superiority of the optimization algorithm proposed in this paper.

Traditional PSO, GA, and DE algorithms are used as comparison algorithms for a performance comparison test to prove the superiority of the proposed IPSO algorithm. The dimension of the standard test function is D = 30. For all test functions and algorithms, the initial value of the algorithm is set as follows: population size NP = 100; the maximum number of iterations g_{max} = 30; threshold stg_{max} = 20; and the maximum number of fitness calculations $FE_{max} = 3 \times 10^5$ and the convergence error $\varepsilon = 10^{-5}$ is set as the termination condition of the algorithm. The simulation results are shown in Figure 20.

In addition to the function speed of IPSO algorithm is slightly slower than DE algorithm except for $(a)f_1$ function, compared with the other three algorithms, the convergence speed is faster. Among them, when solving function $(a)f_1, (c)f_3$, PSO and GA algorithms have not converged to the allowable error range of $\varepsilon = 10^{-5}$ when the maximum computation times of 10,000 D are reached, and the same situation also occurs when solving function $(d)f_4$ for PSO algorithm. Both DE and IPSO algorithms achieve global optimal solutions on $(a)f_1, (b)f_2$ functions. In addition, the GA algorithm achieves global optimal solutions on $(b)f_2$ functions, but its performance in other functions is not as good as the IPSO algorithm. Although the convergence speed of the DE algorithm is faster than that of the IPSO algorithm on $(a)f_1$ function, the convergence speed of the DE algorithm is not as fast as that of the IPSO algorithm on other functions, so it can be seen that the overall quality of its solution is not as good as that of the IPSO algorithm.

The above figure show the convergence curve of the tested algorithm. Under different test functions, the IPSO algorithm (the solid red line in the figure) is more competitive than the other three algorithms regarding solution quality and convergence speed. The search speed has also been improved while ensuring the accuracy of solution.

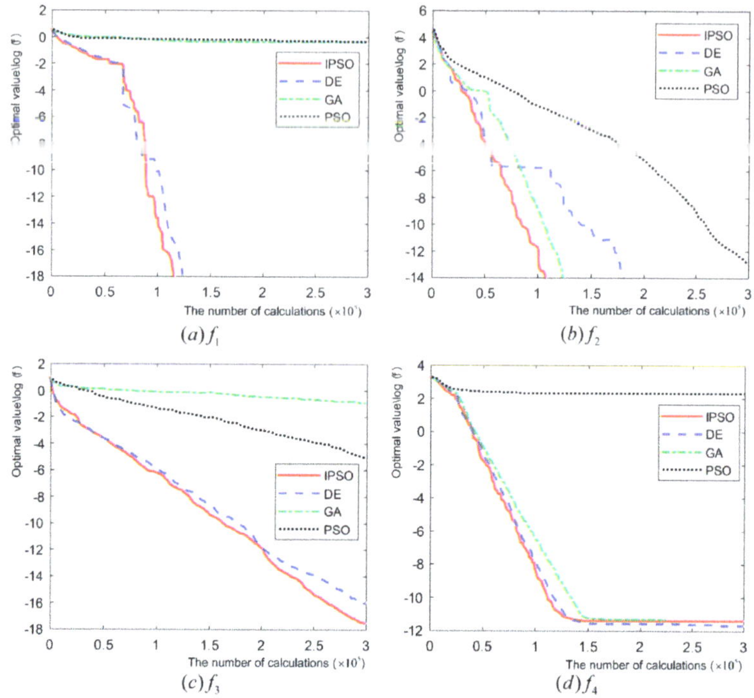

Figure 20. Algorithm comparison graph.

5. Conclusion and Discussion

The energy management strategy of a ship propulsion system based on hierarchical control is studied, including research on the MPPT control strategy for renewable energy and the two-layer coordinated control strategy for distributed power system-connected inverters. The following primary research results were obtained:

- The overall scheme of the layered control system of the ship propulsion system is designed, and a detailed and complete mathematical model is established. An overall simulation model of the ship propulsion system is built, which meets the needs of subsequent related research and simulation tests.
- A P&O algorithm based on dynamic perturbation step size is designed, including oscillation detection mechanism, dynamic perturbation step adjustment strategy, and voltage boundary setting. Through the comparison example simulation test with the traditional algorithm, the results show that the power loss of the MPPT control strategy using the P&O algorithm with a dynamic disturbance step size is reduced by 39.3%, and the overall tracking time is prolonged by 15.4%.
- A three-layer coordinated control strategy of the propulsion system based on the dynamic droop coefficient is designed, which dynamically adjusts the fixed droop coefficient. Realizing the adaptive change of the droop coefficient solves the problem of voltage and frequency deviation. In order to improve the system performance, the power sensitivity factor is designed simultaneously; moreover, a voltage and current double closed-loop controller are further designed to improve the inverter noise immunity and power quality. The simulation results show that the proposed control strategy can effectively suppress the voltage and frequency fluctuations and improve the system-connected security and power quality of the system-connected side.
- A PSO algorithm based on mutation particles is designed, and the collection information of some top-ranked vectors is mixed in the generated mutation vector;

furthermore, the method of the segmental improvement speed formula is adopted in order to improve the accuracy and search speed. Through the standard function performance test with other intelligent optimization algorithms, the results show that the improved algorithm has a faster convergence speed and higher accuracy in solving the load optimization problem. The total running cost of the algorithm is reduced 8.4%, and the total cost was reduced by 8.2%.

In addition, the ship energy management system designed in this paper can give distributed power supply output results through the predicted data. However, there is no relevant operating software and interface display. Despite the lack of visualization-related research, the equipment involved is assumed to be in good operating condition. The built simulation model focuses more on the overall design of the control strategy, ignoring the actual multi-interference situation. Therefore, follow-up work should mainly focus on the above aspects for further study in order to improve the practicability of the simulation model.

Author Contributions: Conceptualization, L.Z. and P.S.; methodology, L.Z. and P.S.; software, Y.R.; validation, Y.R. and Z.Z.; data curation, Y.R.; writing—original draft preparation, Y.R.; writing—review and editing, L.Z. and P.S.; visualization, Y.R. All authors have read and agreed to the published version of the manuscript.

Funding: Supported by the Fundamental Research Funds for the Central Universities. The funding number: 3072022JC0404.

Institutional Review Board Statement: Not applicable.

Informed Consent Statement: Not applicable.

Conflicts of Interest: The authors declare no conflict of interest.

Nomenclature

Category	Symbol	Implication
abbreviations	PSO	Particle Swarm Optimization
	IPSO	Improved Particle Swarm Optimization
	PV	Photovoltaic
	MPPT	Maximum Power Point Tracking
	P&O	Perturbation Observation Algorithm
	GA	Genetic Algorithm
	DE	Differential Evolution
	PWM	Pulse Width Modulation
parameter	K	the damping coefficient
	ω_g	the angular velocity of the diesel engine
	p	the number of pole pairs of the synchronous generator
	M_d	the output torque of the diesel engine
	M_c	resistance torque of the diesel engine
	I_L	the constant current source
	I_D	the current passing through the diode
	α	the temperature coefficient
	β	the pitch angle of the blade
	I_m	the maximum power output of the current
	U_m	the maximum power output of the voltage
	U	the voltage to the photovoltaic array
	U_{oc}	the open circuit voltage
	T_m	the temperature of the photovoltaic array
	T_a	the ambient temperature
	R	the actual solar irradiance
	λ	the tip speed ratio
	ω_T	the rotational speed
	V_0	the wind speed

Category	Symbol	Implication
	C_P	the speed of wind turbines
	V_{BT}	the lithium battery voltage
	E_0	the constant voltage source voltage
	Q_t	the lithium battery charge/discharge capacity
	I^*	the filtered current
	$SOC(t)$	the state of charge of the lithium battery
	i_{oabc}	the three-phase current output by the inverter
	U_{oabc}	the three-phase voltage output by the inverter
	P_0	the reference active power
	Q_0	the reference reactive power
	u^*	the SPWM modulation signal
	ω_0	the given reference frequency
	U_0	the given reference voltage
	k_{f1}	the dynamic droop coefficient
	k_{U1}	the dynamic droop factor
	η_f	the sensitivity factor
	f_{wt}	the operating cost coefficients
	n_{wt}	the numbers of wind power generation units
	$P_{wt,m}(t)$	the output power of wind power generation systems
	$g_{\max}$	the maximum number of iterations
	$f(U_i)$	the test vector function
	$f(X_i)$	the target vector function
	stg	an indicator used to monitor

Appendix A

Table A1. Diesel generator pollutant emission factor (g/kWh).

SO_2	CO_2	CO	NO_X
4.34	2.32	0.47	232.04

Appendix B

Table A2. Environmental pollutant penalty amount (CNY/kg).

SO_2	CO_2	CO	NO_X
0.75	0.0028	0.125	1.00

References

1. Pan, P.; Sun, Y.; Yuan, C.; Yan, X.; Tang, X. Research progress on ship power systems integrated with new energy sources: A review. *Renew. Sust. Energ. Rev.* **2021**, *144*, 111048. [CrossRef]
2. Zhang, R.; Liang, H. Application of solar energy in ship power field. In Proceedings of the 2022 IEEE Asia-Pacific Conference on IPEC, Dalian, China, 14–16 April 2022.
3. Zhu, Y.; Zhou, S.; Feng, Y.; Hu, Z.; Yuan, L. Influences of solar energy on the energy efficiency design index for new building ships. *Int. J. Hydrogen Energy* **2017**, *42*, 19389–19394. [CrossRef]
4. Babarit, A.; Ghani, N.; Brouillette, E. Experiment validation of the energy ship concept for far-offshore wind energy conversion. *Ocean Eng.* **2021**, *239*, 109830. [CrossRef]
5. Altosole, M.; Balsamo, F.; Campora, U.; Mocerino, L. Marine Dual-Fuel Engines Power Smart Management by Hybrid Turbocharging Systems. *J. Mar. Sci. Eng.* **2021**, *9*, 663. [CrossRef]
6. Yuan, Y.; Wang, J.; Yan, X.; Shen, B.; Long, T. A review of multi-energy hybrid power system for ships. *Renew. Sust. Energ. Rev.* **2020**, *132*, 110081. [CrossRef]
7. Huang, Y.; Wang, L.; Zhang, Y.; Wang, L.; Zhang, Z. An Overview of Multi-Energy Microsystem in All-Electric Ships. *Front. Energy Res.* **2022**, *10*, 881548. [CrossRef]
8. Fang, S.; Xu, Y.; Li, Z. Two-Step Multi-Objective Management of Hybrid Energy Storage System in All-Electric Ship Microgrds. *IEEE Trans. Veh. Technol.* **2019**, *68*, 3361–3373. [CrossRef]

9. Anvari, M.; Dragicevic, T.; Lexuan, M. Optimal planning and operation management of a ship electrical power system with energy storage system. In Proceedings of the IECON2016-42nd Annual Conference of the IEEE industrial Electronics Society, Florence, Italy, 23–26 October 2016.
10. Misyris, G.; Marinopoulos, A.; Doukas, D. On battery state estimation algorithms for electric ship applications. *Electr. Power Syst. Res* **2017**, *151*, 115–124. [CrossRef]
11. Lee, K.; Shin, D.; Yoo, D. Hybrid photovoltaic/diesel green ship operating in standalone and system-connected mode Experimental investigation. *Energy* **2013**, *49*, 475–483. [CrossRef]
12. Accetta, A.; Pucci, M. Energy Management System in DC micro-system of Smart Ships: Main Gen-Set Fuel Consumption Minimization and Fault Compensation. *IEEE Trans. Ind. Appl.* **2019**, *55*, 3097–3113. [CrossRef]
13. Brizuela-Mendoza, J.A.; Sorcia-Vazquez, F.D.J.; Rumbo-Morales, J.Y.; Lozoya-Ponce, R.E.; Rodriguez-Cerda, J.C. Active fault tolerant control based on eigen structure assignment applied to a 3-DOF helicopter. *Asian J. Control* **2021**, *23*, 673–684. [CrossRef]
14. Zhu, L.; Han, J.; Peng, D. Fuzzy Logic based Energy Management Strategy for a Fuel Cell/Battery/Ultra-capacitor Hybrid Ship. In Proceedings of the International Conference on Green Energy, Sfax, Tunisia, 25–27 March 2014.
15. Feng, X.; Butler-Purry, K.L.; Zourntos, T. A Multi-Agent System Framework for Real-Time Electric Load Management in MVAC All-Electric Ship Power Systems. *IEEE Trans. Power Syst.* **2015**, *30*, 1327–1336. [CrossRef]
16. Seenumani, G.; Peng, H.; Jing, S. A Reference Governor-Based Hierarchical Control for Failure Mode Power Management of Hybrid Power Systems for All-electric Ships. *J. Power Sources* **2011**, *196*, 1599–1607. [CrossRef]
17. Nelson, M.; Jordan, P.E. Automatic Reconfiguration of a Ship's Power System Using Graph Theory Principles. *IEEE Trans. Ind. Appl.* **2015**, *51*, 2651–2656. [CrossRef]
18. Mensah, E.; Kwatny, H. Models for Optimal Dynamic Reconfiguration and Simulation of Ship Power Systems in Simlink with Stateflow. In Proceedings of the 2007 IEEE Electric Ship Technologies Symposium, Arlington, VA, USA, 21–23 May 2007.
19. Hou, J.; Sun, J.; Hofmann, H. Interaction Analysis and Integrated Control of Hybrid Energy Storage and Generator Control System for Electric Ship Propulsion. In Proceedings of the American Control Conference (ACC), Chicago, IL, USA, 1–3 July 2015.
20. Kanellos, F.D.; Tsekouras, G.J.; Hatziargyriou, N.D. Optimal Demand-Side Management and Power Generation Scheduling in an All-Electric Ship. *IEEE Trans. Sustain. Energy* **2014**, *5*, 1166–1175. [CrossRef]
21. López, A.R.; Mina, J.D.; Calderón, G.; Aguayo, J.; Calleja, J.H. Combined parameters selection of a proportional integral plus resonant controller for harmonics compensation in a wind energy conversion system. *Electr. Eng.* **2018**, *100*, 2277–2286. [CrossRef]
22. Paran, S.; Vu, T.V.; Mezyani, T. MPC-based Power Management in the Shipboard Power System. In Proceedings of the 2015 IEEE Electric Ship Technologies Symposium (ESTS), Alexandria, VI, USA, 21–24 July 2015.
23. Hatata, A.Y.; Osman, G.; Aladl, M.M. An optimization method for sizing a solar/wind/battery hybrid power system based on the artificial immune system. *Sustain. Energy Technol.* **2018**, *27*, 83–93. [CrossRef]
24. Hadidian, M.J.; Nowdeh, S.A. Optimal sizing and energy management of stand-alone hybrid photovoltaic/wind system based on hydrogen storage considering LOEE and LOLE reliability indices using flower pollination algorithm. *Renew Energy* **2019**, *135*, 1412–1434. [CrossRef]
25. Sanajaoba, S.S.; Fernandez, E. Modeling, size optimization and sensitivity analysis of a remote hybrid renewable energy system. *Energy* **2018**, *143*, 719–731. [CrossRef]
26. De la Cruz, J.; Ramirez, J.M.; Leyva, L. Modification of Geometric Parameters in Outer Rotor Permanent Magnet Generators to Improve THD, Efficiency, and Cogging Torque. *Int. J. Emerg. Electr. Power Syst.* **2014**, *15*, 471–483. [CrossRef]
27. Zhang, W.; Maleki, A. A heuristic-based approach for optimizing a small independent solar and wind hybrid power scheme incorporating load forecasting. *J. Clean. Prod.* **2019**, *241*, 117920. [CrossRef]
28. Maleki, A.; Pourfayaz, F. Optimal sizing of autonomous hybrid photovoltaic/wind/battery power system with LPSP technology by using evolutionary algorithms. *Sol. Energy* **2015**, *115*, 471–483. [CrossRef]
29. Borunda, M.; Garduno, R.; Nicholson, A.E.; De la Cruz, J. Assessment of Small-Scale Wind Turbines to Meet High-Energy Demand in Mexico with Bayesian Decision Networks. In Proceedings of the 18th Mexican International Conference on Artificial Intelligence (MICAI), Xalapa, Mexico, 27 November–2 November 2019.
30. Xu, L.; Wei, B.; Yun, Y. Coordinated Control of Diesel Generators and Batteries in DC Hybrid Electric Shipboard Power System. *J. Mar. Sci. Eng.* **2021**, *14*, 6246. [CrossRef]
31. Lamichhane, A.; Zhou, L.; Yao, G. Modeling, Control and Power Management of Six-Phase PMSM Based Shipboard MVDC Distribution System. *J. Mar. Sci. Eng.* **2020**, *13*, 4229. [CrossRef]
32. Nair, R.; Narayanan, G. Emulation of Wind Turbine System Using Vector Controlled Induction Motor Drive. *IEEE Trans. Ind. Appl.* **2020**, *56*, 4124–4133. [CrossRef]
33. Puchalapalli, S.; Tiwari, S.; Singh, B.; Goel, P. A Microsystem Based on Wind-Driven DFIG, DG, and Solar PV Array for Optimal Fuel Consumption. *IEEE Trans. Ind. Appl.* **2020**, *56*, 4689–4699. [CrossRef]
34. He, W.; Mo, O. Accelerating Efficient Installation and Optimization of Battery Energy Storage System Operations Onboard Vessels. *J. Mar. Sci. Eng.* **2022**, *15*, 4098. [CrossRef]
35. Kim, S.; Kim, J. Optimal Energy Control of Battery Hybrid System for Marine Vessels by Applying Neural Network Based on Equivalent Consumption Minimization Strategy. *J. Mar. Sci. Eng.* **2021**, *9*, 1228. [CrossRef]
36. Antonopoulos, S.; Vissser, K.; Kalikatzarakis, M.; Reppa, V. MPC Framework for the Energy Management of Hybrid Ships with an Energy Storage System. *J. Mar. Sci. Eng.* **2021**, *9*, 993. [CrossRef]

37. Jin, Z.; Meng, L.; Guerrero, J.; Han, R. Hierarchical Control Design for Shipboard Power System with DC Distribution and Energy Storage Aboard Future More-Electric Ships. *IEEE Trans. Industr. Inform.* **2018**, *14*, 703–714. [CrossRef]
38. Deng, C.; Wen, C.; Zou, Y.; Wang, W.; Li, X. A Hierarchical Security Control Framework of Nonlinear CPS Agent DoS Attacks With Application To Power Sharing of AC Microgrid. *IEEE Trans. Cybern.* **2022**, *52*, 5255–5266. [CrossRef] [PubMed]
39. Xiao, S.; Dong, J. Distributed Fault-Tolerant Containment Control for Linear Heterogeneous Multiagent System: A Hierarchical Design Approach. *IEEE Trans. Cybern.* **2022**, *52*, 971–981. [CrossRef] [PubMed]
40. Cortajarena, J.; Barambones, O.; Alkorta, P.; Cortajarena, J. Sliding mode control of an active power filter with photovoltaic maximum power tracking. *Int. J. Electr. Power Energy Syst.* **2019**, *110*, 747–758. [CrossRef]
41. Mousa, H.; Youssef, A.; Mohamed, E. State of the art perturb and observe MPPT algorithms based wind energy conversion systems: A technology review. *Int. J. Electr. Power Energy Syst.* **2021**, *126*, 106598. [CrossRef]
42. Farhat, M.; Barambones, O.; Sbita, L. A Real-Time Implementation of Novel and Stable Variable Step Size MTTP. *Energies* **2020**, *13*, 4668. [CrossRef]
43. Abo Gabl, O.M.; Shaaban, M.F.; Zeineldin, H.H.; Ammar, M.E. A Multi-objective Secondary Control Approach for Optimal Design of DG Droop Characteristic and Control Mode for Autonomous Micrsystems. *IEEE Syst. J.* **2021**, *8*, 1–10. [CrossRef]
44. Shivam; Dahiya, R. Distributed Control for DC Microsystem Based on Optimized Droop Parameters. *IETE J. Res.* **2020**, *66*, 192–203. [CrossRef]
45. Mo, S.; Chen, W.-H.; Lu, X. Hierarchical Hybrid Control for Scaled Consensus and Its Application to Secondary Control for DC Microgri. *IEEE Trans. Cybern.* **2022**, 1–13. [CrossRef]
46. Hou, P.; Hu, W.; Soltani, M.; Chen, Z. Optimized Placement of Wind Turbines in Large-Scale Offshore Wind Farm Using Partical Swarm Optimization Algorithm. *IEEE Trans. Sustain. Energy* **2015**, *6*, 1272. [CrossRef]
47. Yang, R.; Wei, H.; Wang, L. Research on Energy Regulation and Optimal Operation Strategy of Multi-energy Ship Power Station Based on Improved Particle Swarm Algorithm. In Proceedings of the 2021 IEEE 5th Conference on IAEAC, Chongqing, China, 12–14 March 2021.
48. Jasim, A.; Jasim, B.; Kraiem, H.; Flah, A. A Multi-Objective Demand/Generation Scheduling Model-Based Microsystem Energy Management System. *J. Mar. Sci. Eng.* **2022**, *14*, 10158.

Journal of
Marine Science and Engineering

MDPI

Article

Hybrid Modeling and Simulation for Shipboard Power System Considering High-Power Pulse Loads Integration

Wanlu Zhu *, Chunpeng Jin and Zhengzhuo Liang

Department of Automation, Jiangsu University of Science and Technology, Zhenjiang 212100, China
* Correspondence: zhuwanlu@just.edu.cn

Abstract: The complex dynamic characteristics of a shipboard power system (SPS) are not only related to its continuous dynamics but also influenced by discrete control behavior. Especially, during combat mission execution of high-power pulse loads (HPPLs), their operation plan as a sequence of discrete control events will cause successive abrupt changes in the continuous dynamics of SPS due to the sudden and intermittent nature of the external attacks, which requires overall comprehension of the hybrid dynamics evolution process driven by discrete events. In this paper, considering the zonal distribution structure of SPS and the influences of extreme events on the discrete dynamics of each zone, the extended hybrid models for each zone, including normal operation configuration and fault configuration, are obtained based on the hybrid automata theory. Then, the global hybrid model of SPS is developed. The mapping relationship of discrete state transition to the continuously controlled system is analyzed to reconstruct the set of differential equations model of the continuous system for the purpose of simulation. Two case studies are carried out to perform the simulation under the proposed hybrid model. It is demonstrated that this proposed method can reveal the operating characteristics of the hybrid dynamic evolution process driven by discrete events, both in normal operation and pulse loads operation. Although the precise measure of discrete states of SPS can be challenging to obtain, especially during the confrontation phase, the proposed method still provides valuable insights on evaluating the sophisticated dynamics of an SPS.

Keywords: shipboard power system; hybrid model; simulation; high-power pulse loads

Citation: Zhu, W.; Jin, C.; Liang, Z. Hybrid Modeling and Simulation for Shipboard Power System Considering High-Power Pulse Loads Integration. *J. Mar. Sci. Eng.* **2022**, *10*, 1507. https://doi.org/10.3390/jmse10101507

Academic Editor: Rosemary Norman

Received: 24 August 2022
Accepted: 14 October 2022
Published: 16 October 2022

1. Introduction

The integrated power system of ships has become the main development direction of future ships, which can supply power to various loads on board, including daily electrical loads, maneuvering systems, communication, navigation, and propulsion systems [1,2]. Compared with conventional ships, a shipboard power system (SPS) can significantly reduce life-cycle costs, support more payloads, and provide greater survivability. By centralizing the propulsion system and other electrical equipment into a unified grid, SPS can rationally distribute energy throughout the ship. This will enable "unlocking" of propulsion energy and provide opportunities for high-powered pulsed loads (HPPLs, e.g., electromagnetic catapults, high-energy radars, and laser weapons) to be used on board, which will fundamentally transform the weaponry and significantly improve the maritime countermeasures capability of naval vessels [3–5].

The main purpose of HPPLs on board is to perform countermeasures [6,7]. External attacks are sudden and intermittent, and SPS also has the capability to actively combat extreme events. This means that the entire countermeasure process does not stop until the end of the event or the system is compromised. Random arrival of multiple concurrent attacks and fluctuating pulse loads during intense attack and defense will have a dramatic impact on SPS dynamics, making it difficult to build a system model that can characterize the entire operation of the confrontation.

Taking the electromagnetic rail gun (EMRG) as an example, the single launch duration is about 9ms, and the power demand is 160 MJ [8]. With the continuous demand (12/min)

that must occur in the fierce confrontation process, such significant power must be provided by SPS. To reduce the impact of instantaneous ultra-high-power demand, energy storage devices should be used before HPPLs integration, such as batteries, supercapacitors, flywheels, and other hybrid energy storage devices [9–11].

However, the uncertainty of the countermeasure process and the complexity of the countermeasure target will cause the dynamic characteristics of SPS to change drastically in a short period of time, and the traditional system model can no longer meet the demand of real-time countermeasure decisions. In order to deal with multiple concurrent attacks arriving randomly and pulse-load operation plans for combating external targets, a model that can describe the system's mixed dynamic evolution process driven by random discrete events in a holistic manner is required.

In addition, the zonal power distribution structure of SPS allows each zone to handle an external attack autonomously, and the damage caused by the attack will not spread beyond the damaged zone, which also determines that the system hybrid dynamic model must take into account the need for distributed control and provide a model basis for subsequent defense and recovery strategies. These are difficult to achieve by traditional modeling theory, simulation, and controller design.

The current research on modeling and simulation of SPS mainly focuses on continuous dynamic electromagnetic transient modeling and simulation methods [12–14]. High-accuracy models for medium-voltage DC (MVDC) SPS were developed to accurately capture the complex and nonlinear system continuous dynamics in physical test systems [15–17]. Attempts to use reduced-order models have also appeared in the study of AC ship integrated power systems [18,19] and in the authors' previous work for MVDC SPS [20]. To minimize the computational time, a mixed SPS model using different component models with varying fidelity was established based on the bond graph method [21]. However, modeling and simulation studies that comprehensively consider the evolution of the mixed dynamics of SPS are still rarely addressed.

Some studies of power electronic systems proposed discrete-state event-driven hybrid models and simulation methods to analyze the multi-timescale mixing characteristics of power electronic systems [22–24]. However, these studies consider discrete events triggered by the evolution of the system's own state, which is quite different from the external discrete attack events that SPS must actively respond to during a countermeasure.

Basically, hybrid modeling methods can be divided into two categories:

The first type considers the system as a continuous dynamic system and describes the whole system using differential equations. Events that make system equations change are used as discrete inputs, which means embedding the discrete events into the continuous dynamic system. This kind of method focuses on the stability and controllability of the system, and its main modeling methods are switching system model, hybrid logic dynamic model, etc. For example, a microgrid switching system model was established to ensure stability of microgrids in different operating states [25]. An AC/DC hybrid system switching system model was established and applied to the field of power system transient stability assessment [26]. The hybrid logic dynamic model was used to study the economic dispatch problem [27].

The second type is to embed continuous dynamic behavior into discrete event dynamic systems, focusing on performance verification and comprehensive design of controllers, whose main modeling methods are the hybrid automaton model, hybrid Petri net model, etc. The Petri net model was used to deal with the discrete switching control events in smart microgrids [28], and this model was also used to analyze the reachability of microgrids consisting of green energy sources [29]. A discrete hybrid automaton was used to model the storage system and controllable electrical loads of the UFES pilot microgrid [30].

While research efforts have been undertaken to study SPS as a hybrid system, the focus was prioritized to studying operations such as load shedding [31] and supervisory control [32]. For example, an equation of state model with integer inequality constraints was developed by building a hybrid logic dynamic model to analyze the optimal control behavior of SPS after loss of a generator [33]. However, this method always requires coarse discretization of the continuous state, and a large set of differential equations must be solved in the search space formed by the discretized continuous and discrete states at each moment, which is not suitable for describing the dynamic characteristics of SPS driven by a sequence of discrete events with drastic sudden changes in a short period of time. Moreover, it also cannot meet the needs of the subsequent defense and recovery strategy optimization considering accuracy and computational efficiency.

In the authors' previous work, a centralized hybrid automaton model for the whole SPS was established for reconfiguration [34], and it was further extended to the model under partial observation in Ref. [35]. In this paper, we would like to continue the previous research by developing a distributed, hybrid model for SPS considering HPPLS integration, which not only provides a complete description of the system dynamics but also accommodates the requirements for zonal autonomy. A hybrid automaton model of SPS including both continuous and discrete dynamics of the system is established to completely describe the system dynamics. The simulation process of the hybrid model is divided into an iterative forward process: differential equations solving of the continuous controlled system in the discrete state, differential equations reconstruction of the continuous controlled system after discrete state transition and initial status resetting, and differential equations solving after the transfer. The implementation steps and algorithms of this process are provided. Further, two case studies are provided to show the SPS hybrid dynamic evolution process driven by discrete events.

Compared with the existing SPS modeling and simulation approaches, the novelty and intellectual merits of this paper can be summarized as follows:

(1) This paper seeks a hybrid model for shipboard power systems considering high-power pulse loads integration, which can deal with multiple concurrent extreme events arriving randomly and fluctuating pulse-load operation, to illustrate the hybrid dynamic evolution process of SPS.
(2) This paper presents the hybrid model in a distributed manner. The proposed model takes full advantage of the zonal distribution structure of SPS and allows each zone in SPS to handle the external events autonomously.
(3) The proposed hybrid model provides a direct input/output interface to be integrated with different-scale continuous models, which allows easy refinement and adjustment to adapt to different ship design intentions and control requirements.

The rest of this paper is organized as follows. Section 2 introduces a representative MVDC SPS and derives its hybrid model description, while Section 3 illustrates the simulation process for this hybrid model. The simulation results are presented in Section 4, followed by the conclusions.

2. Hybrid Model Description for SPS

Based on the theory of hybrid automata, a hybrid model of SPS consisting of four zones is proposed in this section. The continuous dynamics of each zone are represented in the time domain by differential equations, and the continuous state reset relations caused by discrete state transition are provided to realize simulation of this hybrid model.

2.1. Representative SPS

The next-generation ship integrated power system proposed in Ref. [4], as shown in Figure 1, is provided as the baseline topology for this paper. This system consists of two main gas-turbine generator sets (MTGs) and two auxiliary gas-turbine generator sets (ATGs). These generator sets are connected to the port and starboard buses to form a zonal power distribution network connected at the bow and stern of the ship. This zonal distribution network supplies power to four kinds of loads: propulsion loads on each port and starboard side, consisting of the propulsion motor with its drive (variable speed drive, VSD) and propeller; inside the zonal network, there are three pulse loads connected to the port and starboard through energy storages; there are also four zonal loads representing the daily electrical loads of the ship, each of which is internally a small distribution network with a different internal structure depending on the actual needs of the ship. All generators and loads are connected to the distribution network through power conversion modules (PCM) and circuit breakers (black squares in the figure). It should be noted that, although there are connection switches for the port and starboard DC buses at the bow and stern of the ship, the port and starboard are normally operated separately.

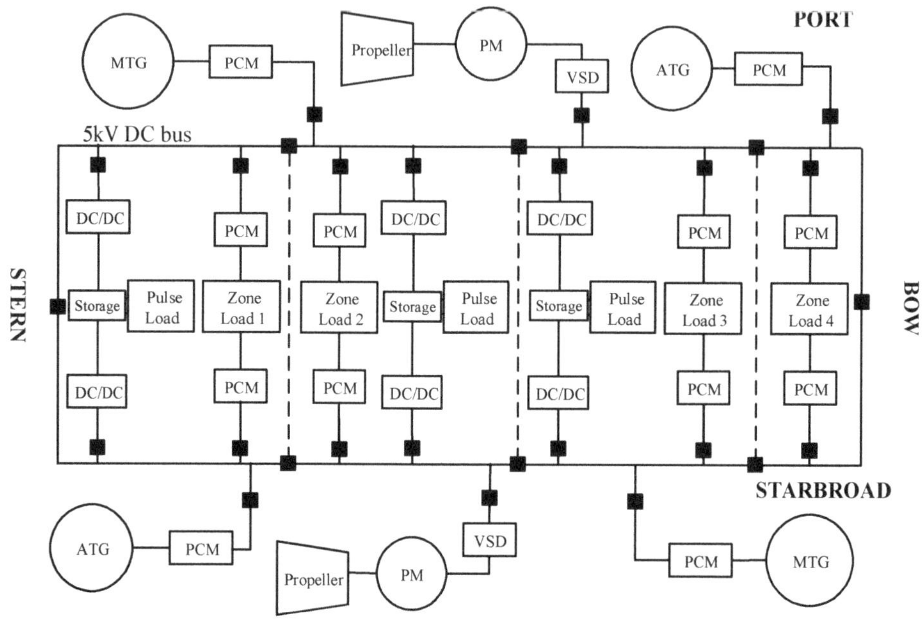

Figure 1. Schematic diagram of the four-zone SPS.

The dotted lines connecting port and starboard breakers divide the entire SPS into four zones. The power supply devices (generator sets and energy storage equipment) are distributed among the zones and can supply power to the loads across the zones. The zonal distribution network allows each zone to be autonomous, and the damage caused by attacks does not spread beyond the damaged zone. Establishing separate hybrid models for each zone will facilitate implementation of subsequent distributed control schemes.

2.2. Automata-Based Hybrid Model for Each Zone

For each zone of SPS, its basic hybrid structure can include discrete controllers and continuous controlled systems and their interfaces, describing the discrete event dynamics and continuous dynamics, respectively. Its basic hybrid model can be represented by a hybrid automaton:

$$H = (Q, X, Y, U, Init, f, g, \Sigma, EG, T, R) \tag{1}$$

where $Q \cup X$ is the state space, Q is finite, Y is the set of output variables, U is the set of continuous control inputs, $Init \subseteq Q \times X \times U$ is the set of initial conditions, $f : Q \times X \times U \to X$ is the set of continuous state evolution laws describing the continuous state corresponding to each $q \in Q$, $g : Q \times X \times Y \times U \to Y$ is the set of algebraic equations for each $q \in Q$, $\Sigma = \Sigma_c \cup \Sigma_u$ is the set of discrete events, where Σ_c is the set of controllable events, Σ_u is the set of uncontrollable events, $EG : X \times U \to \Sigma$ is the event generator function, $T : \Sigma \times Q \to 2^Q$ is the discrete state transition relation, $R : Q \times X \times U \to 2^{X \times U}$ is the reset relation, that is, the control behavior generator function.

The impact of extreme events on the discrete dynamics of the zones is considered to develop an extended zonal hybrid model. The extreme events, i.e., external attacks, have two effects on the discrete dynamics of the zones: one is the intra-zone initiation of pulse loads, and the second is the fault caused by the attacks on the damage of each zone (including communication and cables, etc.). The former can be considered as the normal operation configuration of the zone, while the latter is the fault configuration, according to which the discrete event transfer of each zone is classified. The operational configuration of each zone is regarded as the set of normal operational configuration and fault configuration, and the set of transition events between the two configurations is defined. Then, we can obtain the extended hybrid model of each zone:

$$S = (\mathcal{H}, FT, RE, H_0) \tag{2}$$

where $\mathcal{H}$ is the set of all possible configurations of the subsystem, S and H_0 is the initial configuration of the subsystem S. FT and RE are the set of transition events between the normal operation configuration and the failure configuration:

$$FT = \left\{ft_j^i : i = 1, 2, \cdots, n; j = 1, 2, \cdots, m\right\} \tag{3}$$

$$RE = \left\{r_r^j : r = 1, 2, \cdots, n; j = 1, 2, \cdots, m\right\} \tag{4}$$

where ft_j^i is the fault events, and r_r^j is the recovery events.

In this hybrid model, each discrete state of a zone has its corresponding differential equations to describe the continuous dynamics. Take zone 1 as an example: its continuous controlled dynamics in the discrete state q_i can be expressed as:

$$\begin{aligned} \dot{x} &= f_{q_i}\left(x_{q_i}, y_{q_i}, u_{q_i}\right) \\ 0 &= g_{q_i}\left(x_{q_i}, y_{q_i}, u_{q_i}\right) \end{aligned} \tag{5}$$

The continuous dynamic modeling process for SPS can be found in a previous publication by the authors [28]. Here, we only provide the basic description of variables in (5), where x_{q_i} is the continuous state variables (including state variables of MTG, pulse load, DC bus voltage, and current, etc.), y_{q_i} is the output variables (including variables we want to monitor), and u_{q_i} is the continuous control inputs (including preset values for MTG and other device controllers inside zone 1). f_{q_i} and g_{q_i} are the differential functions that can be derived from Ref. [20].

For the discrete states in this model, let us express the online status of the MTG, ATG, PM, pulse load, and zonal load as integer values, as well as the connection status of breakers in the zone. The set of current states of all these devices is a discrete state in this hybrid model. In addition, take zone 1 as an example: it includes an ATG, two types of load, and a breaker connecting the port and starboard. We use "0", "1", "2", and "3" to indicate the connection status of loads, with "0" indicating no connection, "1" indicating the connection to the starboard bus, and "2" indicating the connection to zone 2, and "3" indicating a simultaneous connection to port and starboard. Then, for ATG, indicators of "0" and "1" can be used to represent their online status, with "0" indicating offline and "1" indicating online. Moreover, for the breaker, "0" indicates open, and "1" indicates closed. Combining these indicators, we can use a string to express the discrete states of zone 1. For example, "1310" means that the ATG is online, the pulse load is connecting to the port and starboard simultaneously, the zone load is connecting to the starboard bus, and the breaker is open.

The discrete dynamics of a zone are represented by event-driven discrete state transitions, such as the discrete state transition of zone 1 from "1310" to "0130" is caused by the event "ATG shutdown". With the transition of discrete states, not only the differential equations describing the continuous dynamics change but also the initial conditions after the transition must be recalculated. This will be discussed in Section 4.

2.3. Global Hybrid Model for SPS

Assume that there are s zones in the SPS (in this paper, it is 4), and the zone $S_k(k = 1, 2, \ldots, 4)$ has $(n + m)_k$ configurations. Therefore, the distributed configuration of SPS can then be denoted by

$$\mathfrak{C} = \left(H_{1,l_1}, H_{2,l_2}, \ldots, H_{s,l_s}\right) \quad (6)$$

where H_{k,l_k} is the zone configuration in $\mathfrak{C}$, $l_k = 0, 1, \ldots, (n + m)_k$. The set of all possible distributed configurations of SPS can be denoted by $\mathcal{C}$. Therefore, the global model for SPS can be represented as:

$$\mathcal{S} = (\mathcal{C}, \mathfrak{C}_0) \quad (7)$$

where $\mathfrak{C}_0$ is the initial distributed configuration of SPS.

This global hybrid model fully reflects the autonomous nature of each zone, where each zone can handle its own internal operation and management as well as respond to external attacks.

3. Hybrid Simulation Method for SPS

Simulation of the hybrid model involves the continuous dynamic mutation caused by a discrete state transition. Therefore, the simulation necessarily includes two steps: (1) calculating the initial conditions of the system after discrete state transition; (2) continuous dynamics simulation of the system in the discrete state after the transition.

3.1. Continuous Dynamic Mutation Due to Discrete State Transition

Suppose that system is first in a discrete state q_i with differential equations as (8); after the discrete state transition occurs, the system is in q_j and differential equations as (9):

$$\begin{aligned} x(k+1) &= f_{q_i}(x_{q_i}(\tau'), y_{q_i}(\tau'), u_{q_i}(\tau')) \\ 0 &= g_{q_i}(x_{q_i}(\tau'), y_{q_i}(\tau'), u_{q_i}(\tau')) \end{aligned} \quad (8)$$

$$\begin{aligned} x(k+1) &= f_{q_j}\left(x_{q_j}(\tau^+), y_{q_j}(\tau^+), u_{q_j}(\tau^+)\right) \\ 0 &= g_{q_j}\left(x_{q_j}(\tau^+), y_{q_j}(\tau^+), u_{q_j}(\tau^+)\right) \end{aligned} \quad (9)$$

where $x(\tau^+), y(\tau^+), u(\tau^+)$ are continuous state variables, output variables, and continuous control inputs after discrete state transition, respectively.

The change in output variables and continuous state variables caused by discrete state transition, which is also the reset relationship of continuous state after discrete state transition, has the following initial condition calculation flow chart:

It should be noted that the numbers of output variables, continuous state variables, and control inputs may change after the discrete state transition. During the flow chart in Figure 2, the output variables are solved first, followed by the continuous state variables; after that, the continuous state variables at the end of time step k are calculated in normal order by using the following backward differencing method:

$$x_{q_j}(k) - x_{q_j}(\tau^+) = \frac{h}{2}\left(\begin{array}{l} f_{q_j}\left(x_{q_j}(k), y_{q_j}(\tau^+), u_{q_j}(k)\right) \\ +f_{q_j}\left(x_{q_j}(\tau^+), y_{q_j}(\tau^+), u_{q_j}(\tau^+)\right) \end{array}\right) \quad (10)$$

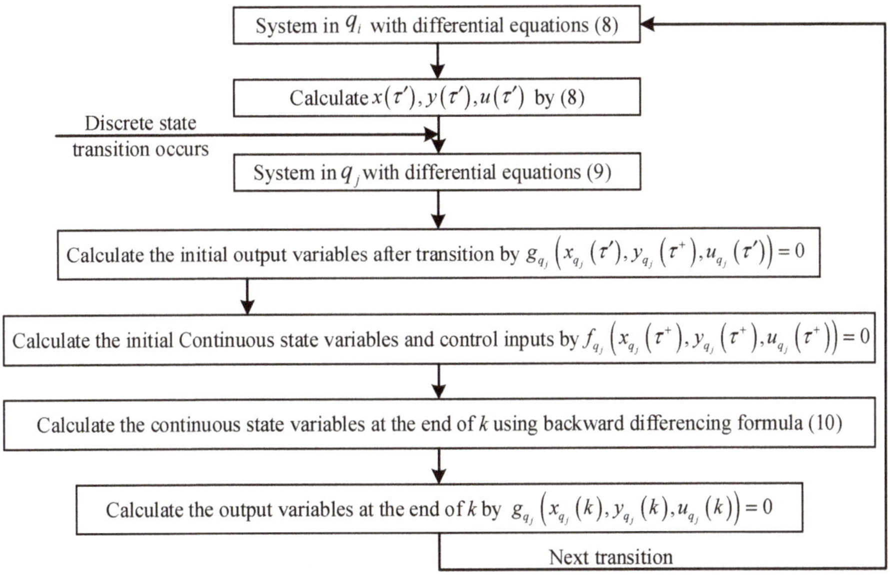

Figure 2. Initial condition calculation flow chart when the discrete state transition occurs.

After the above steps, the system simulation can continue until another discrete state transition occurs.

3.2. Continuous Dynamic Model Reconstruction and Solving after Discrete State Transition

Basically, besides reformulating the differential equations of the system and resetting the initial conditions when the discrete state transition occurs, simulation of hybrid systems only requires solving the differential equations under the current discrete state.

First, differential equations set for each zone are formed by differential equations for all components in that zone with their input–output relationships. Detailed differential equations for each component can be found in Ref. [20], including the power generation module, propulsion module, power conversion module, energy storage module, loads, and power distribution module. These equations cover the parameters, state variables, control inputs, and outputs of each component. Based on the current discrete state of the zone, the operating status of each component is determined. Then, we can construct the differential equations for the zone by connecting the input–output relationships of these components based on the internal distribution network, as shown by $f_Z(x, y, u)$ in Figure 3.

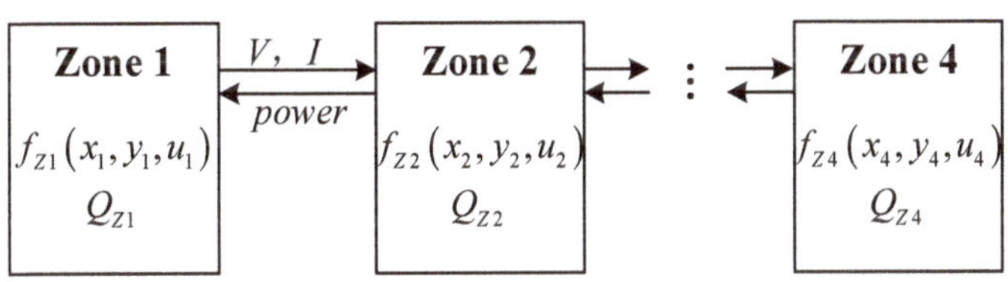

Figure 3. Inter-dependency of the zones.

Then, the global hybrid model can be formed by (7). In this paper, the proposed global hybrid model is a loosely distributed structure under weak restrictions. Each zone can execute its internal operations and calculate the continuous dynamics driven by discrete events. Figure 3 provides the inter-zone power flow relationships under the zonal distribution network. In this way, the hybrid model proposed here can be easily applied to further distributed control and management studies.

Finally, the generalized discretization method (fourth-order Runge–Kutta method) is used to solve the differential equations. The whole simulation is implemented in MATLAB platform.

4. Simulation and Results Analysis

In order to verify the hybrid modeling and simulation method proposed in this paper, a typical SPS shown in Figure 1 is used here to build a hybrid model and perform simulation cases. Table 1 displays the component specifications of the SPS under study, whose detailed parameters can be found in Ref. [28].

Table 1. Components specification in benchmark SPS.

Components	Zone 1	Zone 2	Zone 3	Zone 4
PGM	ATG 4MW	MTG 36MW	MTG 36MW	ATG 4MW
PM		36MW	36MW	
Pulse Load	20MW	1MW	1MW	
Zonal Load	2MW	2MW	2MW	2MW

The discrete states of SPS are determined by the working status of all components and the connection status of power cables. Due to the large number of power cables in the system, here, we only provide the possible discrete states of all components in Table 2. The discrete status of the generator set indicates whether it is online or not. The propulsion motor can be offline or operating at 50% or rated power. The zonal load and pulse load can be disconnected or supplied by port, starboard, or both port and starboard equally. The port and starboard connecting breakers can be open or closed.

Table 2. Discrete status of components in benchmark SPS.

Devices/Components		Status			
		0	1	2	3
Z1	ATG1	Offline	Online		
	PL1	Offline	Port Supply	Stbd Supply	Both Supply
	ZL1	Offline	Port Supply	Stbd Supply	Both Supply
	Brk1	Open	Close		
Z2	MTG1	Offline	Online		
	PL2	Offline	Port Supply	Stbd Supply	Both Supply
	ZL2	Offline	Port Supply	Stbd Supply	Both Supply
	PM1	Offline	50% power	Rated power	
Z3	MTG3	Offline	Online		
	PL3	Offline	Port Supply	Stbd Supply	Both Supply
	ZL3	Offline	Port Supply	Stbd Supply	Both Supply
	PM2	Offline	50% power	Rated power	
Z4	ATG2	Offline	Online		
	ZL4	Offline	Port Supply	Stbd Supply	Both Supply
	Brk2	Open	Close		

4.1. SPS Start-Up Scenario

First, consider the scenario where the ship starts up to full speed, which means that the final propulsion motor will run at its maximum power of 37 MW, and, with four zone loads, the full system load power is close to 80 MW. The discrete event sequences and their occurrence moments during the whole simulation are shown in Table 3. In addition, the operation of the pulsed load is not considered in this condition, and the EMRG will remain disconnected during the whole process.

Table 3. Discrete event sequences during startup.

ID	Time (s)	Event Description
1	0	Simulation starts; all generator sets running at rated speed
2	5	All generator sets connect to the DC bus
3	20	All zone loads online
4	30	Propulsion subsystems connect to DC bus and ship speed gradually increases to 32 knots
5	300	Simulation ends and record the results

The simulation results are shown in Figures 4–9, which provide the full-system simulation results under the starting condition, including the mechanical torque and speed of the gas turbine, DC bus voltage, output current of the generator set, electromagnetic torque and speed of the propulsion motor output, ship's speed and resistance, mechanical power output of the gas turbine, power output of the generator set, mechanical power output of the propulsion motor, and zonal loads power curves.

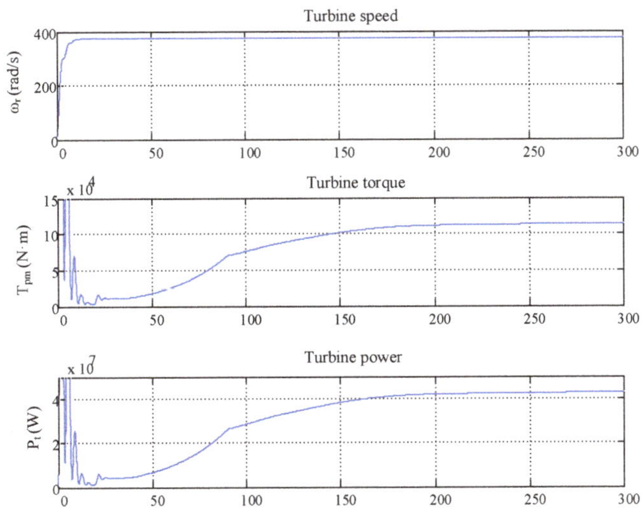

Figure 4. Case 1 gas turbine speed, torque, and power.

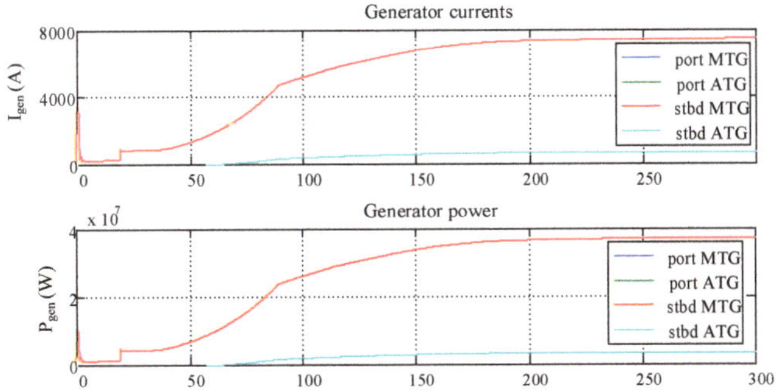

Figure 5. Case 1 generator current and power.

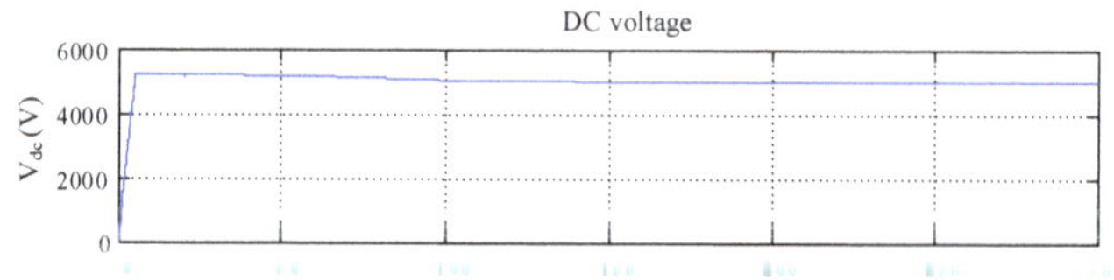

Figure 6. Case 1 DC bus voltage.

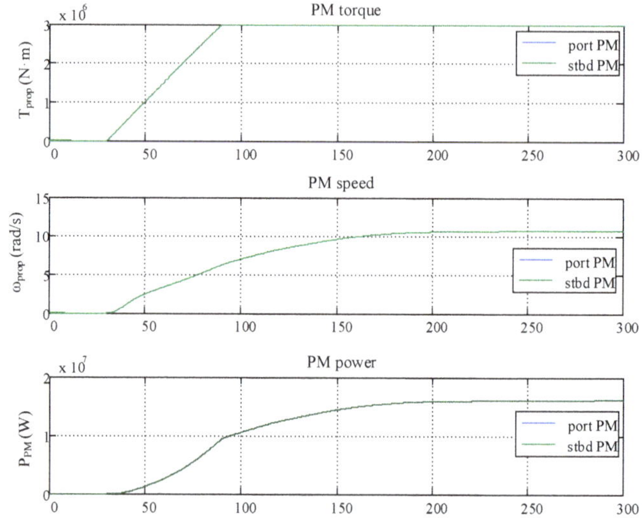

Figure 7. Case 1 PM torque, speed, and power.

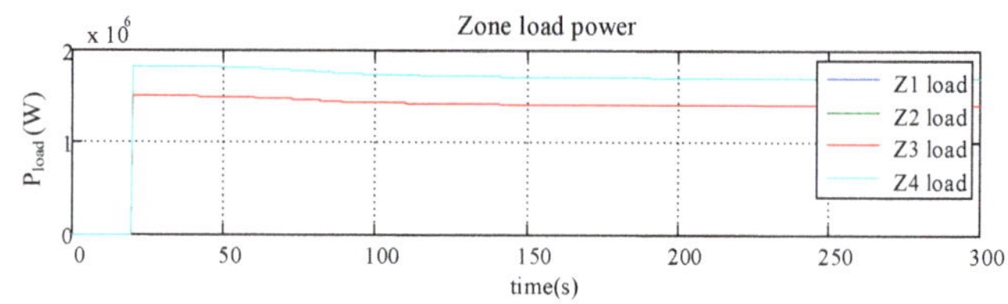

Figure 8. Case 1 zonal loads power.

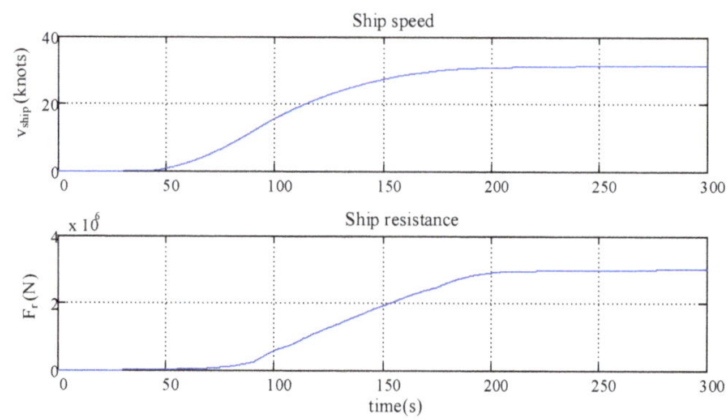

Figure 9. Case 1 ship speed and resistance.

The simulation result curve shows that, from the initial moment to 20 s, the gas turbine is first started under the action of the governor, and its speed is gradually stabilized at the rated value of 377.0 rad/s. Meanwhile, the DC bus voltage rises continuously under the action of the generator excitation controller and remains stable when it reaches the rated value of 5000 V. At 20 s, the zonal loads are online; at this time, the power of the generator sets undergoes a small increase. At 30 s, the propulsion system is connected to the DC bus, and a given electromagnetic torque is added at a speed of 50,000 N·m/s. When the torque reaches 3MN·m, it remains stable. At this time, there is a significant increase in the output current of MTG, which causes a slight fluctuation in the DC bus voltage. The electromagnetic torque of the propulsion motor rises gradually with the given value, driving the propeller to rotate, and the propeller starts to produce resistance torque and thrust for the ship to sail forward. After the electromagnetic torque of the propulsion motor reaches stability, the rotational speed also gradually reaches stability, while the ship's speed takes a longer time to stabilize due to the large inertia time constant, and it finally reaches 32 knots in about 200 s.

4.2. Pulse Load Launch Scenario

Next, consider the pulse load launch scenario. It should be noted that the pulse load of zone 1 in Figure 1 includes the energy storage module, which rises to 15 MW in 5 s and then performs pulse load firing with the whole pulse period of 6 s. This pulse load is powered by the port and starboard together. The pulse power curve is shown in Figure 10, whose load demand will rise to 15 MW in 3.33 s and fall at a rate of 50 MW/s. The discrete event sequence and its occurrence moment during the whole simulation are shown in the Table 4. The gas turbine is controlled by the governor to drive the synchronous generator to always run at a constant speed, and the zonal loads and radar are always kept online with an initial ship speed of 8 knots and accelerated to 25 knots after two launches. The simulation time is 200 s.

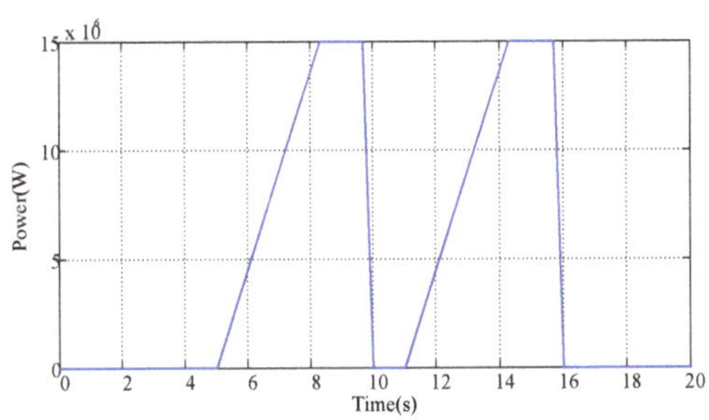

Figure 10. EMRG operation power profile.

Table 4. Discrete event sequences during pulse load launch.

ID	Time (s)	Event Description
1	0	Simulation starts, ship speed 8 knots
2	5	EMRG starts charging
3	10	EMRG launches
4	11	EMRG charging again
5	16	EMRG launches again
6	17	EMRG disconnects, ship accelerates to 25 knots
7	200	Simulation ends and record the results

The simulation results are shown in Figures 11–16. They also provide the mechanical torque and speed of the gas turbine, DC bus voltage, output current of the generator set, electromagnetic torque and speed of the propulsion motor output, ship's speed and resistance, mechanical power output of the gas turbine, power output of the generator set, mechanical power output of the propulsion motor, and zonal loads power curves.

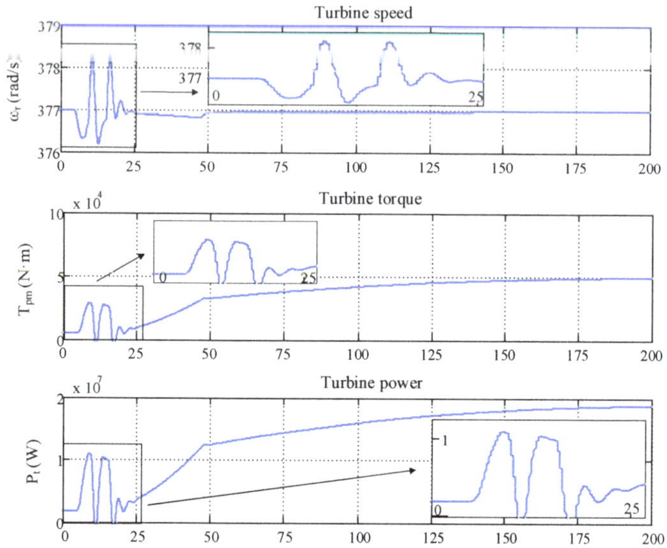

Figure 11. Case 2 gas turbine speed, torque, and power.

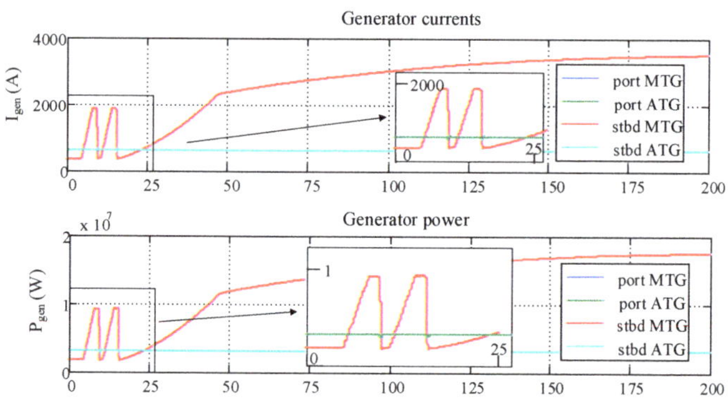

Figure 12. Case 2 generator current and power.

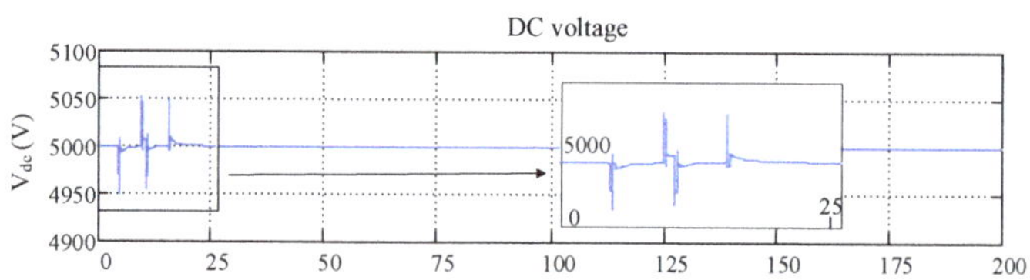

Figure 13. Case 2 DC bus voltage.

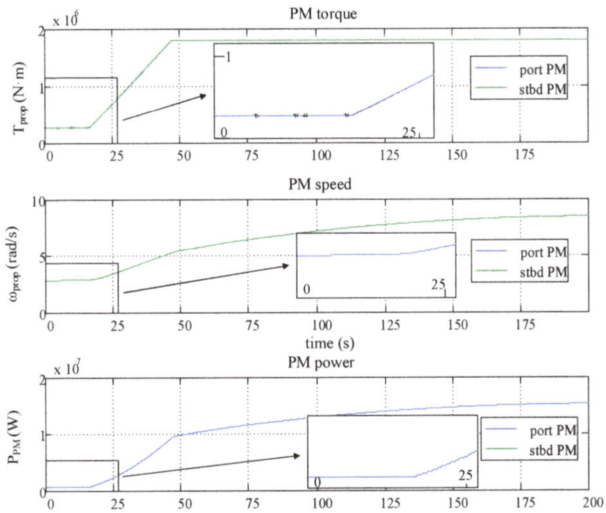

Figure 14. Case 2 PM torque, speed, and power.

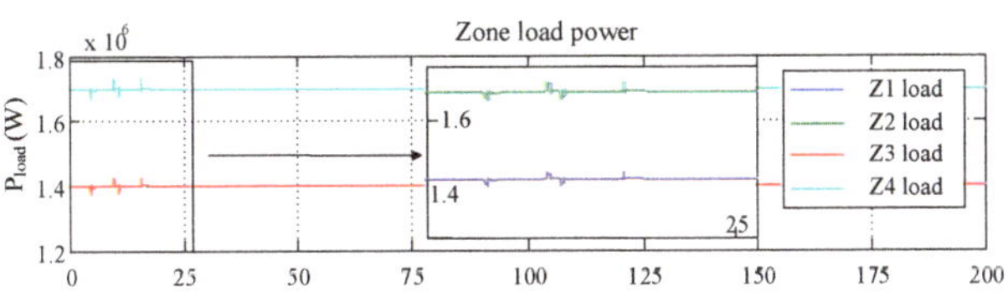

Figure 15. Case 2 zonal load power.

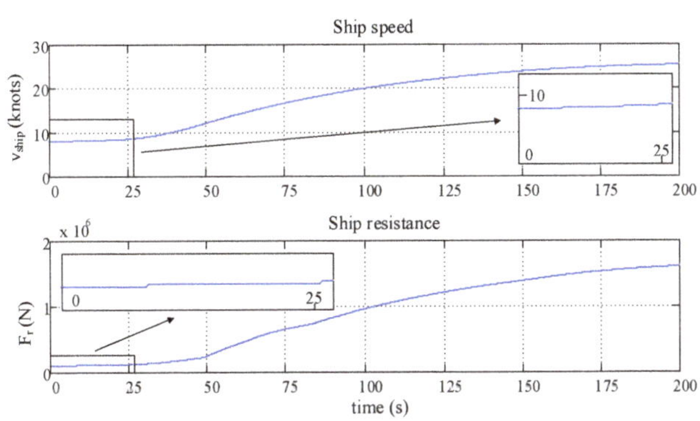

Figure 16. Case 2 ship speed and resistance.

The simulation result curves show that the system is in a steady state at the beginning, at a speed of 8 knots. Before the launch of EMRG, the energy storage device absorbs and accumulates energy from SPS, and this energy is released for pulse load during the mission. In this case, during each pulse duration, the energy storage device is charged at 15 MW for 5 s and then discharged for 1 s as the EMRG is fired. With the launch of the pulse load in 5 s, there is a small oscillation in the speed of the gas turbine. After the pulse load is put into use, its power rises to 15 MW in 3.3 s. As seen from the output power curve of the generator set, there is a sharp rise in the output power of both MTGs to meet the load power. At the same time, the DC bus voltage drops. During the charging and discharging process of the pulse load, there are four oscillations in the DC bus voltage curve, but they all return to the set value quickly. Although it is not obvious in the simulation curve, the output power of the propulsion motor also shows a slight oscillation, which is directly related to the oscillation of the DC bus voltage. After the pulse load stops working, the ship enters an acceleration phase by adding a given electromagnetic torque at 25,000 N·m/s to reach 0.9 MW. The ship speed reaches 25 knots in about 300 s.

4.3. Comparison and Discussion

Our work presents a hybrid model architecture for SPS that focuses on the continuous dynamics triggered by discrete events transition. Benefiting from the properties of the hybrid model, SPS can be treated as a discrete event system by event and control action generation. Thus, this hybrid model can accept pulse loads operation plans as discrete event sequences and perform simulation and analysis. Other modeling methods can also be used for SPS simulation but may not be suitable for the following control and energy management requirements of SPS. As we emphasized in the introduction, multiple concurrent attacks arriving randomly and the pulse loads operation plan will cause the dynamic characteristics of SPS to change drastically in a short period of time, making it important to focus on the discrete event sequences and their impact on the dynamics in each zone. This is exactly the reason why we propose this hybrid model.

In the following discussion, different modeling methods are compared from five different aspects: application SPS topology, model type, configuration, method, and case studies. The results are shown in Table 5. It can be seen that most modeling studies do not take into account the distributed feature of SPS, and they are more concerned with computational efficiency. However, it is obvious that a distributed solution will be more economical in terms of computational resources. Additionally, previous hybrid model studies have tended to focus on system reconfiguration after components failed but have not considered the impact of external attacks. As we mentioned in the introduction, pulse loads operation plans and faults caused by external attacks will have a significant impact on system dynamics, which requires attention to system dynamics affected by discrete event sequences. Among the few hybrid modeling pieces of research for SPS in the literature, our work shows its advantages in distributed and pulse loads adaptability.

Table 5. Comparison of different modeling methods.

Research	SPS Topology	Model Type	Model Configuration	Modeling Method	Case Studies
[12]	DC + Zonal	Continuous	Centralized	Simulink	Pulse load
[13]	DC + Radial	Continuous	Centralized	Simulink	Ship acceleration and deceleration
[14]	DC + Radial	Continuous	Centralized	Simulink	Ship acceleration + Overloaded + Restoration
[15]	DC + Zonal	Continuous	Centralized	RTDS	N/A
[20]	DC + Zonal	Continuous	Centralized	Simulink	Ship acceleration + Pulse load
[31]	DC + Radial	Hybrid	Centralized	Simulink	Load shedding
[32]	AC + Ring	Hybrid	Centralized	Simulink	Supervisory control
[33]	AC + Ring	Hybrid	Centralized	Simulink	Reconfiguration
Our work	DC + Zonal	Hybrid	Distributed	Simulink	Ship acceleration + Pulse load

5. Conclusions

This paper proposes a hybrid modeling method for SPS that takes into account discrete control events and continuous system dynamics. In particular, considering HPPLs integration with discrete and random attack event sequences, the hybrid model of each zone is established separately according to the zonal power distribution structure, which enables dealing with external attack events and actual damage from the perspective of zone autonomy. Through continuous states reset after discrete state transition and reconstruction and solving of continuous differential equations, simulation of the SPS hybrid model is realized. The simulation results show that the hybrid model developed in this paper can specifically describe the hybrid dynamic evolution process of SPS driven by discrete events, as well as the operating characteristics of the whole process into the confrontation phase. This research can improve the modeling theory of SPS and enrich system analysis.

As an extension of our work, one can systematically investigate the effect of cyber failures on the onboard control and communication systems, which will cause abnormal operation of SPS and result in the unobservable state of some equipment. Especially during the confrontation phase, it is probable that the system state is partially observable. State estimates, as with our previous research in Ref. [35], can be further introduced to provide a prediction of system dynamics.

Author Contributions: Conceptualization, W.Z. and C.J.; methodology, W.Z.; validation, W.Z., C.J. and Z.L.; formal analysis, C.J.; writing—original draft preparation, W.Z.; writing—review and editing, W.Z.; project administration, W.Z. All authors have read and agreed to the published version of the manuscript.

Funding: This research was funded by Foundation of National Defense Science and Technology Key Laboratory, grant number 6142217200307.

Institutional Review Board Statement: Not applicable.

Informed Consent Statement: Not applicable.

Data Availability Statement: Not applicable.

Conflicts of Interest: The authors declare no conflict of interest.

References

1. Chalfant, J.S.; Chryssostomidis, C. Analysis of various all-electric-ship electrical distribution system topologies. In Proceedings of the 2011 IEEE Electric Ship Technologies Symposium, Alexandria, VA, USA, 10–13 April 2011; pp. 72–77. [CrossRef]
2. Doerry, N.; McCoy, K. *Next Generation Integrated Power System: NGIPS Technology Development Roadmap*; No. SER-05D/349; Naval Sea Systems Command: Washington, DC, USA, 2007.
3. Doerry, N. Naval Power Systems: Integrated power systems for the continuity of the electrical power supply. *IEEE Electrif. Mag.* **2015**, *3*, 12–21. [CrossRef]
4. Doerry, N.; Amy, J. Design considerations for a reference MVDC power system. *Proc. SNAME Marit. Conv.* **2016**, *124*, 40–59.
5. Zohrabi, N.; Shi, J.; Abdelwahed, S. An overview of design specifications and requirements for the MVDC shipboard power system. *Int. J. Electr. Power Energy Syst.* **2019**, *104*, 680–693. [CrossRef]
6. Domaschk, L.N.; Ouroua, A.; Hebner, R.E.; Bowlin, O.E.; Colson, W.B. Coordination of Large Pulsed Loads on Future Electric Ships. *IEEE Trans. Magn.* **2007**, *43*, 450–455. [CrossRef]
7. Clayton, D.H.; Jebsen, G.M.; Sofia, J.M. The All Electric Warship from Vision to Total Ship System Integration. *Nav. Surf. Warf. Cent. Dahlgren Div.* **2002**, *86*, 1–9.
8. Smart, R.; Chalfant, J.; Herbst, J.; Langland, B.; Card, A.; Leonard, R.; Gattozzi, A. Using S3D to analyze ship system alternatives for a 100 MW 10,000 ton surface combatant. In Proceedings of the 2017 IEEE Electric Ship Technologies Symposium (ESTS), Arlington, VA, USA, 15–17 August 2017; pp. 96–103. [CrossRef]
9. McCoy Timothy, J.; John, V. The state-of-the-art of integrated electric power and propulsion systems and technologies on ships. In Proceedings of the 2009 IEEE Electric Ship Technologies Symposium, Baltimore, MD, USA, 20–22 April 2009.
10. Trinklein, E.H.; Parker, G.G.; McCoy, T.J. Modeling, optimization, and control of ship energy systems using exergy methods. *Energy* **2020**, *191*, 116542. [CrossRef]
11. Weaver, W.W.; Robinett, R.D.; Wilson, D.G.; Matthews, R.C. Metastability of pulse power loads using the hamiltonian surface shaping method. *IEEE Trans. Energy Convers.* **2017**, *32*, 820–828. [CrossRef]
12. Shi, J.; Amgai, R.; Abdelwahed, S. Modelling of shipboard medium-voltage direct current system for system level dynamic analysis. *IET Electr. Syst. Transp.* **2015**, *5*, 156–165. [CrossRef]
13. Zahedi, B.; Norum, L.E. Modeling and Simulation of All-Electric Ships With Low-Voltage DC Hybrid Power Systems. *IEEE Trans. Power Electron.* **2013**, *28*, 4525–4537. [CrossRef]
14. Shajari, Z.; Savaghebi, M.; Guerrero, J.M.; Javidi, M.H. Dynamic Performance Assessment of NG-MVDC Shipboard Power System with Distributed Electric Propulsions. In Proceedings of the 2020 IEEE Electric Power and Energy Conference (EPEC), Edmonton, AB, Canada, 14–16 October 2020; pp. 1–7. [CrossRef]
15. Andrus, M. NGIPS MVDC Baseline Architecture Definition-RTDS Implementation. Center for Advanced Power Systems, Florida State University. 2010. Available online: https://www.esrdc.com/library (accessed on 23 August 2022).
16. Lahiri, S. *Modeling and Simulation of Shipboard Integrated Power Systems*; Drexel University: Philadelphia, PA, USA, 2011.

17. Skjong, S.; Pedersen, E. A real-time simulator framework for marine power plants with weak power grids. *Mechatronics* **2017**, *47*, 24–36. [CrossRef]
18. Abdelwahed, S.; Asrari, A.; Crider, J.; Dougal, R.A.; Faruque, M.O.; Fu, Y.; Langston, J.; Lee, Y.; Mohammadpour, H.A.; Ouroua, A.; et al. Reduced order modeling of a shipboard power system. In Proceedings of the 2013 IEEE Electric Ship Technologies Sympo-sium (ESTS), Arlington, VA, USA, 22–24 April 2013; pp. 256–263. [CrossRef]
19. Park, D.; Zadeh, M.K. Dynamic Modeling and Stability Analysis of Onboard DC Power System for Hybrid Electric Ships. In Proceedings of the 2019 IEEE Transportation Electrification Conference and Expo (ITEC), Novi, MI, USA, 19–21 June 2019; pp. 1–6. [CrossRef]
20. Zhu, W.; Shi, J.; Abdelwahed, S. End-to-end system level modeling and simulation for medium-voltage DC electric ship power systems. *Int. J. Nav. Archit. Ocean. Eng.* **2018**, *10*, 37–47. [CrossRef]
21. Ghimire, P.; Reddy, N.P.; Zadeh, M.K.; Pedersen, E.; Thorstensen, J. Dynamic Modeling and Real-Time Simulation of a Ship Hybrid Power System Using a Mixed-Modeling Approach. In Proceedings of the 2020 IEEE Transportation Electrification Conference & Expo (ITEC), Chicago, IL, USA, 22–26 June 2020; pp. 1–6. [CrossRef]
22. Yang, Y.; Zhengming, Z.; Tan, T.; Boyang, L.; Liqiang, Y. Discrete State Event Driven Method and Self-Adapted Pre-dictor-Corrector Algorithm. *Trans. China Electrotech. Soc.* **2017**, *32*, 33–41.
23. Bochen, S.; Zhengming, Z.; Yicheng, Z.; Zhujun, J. Discrete-state Event-driven Simulation Approach for Multi-time-scale Power Electronic Hybrid System. *Proc. CSEE* **2021**, *41*, 2980–2989.
24. Khan, M.S.; Iravani, M.R. Supervisory Hybrid Control of a Micro Grid System. In Proceedings of the 2007 IEEE Canada Electrical Power Conference, Montreal, QC, Canada, 25–26 October 2007; pp. 20–24. [CrossRef]
25. Sun, L.Y.; Zhang, P.F.; Lv, J.J. Design of Dual-mode Control for Microgrid Inverter Based on Switching System. *Control. Eng. China* **2021**, *28*, 1567–1579.
26. Jiawei, H.; Tong, W.; Zengping, W.; Jiuliang, L.; Jingtian, B. Switching System's MLE Based Transient Stability Assessment of AC/DC Hybrid System Considering Continuous Commutation Failure. *IEEE Trans. Power Syst.* **2021**, *36*, 757–768.
27. Ke, C.; Yiguo, L.; Junli, Z.; Songlin, C.; Honghai, N.; Bing, L. Dynamic economic dispatch for CHP-MG system based on mixed logical dynamic model and MPC method. In Proceedings of the IEEE 2020 39th Chinese Control Conference (CCC), Shenyang, China, 27–29 July 2020; pp. 1630–1636.
28. Chamorro, H.R.; Pazmino, C.; Paez, D.; Jimenez, F.; Guerrero, J.M.; Sood, V.K.; Martinez, W. Multi-agent Control Strategy for Microgrids using Petri Nets. In Proceedings of the 2020 IEEE 29th International Symposium on Industrial Electronics (ISIE), Delft, Netherlands, 17–19 June 2020; pp. 1141–1146. [CrossRef]
29. Lu, X.; Zhou, M.; Ammari, A.C.; Ji, J. Hybrid Petri nets for modeling and analysis of microgrid systems. *IEEE/CAA J. Autom. Sin.* **2016**, *3*, 349–356. [CrossRef]
30. Silva, D.P.; Queiroz, M.D.; Fardin, J.F.; Sales, J.L.F.; Orlando, M.T.D. Hybrid modeling of energy storage system and electrical loads in a pilot-microgrid. In Proceedings of the 2018 13th IEEE International Conference on Industry Applications (INDUSCON), São Paulo, Brazil, 11–14 November 2018; pp. 433–438. [CrossRef]
31. Babaei, M.; Shi, J.; Zohrabi, N.; Abdelwahed, S. Development of a hybrid model for shipboard power systems. In Proceedings of the 2015 IEEE Electric Ship Technologies Symposium (ESTS), Alexandria, VA, USA, 21–24 June 2015; pp. 145–149. [CrossRef]
32. Lahiri, S.; Niebur, D.; Kwatny, H.; Bajpai, G.; Beytin, A.; Patel, J.; Kang, R. A software tool for automated management and supervisory control of shipboard Integrated Power Systems. In Proceedings of the 2012 IEEE Power and Energy Society General Meeting, 22–26 July 2012; pp. 1–7. [CrossRef]
33. Yasar, M.; Beytin, A.; Bajpai, G.; Kwatny, H.G. Integrated Electric Power System supervision for reconfiguration and damage mitigation. In Proceedings of the 2009 IEEE Electric Ship Technologies Symposium, Baltimore, MD, USA, 20–22 April 2009; pp. 345–352. [CrossRef]
34. Zhu, W.; Liang, Z.; Zhu, Z.; Zhi, P. A Distributed Hybrid Control Framework for Shipboard Power System Reconfiguration. In Proceedings of the 2021 4th IEEE International Conference on Industrial Cyber-Physical Systems (ICPS), Victoria, BC, USA, 10–12 May 2021; pp. 769–774. [CrossRef]
35. Zhengzhuo, L.; Wanlu, Z.; Jian, S.; Zhiyu, Z.; Pengfei, Z. Ship Integrated Power System reconfiguration research under partial observation. *Energy Rep.* **2022**, *8*, 444–452.

Journal of
Marine Science and Engineering

MDPI

Article

Study of the LQRY-SMC Control Method for the Longitudinal Motion of Fully Submerged Hydrofoil Crafts

Hongdan Liu, Yunxing Fu and Bing Li *

College of Intelligent Science and Engineering, Harbin Engineering University, Harbin 150001, China
* Correspondence: libing265@hrbeu.edu.cn

Abstract: The control system is one of the important components of the hydrofoil craft. By adjusting the navigation attitude of the craft, the hydrofoil craft can navigate stably and safely in the turbulent environment. Aiming at the problem that existing control algorithms have poor stability in the longitudinal motion control of hydrofoil craft, the longitudinal motion reduction is limited, and there are excessive requirements for accurate disturbance wave data. Based on the fully submerged hydrofoil craft model, this article proposes a joint control method LQRY-SMC combining linear-quadratic optimal control with output regulation (LQRY) and sliding-mode control (SMC), and adds genetic algorithm to optimize the weighting matrix parameters, get better control-feedback gain, improve the global optimal-control stability, thus improving the comfort of the crew, and prevent the attack of the hull, deck wetness and damage to instruments. The simulation results show that compared with the existing methods, the heave displacement and pitch angle obtained by LQRY-SMC under the turbulent flow of different significant wave heights are reduced by about 50%, and the influence of longitudinal motion on hydrofoil crafts is avoided to a large extent, which proves the effectiveness and superiority of the method proposed.

Keywords: fully submerged hydrofoil craft; longitudinal movement; LQRY-SMC control method; wave disturbance

Citation: Liu, H.; Fu, Y.; Li, B. Study of the LQRY-SMC Control Method for the Longitudinal Motion of Fully Submerged Hydrofoil Crafts. *J. Mar. Sci. Eng.* **2022**, *10*, 1390. https://doi.org/10.3390/jmse10101390

Academic Editors: Marco Altosole, Maria Acanfora, Flavio Balsamo and Bowen Xing

Received: 2 September 2022
Accepted: 27 September 2022
Published: 29 September 2022

1. Introduction

The fully submerged hydrofoil craft is a new type of high-performance ship. When sailing at high speed, the hydrofoil installed at the bottom of the craft body is used to generate hydrodynamic lift to lift part or all of the craft body off the water surface, so as to reduce the resistance and improve the sailing speed. The maximum speed can reach more than 50 knots. Compared with traditional ships, it has the characteristics of low resistance, high speed and good seakeeping. The appearance of hydrofoil crafts enables people to save time at sea and obtain a more comfortable shipping experience. However, under the interference of sea waves, hydrofoil crafts will inevitably produce heave and pitch motion, which seriously affects the comfort and work efficiency of the crew. On the other hand, the intense longitudinal movement will cause damage to the onboard instruments and equipment, increase the probability of the attack of the hull, deck wetness, and even hydrofoil out of the water, bring danger to navigation [1].

In order to make the hydrofoil craft navigate stably and safely in the turbulent environment, an important way is to add a control algorithm to the longitudinal motion system of the hydrofoil craft to reduce the heave displacement and pitch angle to a certain extent. By controlling the control surface of the hydrofoil trailing edge flap of the hydrofoil craft, the lift of the hydrofoil craft can be continuously adjusted to offset the interference of the waves and improve the seakeeping performance of the hydrofoil craft. At present, the most commonly used control methods of hydrofoil crafts are based on the improved PID algorithm or classical robust control, and for this multi-input and multi-output nonlinear strong coupling system with uncertain disturbance and unknown parameter disturbance, the above method has poor application effect, large error and cannot meet the requirements. Therefore, many scholars convert the nonlinear model into a linear model for research [2–4]. The longitudinal motion control system of ships can be traced back to the 1950s. Matdaud Z. summarized the key technologies used by the United States and the Soviet Union to control the longitudinal motion of ground-effect ships to stabilize ships in the past, and classified and summarized the control system [5]. Haiye Hu et al. found that the active hydrofoil structure has better stability effect than the fixed hydrofoil in regular waves, and the stability effect decreases with the increase of sea conditions in irregular waves [6]. The longitudinal motion of the high-speed catamaran stability control model also provides a reference for the study of the automatic control system of hydrofoil crafts. Sang Hyun Kim combines LQR controller and Kalman filter to form a totally submerged hydrofoil model and control system. The final results show that it is very effective in still water, but it is not effective under the influence of sea waves, because it cannot reduce the impact of wave track motion and hydrofoil lift change [7]. An optimized preview servo system is designed for the problem, so that the stability of the control system can be enhanced in regular and irregular waves [8]. Hongli Chen et al. designed a PID controller capable of intelligent adaptive interference compensation by using backstepping and online calculation based on generalized dynamic fuzzy neural network, which can greatly reduce the output error of heave displacement and pitch angle, and have a certain control effect on longitudinal attitude [9], and can even use reinforcement learning to make the effect comparable to that of a PID controller [10]. Inspired by the aerospace vehicle [11], the unmanned aerial vehicle [12,13] and the underwater vehicle [14], Sheng Liu et al. proposed an improved adaptive complementary sliding mode controller with disturbance observer, and proved the stability of the system with the Lyapunov stability theory. The improved sliding surface can attenuate the switching gain and maintain the interference reflection performance, and can stabilize the longitudinal motion of the hydrofoil craft with small stability error and fast response [15–17]. Jangwhan Bai compared the advantages and disadvantages of the above three control methods under the same state space equation and gave a conclusion [18].

Based on the analysis of the above documents, the following problems exist: PID control is sensitive to the wave height in irregular waves, and there is even a large instability factor. LQR control is applicable in various wave environments. The algorithm uses small control inputs and can attenuate the motion, but it still cannot make the heave displacement and pitch angle reach the longitudinal stable state. Sliding-mode control provides the maximum reduction of quasi-static motion. However, when the disturbance conditions are uncertain, the performance of sliding mode control is greatly affected by the wave environment. In addition, it also needs a larger flap angle of front and rear hydrofoil than LQR control.

In the field of control, not only can separate control methods be used, but also they can be combined.Subsequently, some scholars have proposed a sliding mode controller based on LQR sliding surface for the balance control of the rotating two-stage inverted pendulum (RDIP) system. The sliding surface is designed based on LQR optimal gain. Under external interference and model and parameter uncertainty, LQR-SMC can maintain the stability of the system and obtain better performance than using them alone [19–21], wwhereas the LQR system considers only the size of the system state and the control quantity. For the model of hydrofoil craft, the output quantity needs to be taken as one of the performance indexes to form a linear quadratic state feedback regulator (LQRY). In document [22], it is also mentioned that LQRY has smaller overshoot and shorter regulation time.

It is found from the above documents that under-the-sea conditions with significant wave height less than 3 m, the heaving displacement amplitude of the current best control method is about 0.5 m, and the pitch angle amplitude is about 4°, which can maintain the basic safety of navigation, but still cannot meet the requirements of stability. Therefore, this article will use LQRY-SMC joint control to reduce the heave displacement and pitch angle of the hydrofoil craft, which can enhance the robustness of the system, greatly reduce the longitudinal motion amplitude and maintain the longitudinal motion stability.

The structure of this article is as follows: In Section 2, the disturbance force and moment of the submerged hydrofoil boat and irregular waves are modeled. Section 3 introduces the control methods used in this article and the algorithm improvement for LQRY optimization. In Section 4, the software is used to establish the simulation model and get the corresponding data for comparison. Finally, Section 5 summarizes the research work of this article. The flow chart of this article is shown in Figure 1.

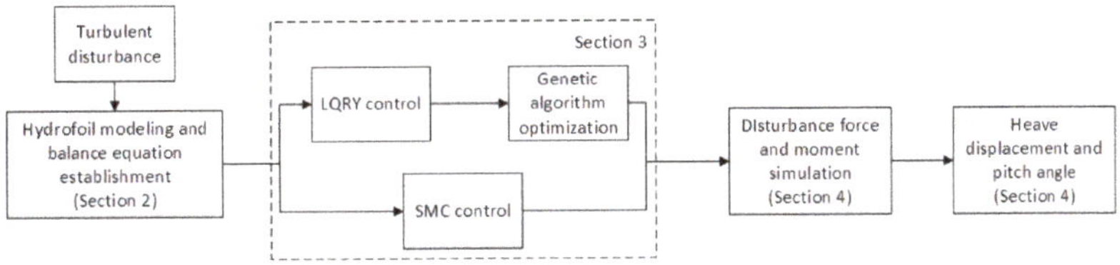

Figure 1. The flow chart of this article.

2. Mathematical Model of Longitudinal Motion

2.1. Construction of Longitudinal Model of Hydrofoil Craft

Hydrofoil craft is composed of four parts: hull, hydrofoil, propulsion system and control system. This article is based on PCH-1 made in the United States [23,24]. It is a fully submersible seaplane designed by the United States, which is suitable for marine conditions and meets military standards. The duckbill configuration is adopted and a control flap is used to control the heave of the fully submerged hydrofoil craft. A hydrofoil with a larger aspect ratio is used as the rear hydrofoil to provide more lift. The control flap is connected to it to control the pitch and heave attitude together with the front flap. The three views of the hydrofoil craft model and the hydrofoil-flap assembly drawing used are shown in Figures 2 and 3 respectively.

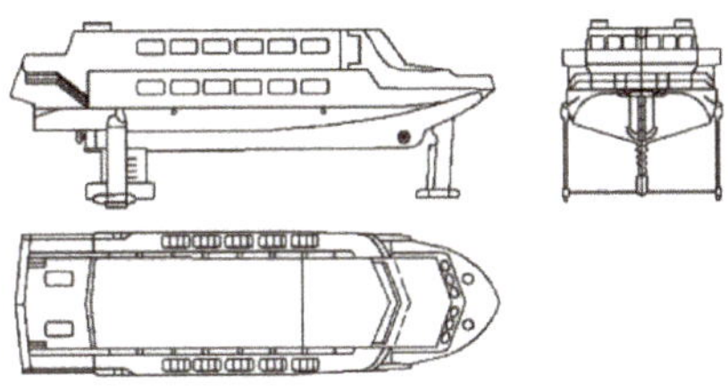

Figure 2. Three views of hydrofoil.

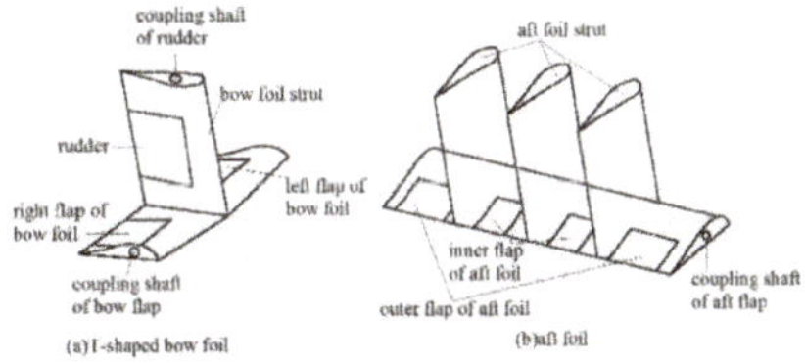

Figure 3. Hydrofoil-flap assembly drawing.

In order to comprehensively study the motion of a ship, it is necessary to establish a fixed coordinate system and a moving coordinate system [25]. The coordinate system fixed on the earth is also called the fixed coordinate system, and O is the origin of the coordinate system fixed on the earth's surface. The X-axis is in the still water plane, and it can usually be selected as the general motion direction of the ship. The Y-axis is selected as the direction in which the X-axis rotates clockwise in the hydrostatic plane. The Z-axis is perpendicular to the hydrostatic plane and points to the earth center. The coordinate system fixed to the ship is also called the moving coordinate system. The origin G of the coordinate system is usually taken at the center of gravity of the ship and moves with the ship. The x-axis is taken as the longitudinal section perpendicular to the midship and pointing to the bow. The y-axis is perpendicular to the midship cross section and points to the right chord. The z-axis is perpendicular to the waterline plane and points to the keel. In ship motion, there are usually six degrees of freedom, of which three degrees of freedom are longitudinal motions, namely, surge, pitch and heave. This article mainly studies pitch and heave, and does not consider surge. As shown in Figure 4, it is the coordinate system of fully submerged hydrofoil craft.

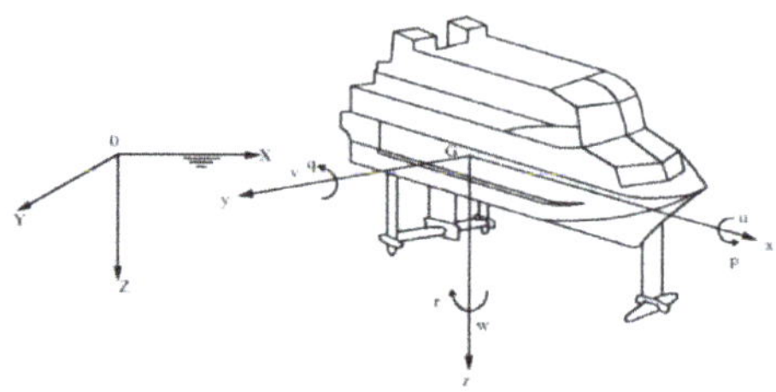

Figure 4. Definition of hydrofoil coordinate system.

Because of the irregularity of the hull shape and the randomness of the sea state, the hydrofoil craft is assumed to move in an infinite uniform flow field. The mathematical expressions of heave and pitch are as follows [26]:

$$\begin{cases} m(\ddot{h}+U_e\dot{\theta}) = Z_f + Z_c + Z_w \\ I_{yy}\ddot{\theta} = M_f + M_c + M_w \end{cases} \tag{1}$$

where m is the mass of the hydrofoil craft, h is the heave displacement, U_e is the ship speed, θ is the pitch angle, Z_f, M_f is the lift and moment of the hydrofoil, Z_c, M_c is the control force and control moment, and Z_w, M_w is the disturbance force and disturbance moment of the sea wave. The expansion of formula (1) can be written as follows:

$$\begin{cases} m(\ddot{h}+U_e\dot{\theta}) = \sum\limits_{i=1}^{2}(F_{fi}+F_{fpi}) + mg\cos\theta + Z_w \\ I_{yy}\ddot{\theta} = -\sum\limits_{i=1}^{2}(x_{fi}-x_G)(F_{fi}+F_{fpi}) + M_w \end{cases} \tag{2}$$

where F_{fi} is the force generated by the hydrofoil, I_{yy} is the moment of inertia of the hull, and x_{fi}, x_G is the distance from the hydrofoil and the center of gravity to the center of the ship. The symbol of x_{fi}, x_G is determined by the position of the relative stress action point in the ship. The "+" sign is taken before and "−" sign is taken after. When $i = 1$, it is related to the front wing, and $i = 2$, it is related to the rear wing.

Then rewrite the above formula into the following form to obtain:

$$\begin{cases} Z(\ddot{h},\dot{h},h,\ddot{\theta},\dot{\theta},\theta) = m(\ddot{h}+U_e\dot{\theta}) - \sum\limits_{i=1}^{2}(F_{fi}+F_{fpi}) - mg\cos\theta - Z_w = 0 \\ M(\ddot{h},\dot{h},h,\ddot{\theta},\dot{\theta},\theta) = I_{yy}\ddot{\theta} + \sum\limits_{i=1}^{2}(x_{fi}-x_G)(F_{fi}+F_{fpi}) - M_w = 0 \end{cases} \tag{3}$$

Linearize the left end of the equation and substitute the parameters of PCH hydrofoil craft to obtain:

$$\begin{cases} Z_{\ddot{h}}\ddot{h} + Z_{\dot{h}}\dot{h} + Z_h h + Z_{\ddot{\theta}}\ddot{\theta} + Z_{\dot{\theta}}\dot{\theta} + Z_\theta\theta = -Z_{\delta_e}\delta_e - Z_{\delta f}\delta_f - Z_w \\ M_{\ddot{h}}\ddot{h} + M_{\dot{h}}\dot{h} + M_h h + M_{\ddot{\theta}}\ddot{\theta} + M_{\dot{\theta}}\dot{\theta} + M_\theta\theta = -M_{\delta_e}\delta_e - M_{\delta f}\delta_f - M_w \end{cases} \tag{4}$$

$$\begin{cases} \ddot{h} + 6.06\dot{h} + 0.338h + 3.41\ddot{\theta} + 42.4\dot{\theta} + 454\theta = -51.5\delta_e - 62.9\delta_f - Z_w \\ 0.016\ddot{h} + 0.069\dot{h} + \ddot{\theta} + 8.45\dot{\theta} + 0.654\theta = 4.58\delta_e - 1.88\delta_f - M_w \end{cases} \quad (5)$$

In modern control theory, state feedback and matrix operation are adopted to convert the above differential equation into the form of state equation, so the continuous state equation of longitudinal motion is:

$$\begin{bmatrix} \ddot{h} \\ \ddot{\theta} \end{bmatrix} = \begin{bmatrix} -0.6079 & -6.412 & -478.024 & -14.3747 \\ 0.0791 & 0.1033 & 7.0452 & -8.2187 \end{bmatrix} \begin{bmatrix} h \\ \dot{h} \\ \theta \\ \dot{\theta} \end{bmatrix} + \begin{bmatrix} -71.016 & -59.7706 \\ 5.7234 & -0.9177 \end{bmatrix} \begin{bmatrix} \delta_e \\ \delta_f \end{bmatrix} + \begin{bmatrix} 1.0577 & -3.6069 \\ 0.017 & 1.0577 \end{bmatrix} \begin{bmatrix} Z_w \\ M_w \end{bmatrix} \quad (6)$$

2.2. Force analysis of Hydrofoil

The hydrofoil installed at the bottom is the biggest difference between hydrofoil crafts and other ships, which is the key to ensure its stable navigation. When traveling at a certain speed, the pressure difference caused by the flow velocity difference between the upper and lower hydrofoil plates will generate buoyancy, which will make the hull come completely out of the water, reduce the resistance, improve the speed, and provide better stability. The plane geometric structure and wing section structure of the hydrofoil are shown in Figures 5 and 6.

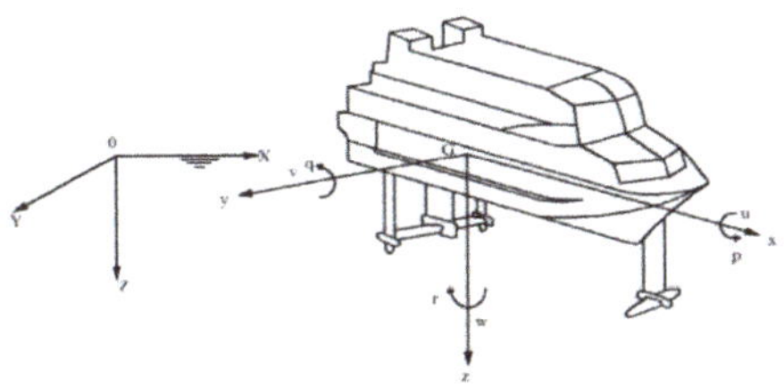

Figure 5. Plane geometry of hydrofoil.

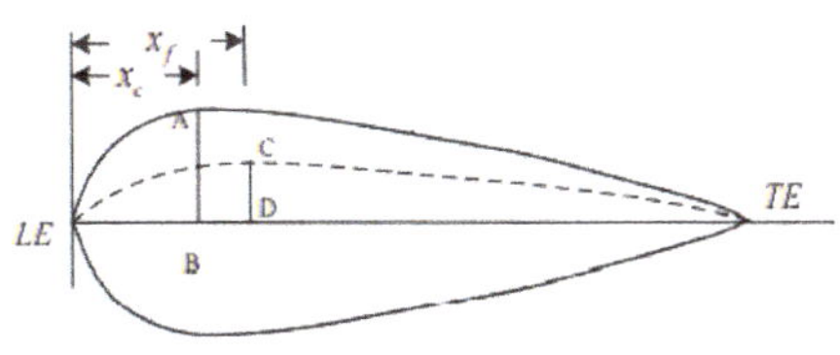

Figure 6. Wing section structure drawing.

In Figure 5, where b is the wingspan, l_r is the tip chord, l_t is the root chord, the average chord length is l, and the aspect ratio is $\lambda = b/l$. In Figure 6, the section along the inflow direction is called the wing section. The foremost point LE on the wing section is the leading edge, and the rearmost point TE is the trailing edge. AB is the maximum thickness of the wing section and the distance from AB to the leading edge is x_c. The longest line segment CD is the maximum camber, and the distance from CD to the leading edge is x_f.

Next, the force analysis is carried out. The characteristic that the hydrofoil can generate lift is also called the hydrodynamic characteristic. The complete force analysis of the hydrofoil during the navigation of the hydrofoil craft is shown in Figure 7 [27].

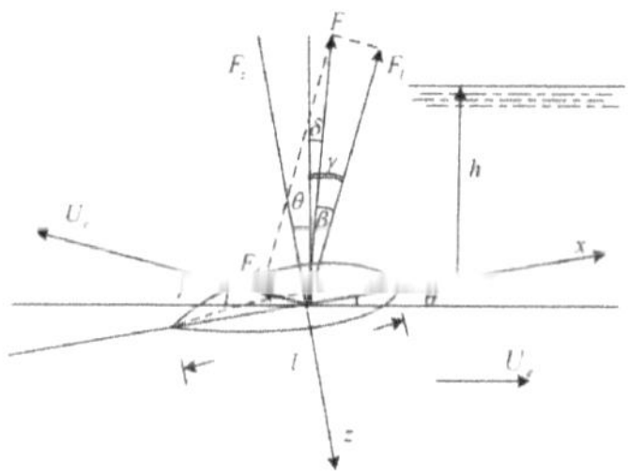

Figure 7. Force analysis of hydrofoil.

In the figure, F is the resultant force of the hydrofoil, F_l is the lift of the hydrofoil, and F_d is the resistance of the hydrofoil. δ is the angle between the resultant force and the vertical line of the water surface, θ is the angle between the hydrofoil plane and the horizontal direction, β is the angle between the lift force and the resultant force, γ is the angle between the lift and the vertical water surface line. Lift, drag and lift drag ratio can be expressed as follows:

$$F_l = \frac{1}{2} C_L \rho {U_e}^2 S \tag{7}$$

$$F_d = \frac{1}{2} C_D \rho {U_e}^2 S \tag{8}$$

$$K = \frac{F_l}{F_d} = \frac{C_L}{C_D} \tag{9}$$

where ρ is the sea water density, U_e is the ship speed, and $S = l \cdot b$ is the area of the hydrofoil. Among them, the lift drag ratio K is one of the important parameters reflecting the performance of the hydrofoil, and it is the ratio of the lift coefficient and the drag coefficient. The lift coefficient can be obtained by two methods of test or calculation according to the hydrofoil theory. Generally, it is assumed that the thickness, camber and angle of attack of the airfoil section are small and the airfoil is deep from the water surface. The formula is as follows:

$$C_L = 2\pi(\alpha + \frac{2f}{l}) \tag{10}$$

where α is the attack angle of the hydrofoil and f is the camber. When the camber coefficient of the thin wing of the plate is approximately 0, the lift coefficient is obtained $C_L = 2\pi\alpha$.

The flap is the part that provides stable control force for the longitudinal movement of the hydrofoil craft. The control force formula generated by the flap is as follows:

$$F_f = \frac{1}{2} C_{Lf} \rho {U_e}^2 S_f \alpha_f \tag{11}$$

where, S_f is the area of the flap, α_f is the angle of attack of the flap, C_{Lf} is the lift coefficient of the flap, and its calculation method is similar to that of the hydrofoil. However, due to the limitation of the mechanical structure, the lift coefficient of the flap is linear within a certain angle, so the flap servo system needs to set the maximum angle to ensure the appropriate control force of the hydrofoil craft.

Since the waves have great randomness and complexity in time and space, they can be considered to be formed by the superposition of multiple regular waves with different wavelengths, frequencies, wave amplitudes, phases and propagation directions, and then the disturbance force and moment are analyzed. The irregular wave model is as follows:

$$\xi = \sum_{i=1}^{n} A_i \cos(\omega_{ei} t + \varepsilon_i + \psi_i) \tag{12}$$

where $n = 50$, ε_i is the ith phase randomly generated on [0, 20], and the formula of wave amplitude A_i and encounter frequency ω_{ei} is as follows:

$$A_i = (2S(\omega_i) \triangle \omega)^{\frac{1}{2}} \tag{13}$$

$$\omega_{ei} = (\frac{2\pi}{\lambda_i})(U_R \cos\chi - c) \tag{14}$$

$$\psi_i = (\frac{2\pi}{\lambda_i})(x\cos\chi + y\sin\chi) \tag{15}$$

$$\omega_i = \frac{2\pi c}{\lambda_i} \tag{16}$$

where, $S(\omega_i)$ is the P-M spectrum, λ_i is the wavelength, c is the wave velocity, χ is the encounter angle, and ω_i is the actual frequency. In this article, the P-M spectrum used in the research of ship hull, which is currently popular internationally, is used for analysis. The spectral density function is as follows [28].

$$S(\omega) = \frac{A}{\omega^5} e^{-\frac{B}{\omega^4}} \tag{17}$$

$A = 8.1 \times 10^{-3} g^2$, $B = 3.11/H_{1/3}{}^2$, $H_{1/3}$ is significant wave height, and Figure 8 is the curve under different significant wave heights.

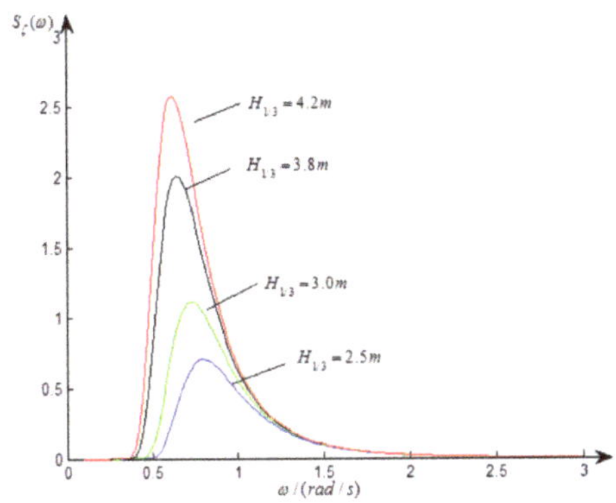

Figure 8. P-M spectra with different significant wave heights.

After analysis and collation, it is concluded that the forces and moments generated by irregular waves on the hydrofoil are:

$$Z_2 = \sum_{i=1}^{n} F_{zi} \cos(\omega_{ei}t + \phi_{zi} + \varepsilon_i) \tag{18}$$

$$M_2 = \sum_{i=1}^{n} M_{ei} \cos(\omega_{ei}t + \phi_{Mi} + \varepsilon_i) \tag{19}$$

where:

$$F_{zi} = 2\pi U_R A_i c/\lambda_i e^{-2\pi z/\lambda_i}[(K_b \cos\psi_{bi} + K_f \cos\psi_{fi})^2 + (K_b \sin\psi_{bi} + K_f \sin\psi_{fi})^2]^{\frac{1}{2}} \tag{20}$$

$$K_b = (1/2)\rho U_R{}^2 A_{fb}(\partial C_L/\partial\alpha)_b \tag{21}$$

$$\psi_{bi} = (2\pi X_b/\lambda_i)\cos\chi \tag{22}$$

$$K_f = (1/2)\rho {U_R}^2 A_{ff}(\partial C_L/\partial\alpha)_f \tag{23}$$

$$\psi_{fi} = (-2\pi/\lambda_i)(L_s - X_b)\cos\chi \tag{24}$$

$$\phi_{zi} = \arctan(-\frac{K_b\cos\psi_{bi} + K_f\cos\psi_{fi}}{K_b\sin\psi_{bi} + K_f\sin\psi_{fi}}) \tag{25}$$

$$\begin{aligned} M_{ei} \quad & = 2\pi U_R A_i c/\lambda_i e^{-2\pi z/\lambda_i}[(-X_bK_b\cos\psi_{bi} + (L_s - X_b)K_f\cos\psi_{fi})^2 \\ & +(-X_bK_b\sin\psi_{bi} + (L_s - X_b)K_f\sin\psi_{fi})^?]^{\frac{1}{2}} \end{aligned} \tag{26}$$

$$\phi_{Mi} = \arctan[-\frac{-X_bK_b\cos\psi_{bi} + (L_s - X_b)K_f\cos\psi_{fi}}{-X_bK_b\sin\psi_{bi} + (L_s - X_b)K_f\sin\psi_{fi}}] \tag{27}$$

where, X_b is the distance from the center of gravity of the hull to the front hydrofoil, and L_s is the distance between the front and rear hydrofoils. It should be noted that when the encounter frequency ω_{ei} is negative, the minus sign before Equations (25) and (27) should be removed to calculate φ_{zi} and φ_{Mi}.

3. Design of Longitudinal Motion Controller of Hydrofoil Based on LQR/LQRY-SMC

The control system has a very important role in the stable navigation of the hydrofoil craft, and is also the focus of this article. According to the changes of the sea conditions, the flap angle is adjusted to generate the required restoring force and restoring moment, reduce the longitudinal motion of the full-submerged hydrofoil craft, and realize the stabilization of the longitudinal motion attitude of the full-submerged hydrofoil craft. Generally, a closed loop is composed of the controlled object (hull), sensor, controller and actuator. The state observer is used to obtain the required state value. The controller gives the flap-command angle signal according to the attitude angle and displacement to reduce the deviation between the actual value and the expected value. The actuator is mainly composed of the servo system, the flap-mechanical structure and the flap-angle feedback measurement device. Figure 9 is a block diagram of the control system.

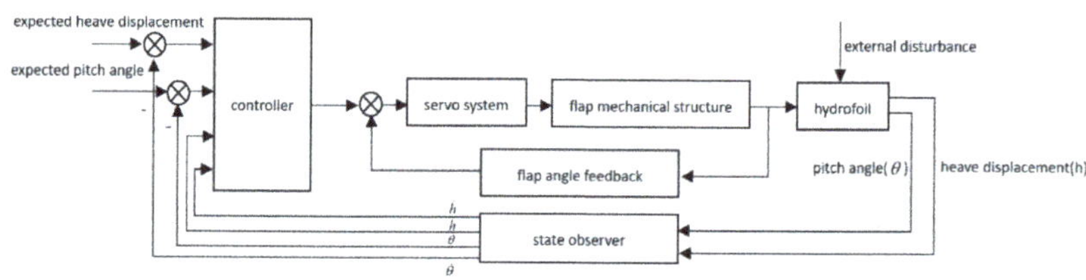

Figure 9. Structure diagram of control system.

The joint control in this article first needs to analyze the LQR control part. This method is easy to realize, and the optimal control law of state-linear feedback can be obtained. At the same time, the original system can achieve a better performance index with low cost. The performance index function *J(u)* needs to be defined to have a minimum value.

$$J(u) = \frac{1}{2}x^T(T)S(T)x(t) + \frac{1}{2}\int_0^T (x^T(t)Q(t)x(t) + u^T(t)R(t)u(t))dt \tag{28}$$

The Q and R matrices are used to adjust the input variables and state variables of the cost function to find the optimal value. If not only the influence of system state *x(t)* and control quantity *u(t)* is considered, but also the output quantity *y(t)* is introduced, it can be called LQ optimal control based on output regulation, also called LQRY control.

$$J(u) = \frac{1}{2}x^T(T)S(T)x(t) + \frac{1}{2}\int_0^T (x^T(t)Q(t)x(t) + u^T(t)R(t)u(t) + y^T(t)F(t)y(t))dt \tag{29}$$

The optimal feedback gain matrix of LQR/LQRY can be expressed as:

$$u(t) = -R^{-1}B^T Px(t) \tag{30}$$

$$K = -R^{-1}B^T P \tag{31}$$

where P is the solution of Riccati equation, the equation is as follows:

$$PA + A^T P + Q - PBR^{-1}B^T P = 0 \tag{32}$$

$$PA + A^T P + Q + C^T FC - PBR^{-1}B^T P = 0 \tag{33}$$

The above control mode can also use some optimization algorithms to further reduce the longitudinal motion of the hydrofoil craft. In this article, the genetic algorithm is used to select the appropriate fitness function to optimize the Q and R matrix [29]. The process diagram is shown in Figure 10.

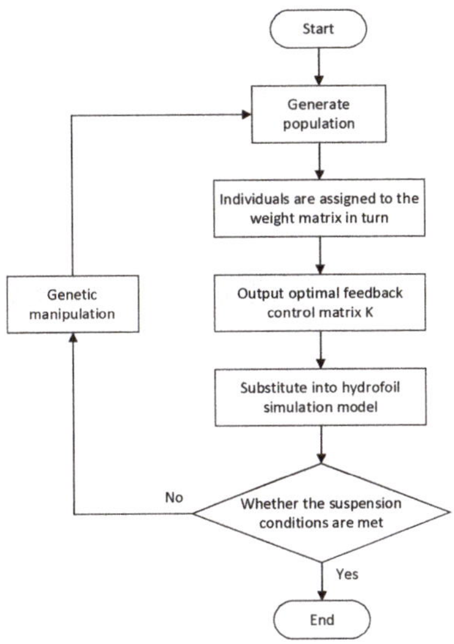

Figure 10. Schematic diagram of genetic algorithm.

Then analyze the sliding mode control part, let a system be: $\dot{x} = f(x), x \in R^n$, there is a plane $s(x) = s(x_1, x_2, \ldots, x_n) = 0$, as shown in Figure 11.

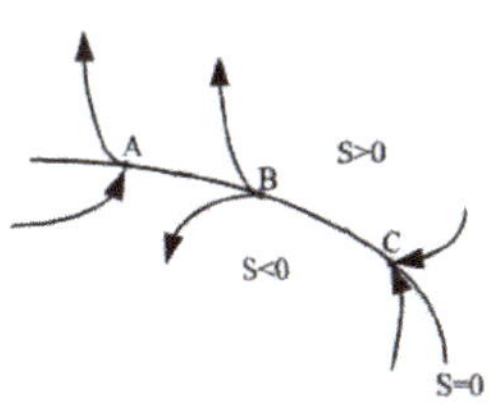

Figure 11. Sliding surface.

As can be seen from the figure, this plane divides the state space into two parts. Once the moving point in the state space enters the plane range during the movement, there are three different states, namely, the normal point A, the starting point B and the ending point C. In order to keep the system stable, it is desirable that the motion point in the plane range be the termination point, because other motion points in the state space will become the constant motion of the termination point in the region when they enter the plane range, so this region is called the sliding-mode region. From the sliding-mode analysis, it needs to meet the following requirements: $\lim_{s \to 0} s\dot{s} \le 0$.

According to formula (6), the hydrofoil craft can be regarded as the following second-order uncertain nonlinear dynamic system:

$$\begin{cases} \dot{x}_1 = x_2 \\ \dot{x}_2 = f(x) + g(x) + b(x)u \end{cases} \tag{34}$$

$f(x)$, $g(x)$ and $b(x)$ can be determined by formula (6) to obtain:

$$f(x) = \begin{bmatrix} -0.6079 & -478.024 & -6.4120 & -14.3747 \\ 0.0791 & 7.0452 & 0.1033 & -8.2187 \end{bmatrix} \begin{bmatrix} h \\ \dot{h} \\ \theta \\ \dot{\theta} \end{bmatrix} \tag{35}$$

$$b(x) = \begin{bmatrix} -71.016 & -59.7706 \\ 5.7234 & -0.9177 \end{bmatrix} \tag{36}$$

$$g(x) = \begin{bmatrix} 1.0577 & -3.6069 \\ 0.017 & 1.0577 \end{bmatrix} \begin{bmatrix} Z_w \\ M_w \end{bmatrix} \tag{37}$$

According to the characteristics of the fully submerged hydrofoil craft, the sliding-mode function selected by the sliding mode variable structure control part in this article is as follows:

$$s = x_1 + \frac{1}{\beta} {x_2}^{p/q} \tag{38}$$

where $\beta > 0$, $p > q$, p and q are positive odd numbers. The sliding-mode controller designed according to the linearization feedback theory is:

$$u = -b^{-1}(x)\left(f(x) + \beta \frac{q}{p} {x_2}^{2-p/q} + 0.1\mathrm{sgn}(s)\right) \tag{39}$$

Its stability is proved as follows:

$$\begin{aligned} \dot{s} &= \dot{x}_1 + \tfrac{p}{\beta q} {x_2}^{p/q-1} \dot{x}_2 \\ &= x_2 + \tfrac{p}{\beta q} {x_2}^{p/q-1} [f(x) + g(x) + b(x)u] \\ &= \tfrac{p}{\beta q} {x_2}^{p/q-1} (g(x) - 0.1\mathrm{sgn}(s)) \end{aligned} \tag{40}$$

Multiply both sides by s:

$$s\dot{s} = \frac{p}{\beta q} {x_2}^{p/q-1} (s g(x) - 0.1|s|) \tag{41}$$

Because $1 < p/q < 2$, then $0 < p/q - 1 < 1$, ${x_2}^{p/q-1} > 0$, it can be proved that

$$s\dot{s} \le -\frac{0.1p}{\beta q} {x_2}^{p/q-1} |s| \tag{42}$$

Because $-0.1p/\beta q * {x_2}^{p/q-1} \le 0$, then $s\dot{s} \le 0$, the controller satisfies the Lyapunov stability condition. In order to further enhance the control effect, it is considered to combine the above two control methods to form LQRY-SMC. The controller formula is as follows:

$$u = -Kx - b^{-1}(x)\left(f(x) + \beta \frac{q}{p} {x_2}^{2-p/q} + 0.1\mathrm{sgn}(s)\right) \tag{43}$$

4. Design of Longitudinal Motion Controller of Hydrofoil Based on LQR/LQRY-SMC

The simulation in this section firstly needs to substitute the disturbance force and disturbance moment under irregular wave waves with different parameters into the formula in Section 2.2, and then substitute the data results into the controller simulation as the external disturbance part. Under different wave heights, the LQRY and sliding-mode control alone and the LQRY-SMC using them together are simulated and the data diagram is obtained.

4.1. Simulation of Disturbance Force and Moment

In order to more intuitively show the disturbance effect of the hydrofoil craft by the sea waves, the ship model is used as a reference to simulate the disturbance force and moment of the ship under the action of random sea waves according to the above formula. The parameters of the hull model used are shown in Table 1. The disturbance force and moment curves of 180° encounter angle and 1.5 m, 2 m, 3 m significant wave heights of the turbulent flow will be given in the article, as shown in Figures 12–14.

Table 1. Hydrofoil craft parameters.

Parameter	Symbolic Representation	Value	Unit
Craft weight	m	26,200	kg
Craft speed	U_e	35	kn
Average immersion depth	Z	1.52	m
Front hydrofoil area	A_{fb}	6.08	m^2
Rear hydrofoil area	A_{ff}	13.90	m^2
Distance from front hydrofoil to center of gravity	X_b	12.68	m
Distance between two hydrofoils	L_s	17.86	m

It can be seen from the figure that the disturbance force of the hydrofoil craft is basically stable at 10^4, and the disturbance moment is basically stable at 10^5. The data results obtained here provide preconditions for the following control system simulation.

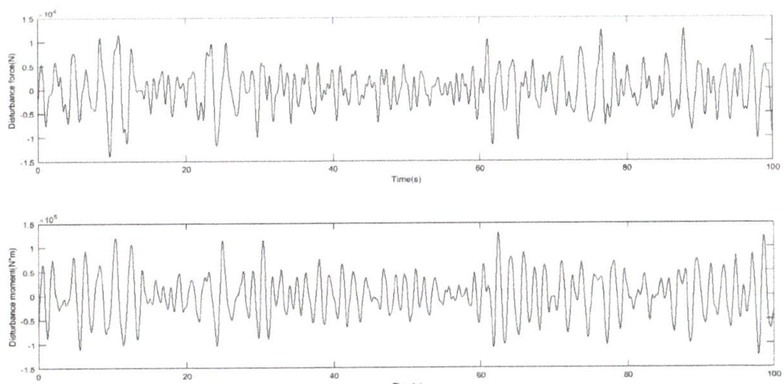

Figure 12. Disturbance force and moment curve (significant wave height 1.5 m).

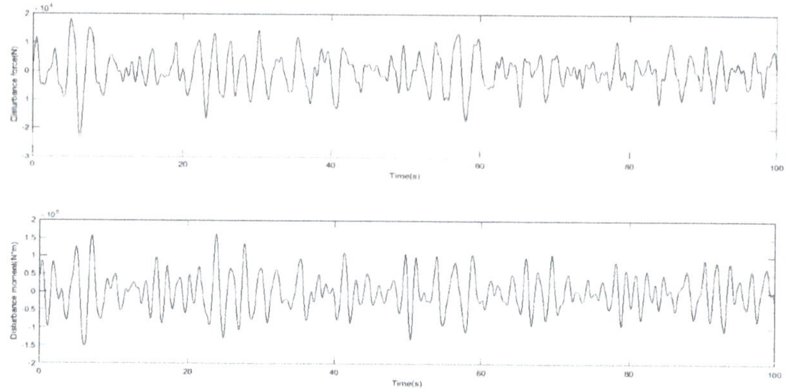

Figure 13. Disturbance force and moment curve (significant wave height 2 m).

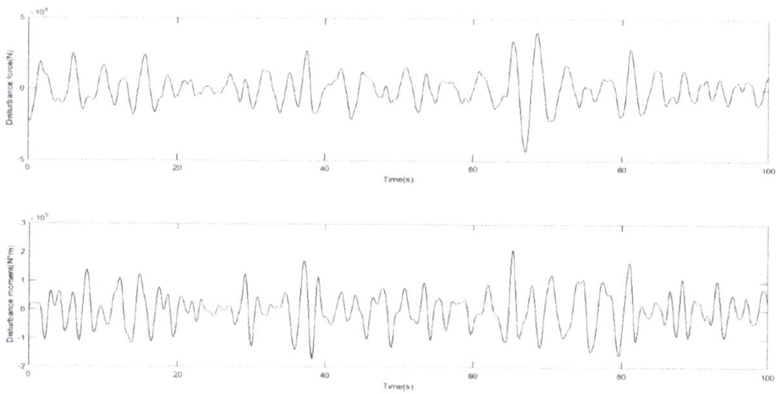

Figure 14. Disturbance force and moment curve (significant wave height 3 m).

4.2. Comparative Simulation of Optimized LQR and LQRY Simulation

After the data of disturbance force and moment are obtained, the control performance of LQR and LQRY will be compared in this section. After comparison, the method more suitable for the next joint control will be selected to obtain better performance. The parameter table of optimization algorithm simulation is shown in Table 2, and the results are shown in Figure 15.

Table 2. Genetic algorithm parameters.

Parameter	Value
Initial population size	100
Number of elite individuals	10
Cross offspring ratio	0.75
Lower limit	[0.1 0.1 0.1 0.1 0.1 0.1]
Upper limit	[1000 1000 1000 1000 500 500]
Evolutionary algebra	30
Fitness function deviation	$1e^{-100}$

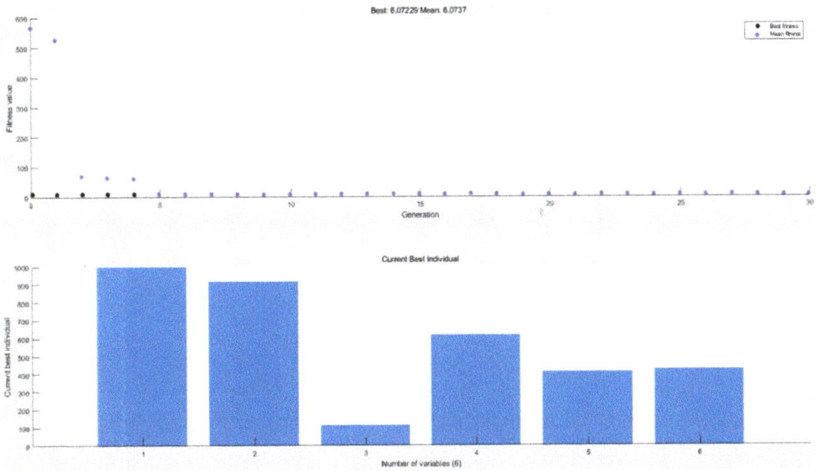

Figure 15. Result graph of LQRY optimized by genetic algorithm.

As can be seen from Figure 14, the result shows that the optimal fitness is 6.072, Q = diag [999.8614 918.9989 116.3479 618.8724], R = diag [415.6694 427.8040], and the value of K is:

$$Klqr = \begin{bmatrix} -1.1878 & -1.0902 & 5.8968 & 0.5587 \\ -0.9730 & -0.9119 & 0.9275 & -0.3399 \end{bmatrix} Klqry = \begin{bmatrix} -1.1884 & -1.0908 & 5.8970 & 0.5590 \\ -0.9735 & -0.9125 & 0.9274 & -0.3403 \end{bmatrix}$$

The simulation results of the longitudinal motion model substituted into the hydrofoil craft are shown in Figure 16 and Table 3 below:

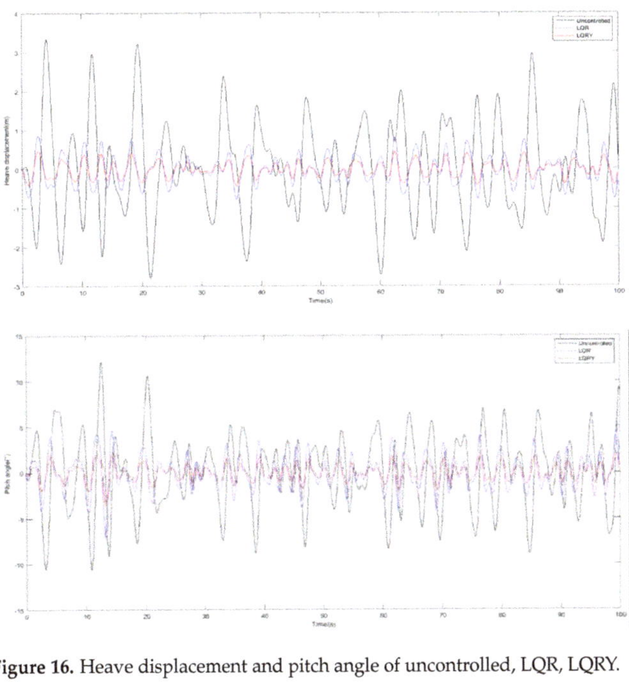

Figure 16. Heave displacement and pitch angle of uncontrolled, LQR, LQRY.

Table 3. Comparison results of uncontrolled, LQR, LQRY.

Control Method	h_{max}	E(h)	STD(h)	θmax	E(θ)	STD(θ)
Uncontrolled	3.3372	0.0036	1.2489	12.1330	0.0049	4.0937
LQR	0.8865	−0.0064	0.3535	4.6508	0.0236	1.8429
LQRY	0.4801	−0.0035	0.1930	2.1788	0.0113	0.8784

In the specific case selected, it can be found that the optimized LQRY algorithm is better than the LQR algorithm in all indicators, in which the maximum-heave displacement and pitch angle are reduced by 45.8% and 53.2%, so the optimized LQRY will be used in the joint control with the sliding-mode algorithm.

4.3. Comparative Simulation of LQRY, SMC and LQRY-SMC

After the superiority of LQRY is proved in the previous section, this section will verify the joint control and compare the heave displacement and pitch angle of LQRY and SMC alone. First of all, the control rate of the sliding mode control part shall be determined. It can be seen from Equation (39) that the parameter β, p, q shall be determined. In this article, β = diag [19.2 7.5], $p = 7$, $q = 5$.

In this section, in order to reflect the universality of joint control, to be applicable to a variety of complex sea conditions and to be more stable, the case of 180° encounter angle and 1.5 m, 2 m, 3 m significant wave heights of the turbulent flow are selected in the simulation. The simulation results are shown in Figures 17–19, and the numerical values are arranged in Table 4.

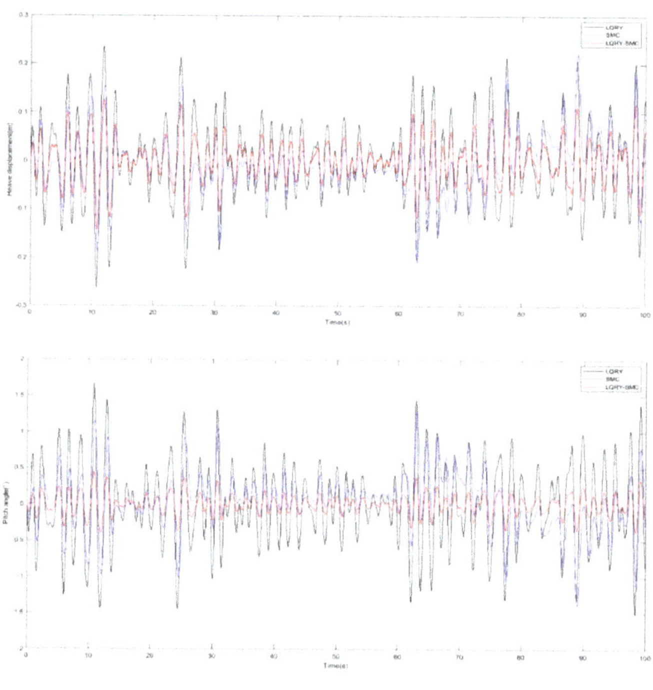

Figure 17. Heave displacement and pitch angle of LQRY, SMC, LQRY-SMC (significant wave height 1.5 m).

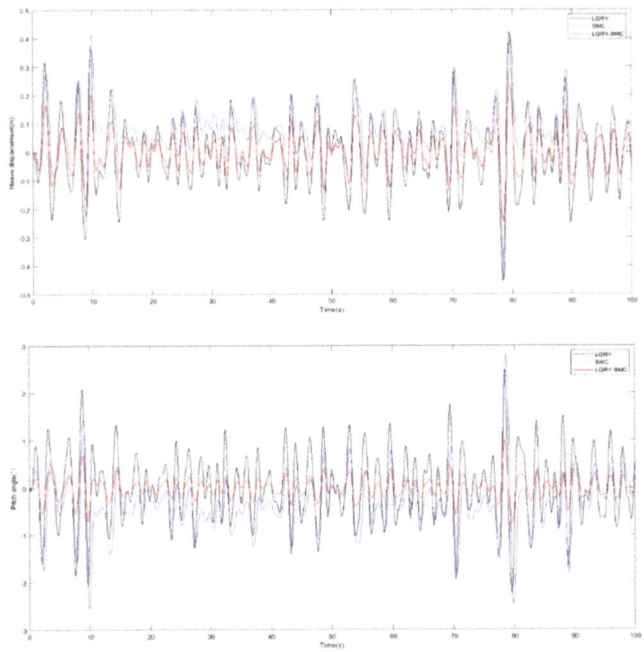

Figure 18. Heave displacement and pitch angle of LQRY, SMC, LQRY-SMC (significant wave height 2 m).

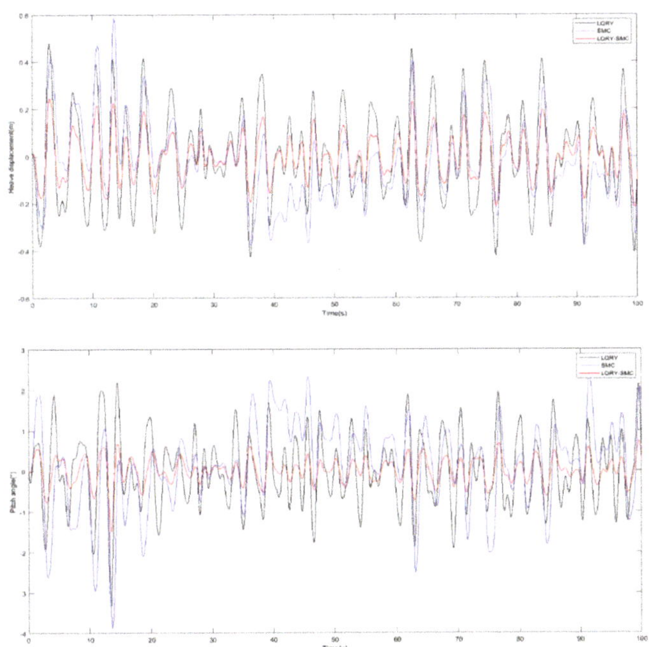

Figure 19. Heave displacement and pitch angle of LQRY, SMC, LQRY-SMC (significant wave height 3 m).

Table 4. Comparison results of LQRY, SMC, LQRY-SMC.

Wave Height	Control Method	h_{max}	E(h)	STD(h)	θmax	E(θ)	STD(θ)
1.5 m	LQRY	0.2625	1.9601×10^{-5}	0.0814	1.6591	3.5457×10^{-4}	0.5508
	SMC	0.2219	0.0057	0.0611	1.3888	0.0355	0.3792
	LQRY-SMC	[illegible]	[illegible]	[illegible]	[illegible]	[illegible]	[illegible]
2 m	LQRY	0.4552	8.6007×10^{-4}	0.1221	2.4925	-8.4116×10^{-4}	0.7204
	SMC	0.4506	0.0538	0.0949	2.8112	−0.3344	0.5907
	LQRY-SMC	0.2453	9.2481×10^{-4}	0.0644	0.9975	−0.0019	0.2132
3 m	LQRY	0.4801	−0.0035	0.1930	3.3249	0.0113	0.8784
	SMC	0.5864	−0.0242	0.1673	3.8698	0.1399	1.0303
	LQRY-SMC	0.2462	−0.0026	0.0942	1.4875	0.0053	0.3173

It can be seen from the simulation curves from Figure 17 to Figure 19 that when the significant wave height is 3 m, the sliding-mode control is greatly affected by the high sea conditions and cannot effectively suppress the longitudinal attitude movement of the hydrofoil craft. Although LQRY control can reduce the longitudinal motion, the index still stays at a large value. After combining the two, it can be seen that the maximum absolute value of the heave displacement curve is about 0.2 m, and the maximum absolute value of the pitch angle curve is about 1.5°, which has a good control effect.

It can be seen from this Table 4 that the joint control of LQRY and SMC is better than each individual control algorithm, and its maximum value and standard deviation are greatly reduced. It is suitable for various significant wave heights of hydrofoil crafts, especially in high sea conditions. Compared with LQRY and SMC alone, the maximum heave displacement and pitch angle are reduced by 48.7% and 58.0% respectively, and 55.3% and 61.6% respectively. Therefore, the controller designed in this article can effectively offset the wave interference, and the amplitude of pitch angle and heave displacement is significantly reduced, which basically meets the requirements of stability and comfort.

Through the research on the control system in this article, we propose the following suggestions: hydrofoil craft can basically reach a stable state in the low speed state of entering and leaving the port, or under the conditions of good sea conditions. As shown in Figure 20, when the significant wave height is 0.5 m, the maximum heave displacement is about 0.2 m, and the maximum pitch angle does not exceed 1.5°, then it is unnecessary to use the control system during navigation, and energy can be saved. Under high sea conditions or severe weather conditions, the control system of hydrofoil craft will gradually saturate, thus losing stability and causing severe shaking. As shown in Figure 21, when the significant wave height is 5 m, the control system can still work normally at first, but the heave displacement can reach more than 2 m, and the pitch angle can reach 10°. Therefore, we suggest that the hydrofoil craft reduce its speed or not go to sea in this case.

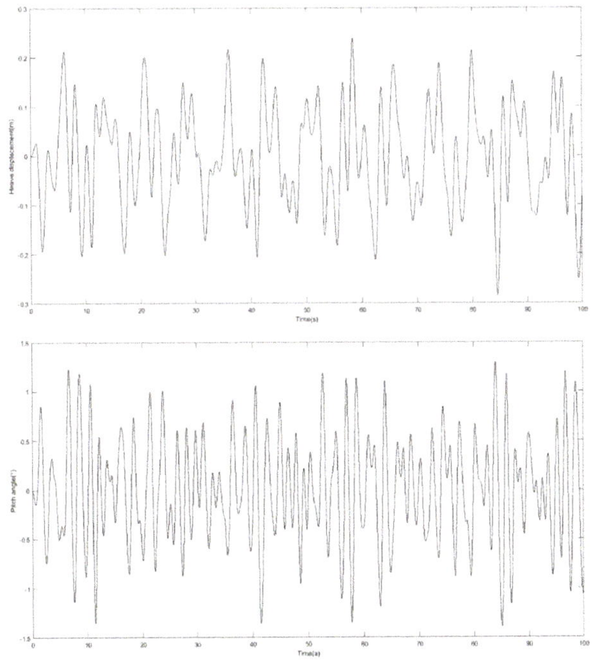

Figure 20. Heave displacement and pitch angle without control (significant wave height 0.5 m).

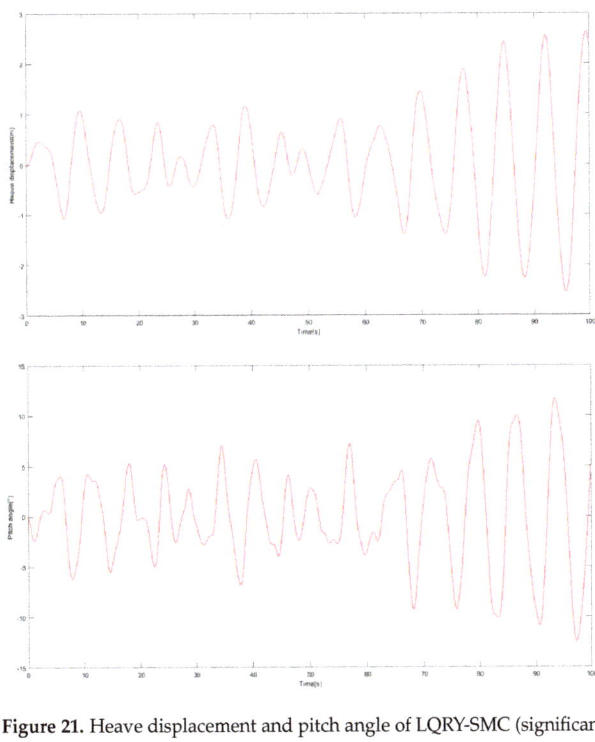

Figure 21. Heave displacement and pitch angle of LQRY-SMC (significant wave height 5 m).

5. Conclusions

In this article, a control method of LQRY-SMC is proposed to reduce the heave displacement and pitch angle in the longitudinal motion of the fully submerged hydrofoil craft. Firstly, the longitudinal-motion mathematical model of the hydrofoil craft is established and the force of the hydrofoil is analyzed. Then the hull and hydrofoil structure are analyzed. The simulation results of the disturbance force and moment of the craft in the irregular wave environment are obtained. In the LQRY control part, the genetic algorithm is used to optimize the controller parameters. After simulation and verification, the joint control method's heave displacement and pitch angle are smaller than LQRY and sliding mode control under different sea conditions, which further reduces the longitudinal motion and makes the hull more stable, and improves the comfort of hydrofoil crafts in sea navigation. Therefore, it has some enlightening effects on practical engineering applications.

Author Contributions: Conceptualization, Y.F. and H.L.; methodology, Y.F. and H.L.; software, Y.F.; validation, Y.F., H.L. and B.L.; formal analysis, B.L.; writing—original draft preparation, Y.F. and H.L.; writing—review and editing, Y.F., H.L. and B.L. All authors have read and agreed to the published version of the manuscript.

Funding: This work was supported by the Natural Science Foundation of Heilongjiang Province (grant number KY10400210217), Central University basic scientific research business fee project funding (grant number 3072020CFT1501) and Foundation strengthening program technical field fund support (grant number 2021-JCJQ-JJ-0026).

Institutional Review Board Statement: Not applicable.

Informed Consent Statement: Not applicable.

Data Availability Statement: Not applicable.

Conflicts of Interest: The authors declare no conflict of interest. The funders had no role in the design of the study.

References

1. Ruggiero, V.; Morace, F. Methodology to study the comfort implementation for a new generation of hydrofoils. *Int. J. Interact. Des. Manuf. (IJIDeM)* **2019**, *13*, 99–110. [CrossRef]
2. Wang, Y.; Liu, S.; Su, X. Hydrofoil catamaran longitudinal motion robust gain scheduling control study. In Proceedings of the 33rd Chinese Control Conference, Nanjing, China, 28–30 July 2014; pp. 1983–1987.
3. Ling, H.; Wang, Z.; Wu, N. On prediction of longitudinal attitude of planing craft based on controllable hydrofoils. *J. Mar. Sci. Appl.* **2013**, *12*, 272–278. [CrossRef]
4. Piene, E.B. Disturbance Rejection of a High Speed Hydrofoil Craft Using a Frequency Weighted H2-Optimal Controller. Master's Thesis, Norwegian University of Science and Technology, Trondheim, Norway, 2018.
5. Matdaud, Z.; Zhahir, A.; Pua'at, A.A.; Hassan, A.; Ahmad, M.T. Stabilizing Attitude Control for Mobility of Wing in Ground (WIG) Craft—A Review. *IOP Conf. Ser. Mater. Sci. Eng.* **2019**, *642*, 012005. [CrossRef]
6. Hu, K.; Ding, Y.; Wang, H. High-speed catamaran's longitudinal motion attenuation with active hydrofoils. *Pol. Marit. Rearch* **2018**, *25*, 56–61. [CrossRef]
7. Kim, S.-H.; Yamato, H. An experimental study of the longitudinal motion control of a fully submerged hydrofoil model in following seas. *Ocean. Eng.* **2004**, *31*, 523–537. [CrossRef]
8. Kim, S.-H.; Yamato, H. On the design of a longitudinal motion control system of a fully-submerged hydrofoil craft based on the optimal preview servo system. *Ocean. Eng.* **2004**, *31*, 1637–1653. [CrossRef]
9. Hongli, C.; Haokai, L.; Xiaojing, X.; Xiaoyue, Z. Design of adaptive sliding mode controller for longitudinal motion of hydrofoil. In Proceedings of the OCEANS 2019, Marseille, France, 17–20 June 2019; pp. 1–9.
10. Schaaf, J. Using Reinforcement Learning to Control Hydrofoils. Bachelor's Thesis, University of Twente, Enschede, The Netherlands, 2022.
11. Gao, H.; Lv, Y.; Ma, G.; Li, C. Backstepping sliding mode control for combined spacecraft with nonlinear disturbance observer. In Proceedings of the 2016 UKACC 11th International Conference on Control, Belfast, UK, 31 August–2 September 2016; pp. 1–6.
12. Li, H.; Wu, Y.j.; Zuo, J.x. Sliding mode controller design for UAV based on backstepping control. In Proceedings of the 2016 IEEE Chinese Guidance, Navigation and Control Conference (CGNCC), Nanjing, China, 12–14 August 2016; pp. 1448–1453.
13. Guo, Y.; Luo, L.; Bao, C. Design of a Fixed-Wing UAV Controller Combined Fuzzy Adaptive Method and Sliding Mode Control. *Math. Probl. Eng.* **2022**, *2022*, 13–21. [CrossRef]

14. Elmokadem, T.; Zribi, M.; Youcef-Toumi, K. Terminal sliding mode control for the trajectory tracking of underactuated Autonomous Underwater Vehicles. *Ocean. Eng.* **2017**, *129*, 613–625. [CrossRef]
15. Liu, S.; Niu, H.; Zhang, L.; Xu, C. Modified adaptive complementary sliding mode control for the longitudinal motion stabilization of the fully-submerged hydrofoil craft. *Int. J. Nav. Archit. Ocean. Eng.* **2019**, *11*, 584–596. [CrossRef]
16. Liu, S.; Niu, H.; Zhang, L.; Xu, C. The Longitudinal Attitude Control of the Fully-Submerged Hydrofoil Vessel Based on the Disturbance Observer. In Proceedings of the 2018 37th Chinese Control Conference (CCC), Wuhan, China, 25–27 July 2018; pp. 397–401.
17. Liu, S.; Niu, H.; Zhang, L.; Guo, X. Adaptive compound second-order terminal sliding mode control for the longitudinal attitude control of the fully submerged hydrofoil vessel. *Adv. Mech. Eng.* **2019**, *11*, 1687814019895637. [CrossRef]
18. Bai, J.; Kim, Y. Control of the vertical motion of a hydrofoil vessel. *Ships Offshore Struct.* **2010**, *5*, 189–198. [CrossRef]
19. Sanjeewa, S.D.; Parnichkun, M. Control of rotary double inverted pendulum system using LQR sliding surface based sliding mode controller. *J. Control. Decis.* **2022**, *9*, 89–101. [CrossRef]
20. Chawla, I.; Singla, A. Real-Time Stabilization Control of a Rotary Inverted Pendulum Using LQR-Based Sliding Mode Controller. *Arab. J. Sci. Eng.* **2021**, *46*, 2589–2596. [CrossRef]
21. Ali, H.; Abdulridha, A.J.; Khaleel, R.; Hussein, K. LQR/Sliding Mode Controller Design Using Particle Swarm Optimization for Crane System. *Al-Nahrain J. Eng. Sci.* **2020**, *23*, 45–50. [CrossRef]
22. Jibril, M.; Tadese, M.; Hassen, N. Position Control of a Three Degree of Freedom Gyroscope using Optimal Control. *Preprints* **2020**, *97*, 5–9.
23. Deng, Y.; Zhang, X.; Im, N.; Liang, C. Compound learning tracking control of a switched fully-submerged hydrofoil craft. *Ocean. Eng.* **2021**, *219*, 108260. [CrossRef]
24. ELLSWORTH, W.M. US Navy Hydrofoil Craft. *J. Hydronautics* **2012**, *1*, 66–73. [CrossRef]
25. Yasukawa, H.; Hirata, N.; Matsumoto, A.; Kuroiwa, R.; Mizokami, S. Evaluations of wave-induced steady forces and turning motion of a full hull ship in waves. *J. Mar. Sci. Technol.* **2019**, *24*, 1–15. [CrossRef]
26. Hongli, C.; Jinghui, S.; Yuwei, C. The applied research of hydrofoil catamaran attitude estimation based on the fusion filtering. In Proceedings of the 2015 34th Chinese Control Conference (CCC), Hangzhou, China, 28–30 July 2015; pp. 1758–1763.
27. Touw, M. Prediction of the Longitudinal Stability and Motions of a Hydrofoil Ship with a Suspension System between the Wings and the Hull Using a State-Space Model. Master's Thesis, Delft University, Delft, The Netherlands, 2020.
28. Lee, D.; Ko, S.; Park, J.; Kwon, Y.C.; Rhee, S.H.; Jeon, M.; Kim, T.H. An Experimental Analysis of Active Pitch Control for an Assault Amphibious Vehicle Considering Waterjet-Hydrofoil Interaction Effect. *J. Mar. Sci. Eng.* **2021**, *9*, 894. [CrossRef]
29. Yu, W.; Li, J.; Yuan, J.; Ji, X. LQR controller design of active suspension based on genetic algorithm. In Proceedings of the 2021 IEEE 5th Information Technology, Networking, Electronic and Automation Control Conference (ITNEC), Xi'an, China, 15–17 October 2021; pp. 1056–1060.

Journal of
Marine Science and Engineering

MDPI

Article

Power Tracking Control of Marine Boiler-Turbine System Based on Fractional Order Model Predictive Control Algorithm

Shiquan Zhao [1], Sizhe Wang [1], Ricardo Cajo [2], Weijie Ren [1,3] and Bing Li [1,*]

1 College of Intelligent Systems Science and Engineering, Harbin Engineering University, Harbin 150001, China
2 Facultad de Ingeniería en Electricidad y Computación, Escuela Superior Politécnica del Litoral, ESPOL, Campus Gustavo Galindo Km 30.5 Vía Perimetral, P.O. Box 09-01-5863, Guayaquil 090150, Ecuador
3 Faculty of Electronic Information and Electrical Engineering, Dalian University of Technology, Dalian 116024, China
* Correspondence: libing265@hrbeu.edu.cn

Abstract: The marine boiler-turbine system is the core part for the steam-powered ships with complicated dynamics. To improve the power tracking performance and fulfill the requirement of high utilization rate of fossil energy, the control performance of the system should be improved. In this paper, a nonlinear model predictive control method is proposed for the boiler-turbine system with fractional order cost functions. Firstly, a nonlinear model of the boiler-turbine system is introduced. Secondly, a nonlinear extended predictive self adaptive control(EPSAC) method is designed to the system. Then, integer order cost function is replaced with a fractional order cost function to improve the control performance, and also the configuration of the cost function is simplified. Finally, the superiority of the proposed method is proved accordring to the comparison experiments between the fractional order model predictive control and the traditional model predictive control.

Keywords: boiler-turbine; nonlinear model predictive control; fractional order calculus; distributed control

Citation: Zha, S.; Wang, S.; Cajo, R.; Ren, W.; Li, B. Power Tracking Control of Marine Boiler-Turbine System Based on Fractional Order Model Predictive Control Algorithm. *J. Mar. Sci. Eng.* **2022**, *10*, 1307. https://doi.org/10.3390/jmse10091307

Academic Editors: Bowen Xing, Marco Altosole, Maria Acanfora and Flavio Balsamo

Received: 7 August 2022
Accepted: 12 September 2022
Published: 15 September 2022

1. Introduction

In order to reduce the waste of fossil energy and CO_2 emissions, many countries have released different policies. China has released a policy document to fulfill its target of reaching peak carbon emissions by 2030. The United States of America released policy to cut carbon emissions in half by 2030. A lot of renewable power technologies were also proposed. However, most applied energy still comes from the combustion of fossil fuels. In addition, the control performance has a close relationship with the utilization of fossil fuels [1]. In this paper, the fossil fuels in ships are focused. Many of the ships are equipped with internal combustion engine, or gas turbine. However, for large scale ships, the power systems based on boiler-turbine still occupy a large proportion [2–4]. For example, many aircraft carriers are powered with boiler-turbine system. Hence, a lot of academics and companies are doing research to improve the control performce of the boiler-turbine system [5–9], of which there are three manipulated variables and three controlled variables with complicated dynamics. For the marine steam power plant, the disturbance and the energy required changes more frequently compared than that on land. However, there is not so much research about the control for the marine boiler-turbine system.

In the boiler-turbine system, the interactions of rvariables and constraints are the mainly reasons which make it difficult to obtain a satisfied control performance. The input variables for the system are the flow rates of fuel, steam to the turbine, and feedwater to the boiler, while the output variables are the steam pressure in the drum of the boiler, power required of the turbine, and the water level of the drum. The constraints are the limitations for the actuator, including the upper and lower bound, and rate limiter. In addition, power requirement changes a lot, which is usually treated as disturbance. In order to compensate unknown disturbances in the boiler-turbine system, a high order sliding mode observer was designed for a baseline exponentially stable feedback controller [9]. To improve the economy of the boiler-turbine system, the economic index was utilized directly in the cost function, and a global economic optimum routine was obtained for the system [10]. In the literature [11], the model of the boiler turbine system was linearized and decoupled with an adaptive feedback linearization method, and a second order sliding mode controller was designed to deal with the disturbances and uncertainties. Ref. [12] applied an online policy iteration integral reinforcement learning method to the boiler-turbine system, and optimal tracking control performance was obtained.

The model predictive control(MPC) has the advantages in dealing with the nonlinear dynamics, interactions and constraints problems [13], hence, MPC is a preferred choice for researchers. The MPC is studied in many fields such as building energy management [14], landscape office lighting regulatory system [15,16], wind turbines [17], tank-system [18], pressure oscillation adsorption process [19,20], permanent-magnet synchronous motor [21], autonomous underwater vehicle [22], and so on. For the boiler-turbine system, there are also some applications of MPC. A nonlinear model predictive control method was designed for boiler-turbine system with a data driven model [23], and the optimal problem was solved by immune genetic algorithm. Ref. [24] presented a zone economic model predictive cotroller to fulfill the economic target of the boiler-turbine system. Fuzzy model predictive control was designed to realize load tracking and economy of the boiler-turbine system, and a fuzzy model was used to approximate the nonlinear dynamics of the system [25]. Other MPC applications on boiler-turbine system can be found in [26–28].

For the MPC method, the weighting factors have a significant different effect on the control performance, and the number of the weighting factors is large, which makes it diffcult to obtain a good choice for these parameters. For example, if the N_{ip} denotes the prediction horizon for the ith output, and N_{jc} for the jth control horizon, the number of weighting factor will be $\sum_{i=1}^{n} N_{ip} + \sum_{j=1}^{m} N_{jc}$, where m and n are the numbers of output and input. The commonly used method to tune the weighting factors is trial and error or choose them empirically. Another alternative way to choose the control parameters is optimization method [29,30]. However, due to the high dimension of the weighting factors, it is difficult to optimize them. So, in this paper, the fractional order model predictive control(FOMPC) for the boiler-turbine system is proposed. By two fractional order papermeters (one for the tracking error, and another for the control effort), the weight factors matrices can be obtained, which reduces the difficulties in weighting factors configuration. The boiler-turbine system is a nonlinear multiple inputs and multiple outputs system, so the nonlinear distributed structure of MPC is studied for the system.

The rest of the paper is structured as follows. The boiler-turbine system is formulated in Section 2; Section 3 presents the nonlinear distributed MPC with EPSAC framework. Fractional order MPC is designed for the boiler-turbine system in the Section 4. The simulation experiments are presented in Section 5. The last section gives the conclusions.

2. Boiler-Turbine System

The boiler-turbine system is a core part in the power plant, and Figure 1 shows the structure of the system [9]. The details elements are indicated in the figure, and they are listed as follows: 1—drum; 2—superheater; 3—water spray desuperheater; 4—valve for the steam to turbine; 5—turbine high-pressure cylinders; 6—foward control valve; 7—backward control valve; 8—turbine middle- and low- pressure cylinders; 9—shafting; 10—condenser; 11—replenish water; 12—condensate pump; 13—low-pressure heater; 14—deaerator; 15—feed water pump; 16—high-pressure heater; 17—feed water valve; 18—economizer; 19—downcomers; 20—water-cooled walls; 21—furnace; 22—heat flow control valve; 23—nozzle; 24—blower; 25—preheater for air; 26—air conditioner; 27—flue gas baffle; 28—induced draft fan; 29—flue; 30—gearbox; 31—turbine.

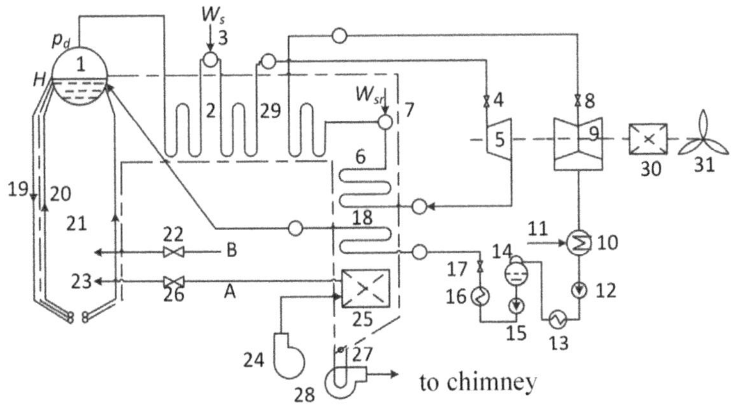

Figure 1. The structure of a boiler-turbine unit.

Due to the lack of data for marine boiler-turbine sytem, it is difficult to obtain its model. The structure is similar to that on land. Hence, a model of on load power plant is chosen for study, and the nonlinear model of the boiler-turbine system is shown as follows according to the literature [31]:

$$\dot{x}(t) = F(x(t)) + G(x(t))u(t), \tag{1}$$

and the $F(x(t))$, $G(x(t))$ are defined as follows:

$$F(x(t)) = \begin{bmatrix} 0 \\ -0.1x_2 - 0.016x_1^{9/8} \\ 0.0022x_1 \end{bmatrix} \tag{2}$$

$$G(x(t)) = \begin{bmatrix} 0.9 & -0.0018x_1^{9/8} & -0.15 \\ 0 & 0.073x_1^{9/8} & 0 \\ 0 & -0.0129x_1 & 1.6588 \end{bmatrix} \tag{3}$$

where the inputs $u = [u_1, u_2, u_3]^T$ for the system are the valve opening of fuel, steam to the turbine and feedwater to the drum. The states are drum steam pressure, power required for the turbine and steam water density denoted by $x = [x_1, x_2, x_3]^T$. The outputs are drum steam pressure, power required for the turbine and water level in the drum. The level of the drum can be calculated as:

$$L = 0.05(0.13073x_3 + 100\alpha_s + q_e/9 - 67.975) \tag{4}$$

and $q_e = (0.854u_2 - 0.147)x_1 + 45.59u_1 - 2.51u_3 - 2.096$; $\alpha_s = \frac{(1-0.001538x_3)(0.8x_1-25.6)}{x_3(1.0394-0.0012304x_1)}$. The q_e and α_s denote the evaporation rate and steam quality, respectively.

The rates and amplitudes limitaion for the inputs are listed as follows:

$$\begin{cases} -0.007 \le \frac{du_1}{dt} \le 0.007 & 0 \le u_1 \le 1 \\ -2.0 \le \frac{du_2}{dt} \le 0.02 & 0 \le u_2 \le 1 \\ -0.05 \le \frac{du_3}{dt} \le 0.05 & 0 \le u_3 \le 1 \end{cases} \tag{5}$$

For the boiler-turbine, there are different operating points. To evaluate the effectiveness of the proposed method, experiments around the following operating points shown in Table 1 are carried out.

Table 1. Operating points for the boiler-turbine system.

Operating Point	Pressure	Power	Density
1	99.3 kg/m^2	80.9 MW	396
2	120 kg/m^2	110 MW	331

The drum water level should always be kept as zero meters.

3. Nonlinear Distributed MPC for the Boiler-Turbine System

According to the introduction of the boiler-turbine, it can be found that this system is a nonlinear multiple inputs multiple outputs system. Hence, the nonlinear MPC is designed for the system with distributed structure.

3.1. The Basic of the EPSAC

This part presents the basic of EPSAC. For more details about the EPSAC, it can be found in the refs. [32–34]. For a discrete system, the system output can be expressed as:

$$y(t) = x(t) + w(t) \tag{6}$$

where $y(t)$ is the system output; $x(t)$ is the model output and the $w(t)$ is the disturbances. $x(t)$ can be calculated according to the model of the system as follow:

$$x(t) = f[x(t-1), x(t-2), \ldots, u(t-1), u(t-2), \ldots] \tag{7}$$

In Equation (7), the $f(x,u)$ denotes the model of the system, $x(t-i)$ and $u(t-i)$ $i = 1,2,\ldots$ indicate the past model outputs and inputs.

In the EPSAC, the input scenario for the future is composed with two parts:

$$u(t+k|t) = u_{base}(t+k|t) + \delta u(t+k|t) \tag{8}$$

where the $u_{base}(t+k|t)$ and $\delta u(t+k|t)$ are the basic and optimized future control actions. Then the future system output can be predicted as:

$$y(t+k|t) = y_{base}(t+k|t) + y_{opt}(t+k|t) \tag{9}$$

where $y_{base}(t+k|t)$ is the result of the basic future control action; $u_{base}(t+k|t)$ and $y_{opt}(t+k|t)$ can be calculated according to the optimized future control action $\delta u(t+k|t)$.

The $y_{opt}(t+k|t)$ can be obtained with:

$$y_{opt}(t+k|t) = h_k \delta u(t|t) + h_{k-1}\delta u(t+1|t) + \ldots + g_{k-N_c+1}\delta u(t+N_c-1|t) \tag{10}$$

In Equation (10), the h_i and g_i are the impulse response and step response coefficients of the system, respectively; N_c and N_p are the control horizon and the prediction horizon, respectively. The system output can be re-written in matrix form:

$$\mathbf{Y} = \bar{\mathbf{Y}} + \mathbf{GU} \tag{11}$$

where $\mathbf{Y} = [y(t+N_1|t)\ldots y(t+N_p|t)]^T$, $\mathbf{U} = [\delta u(t|t)\ldots \delta u(t+N_c-1|t)]^T$, $\bar{\mathbf{Y}} = [y_{base}(t+N_1|t)\ldots y_{base}(t+N_P|t)]^T$; N_1 is the time delay of the system, and

$$\mathbf{G} = \begin{bmatrix} h_{N_1} & h_{N_1-1} & \cdots & g_{N_1-N_c+1} \\ h_{N_1+1} & h_{N_1} & \cdots & \cdots \\ \cdots & \cdots & \cdots & \cdots \\ h_{N_P} & h_{N_P-1} & \cdots & g_{N_P-N_c+1} \end{bmatrix}$$

The disturbance term $w(t)$ in (6) includes all the effects on the system output. It can be modeled by a colored noise process as:

$$w(t+k|t) = \frac{C(q^{-1})}{D(q^{-1})} w_f(t+k|t) \tag{12}$$

where q^{-1} is the backward shift operator.

In this work, the $C(q^{-1})$ and $D(q^{-1})$ are designed as follows:

$$\frac{C(q^{-1})}{D(q^{-1})} = \frac{1}{(1-q^{-1})(1-ae^{+j\alpha}q^{-1})(1-ae^{-j\alpha}q^{-1})} \tag{13}$$

with $\alpha = 2\pi f_0 T_s$ and $a \approx 1$; T_s is the sampling time and $a \leq 1$ for stability.

The cost function for the boiler-turbine system can be defined as:

$$J_{MPC} = \sum_{k=N_1}^{N_2} p_k[r(t+k|t) - y(t+k|t)]^2 + \sum_{k=1}^{N_u} q_k \triangle u(t+k)^2 \tag{14}$$

The p_k and q_k are nonnegative weighting factors, and they are usually kept as constants. The matrix form of Equation (14) can be written as:

$$J_{MPC} = (\mathbf{R}-\mathbf{Y})^T\mathbf{P}(\mathbf{R}-\mathbf{Y}) + \mathbf{U}^T\mathbf{Q}\mathbf{U} = (\mathbf{R}-\bar{\mathbf{Y}}-\mathbf{G}\mathbf{U})^T\mathbf{P}(\mathbf{R}-\bar{\mathbf{Y}}-\mathbf{G}\mathbf{U}) + \mathbf{U}^T\mathbf{Q}\mathbf{U} \tag{15}$$

with $P = diag(p_1, p_2, \ldots, p_{(N_2-N_1+1)})$ and $Q = diag(q_1, q_2, \ldots, q_{N_u})$.

For systems with constraint, the optimization problem can be solved with quadratic programming. Otherwise, the results of the optimal input part, which are indicated by $\delta u(t+k|t)$, can be obtained as:

$$\mathbf{U}^*_{\mathbf{MPC}} = (\mathbf{G}^T\mathbf{P}\mathbf{G}+\mathbf{Q})^{-1}\mathbf{G}^T\mathbf{P}(\mathbf{R}-\bar{\mathbf{Y}}) \tag{16}$$

3.2. The Fractional Order MPC

For the fractional order MPC, the cost function is designed as:

$$J_{FOMPC} = {}^{\gamma}I_{N_1}^{N_2} p_k[r(t+k|t) - y(t+k|t)]^2 + {}^{\lambda}I_1^{N_c} q_k \triangle u(t+k)^2 \tag{17}$$

where ${}^{\gamma}I_{N_1}^{N_2}$ and ${}^{\lambda}I_1^{N_c}$ indicate fractional order integral with fraction order of γ and λ; [N_1, N_2] and [1, N_c] are the integration intervals.

According to [35], the Equation (17) can be written by:

$$\begin{aligned} J_{FOMPC} &= (\mathbf{R}-\mathbf{Y})^T\mathbf{P}\mathbf{\Gamma}(T_s,\gamma)(\mathbf{R}-\mathbf{Y}) + \mathbf{U}^T\mathbf{Q}\mathbf{\Lambda}(T_s,\lambda)\mathbf{U} \\ &= (\mathbf{R}-\bar{\mathbf{Y}}-\mathbf{G}\mathbf{U})^T\mathbf{P}\mathbf{\Gamma}(T_s,\gamma)(\mathbf{R}-\bar{\mathbf{Y}}-\mathbf{G}\mathbf{U}) + \mathbf{U}^T\mathbf{Q}\mathbf{\Lambda}(T_s,\lambda)\mathbf{U} \end{aligned} \tag{18}$$

$$\mathbf{\Gamma}(T_s,\gamma) = T_s^{\gamma} diag(m_{N_2-N_1}, m_{N_2-N_1-1}, \ldots, m_1, m_0) \tag{19}$$

$$\mathbf{\Lambda}(T_s,\lambda) = T_s^{\lambda} diag(m_{N_c}, m_{N_c-1}, \ldots, m_1, m_0) \tag{20}$$

The m_i in Equations (19) and (20) with fractional order α can be calculated as:

$$m_j = \omega_j^{(-\alpha)} - \omega_{j-n}^{(-\alpha)} \tag{21}$$

where n is the number of the m_i, and ω can be calculated with:

$$\omega_j^{-\alpha} = \begin{cases} (1-(1-\alpha)/j)\omega_{j-1}^{(-\alpha)} & j > 0; \\ 1 & j = 0; \\ 0 & j < 0. \end{cases} \tag{22}$$

According to Equation (19) to Equation (22), the weight matrix in the cost function J_{FOMPC} can be easily tuned with fractional order γ and λ. For system with constraints, the optimization for input sequence can be solved with quadratic programming. For the case without constraints, the results for the optimal input $\delta u(t+k|t)$ can be calculated as:

$$U^*_{FOMPC} = (G^T P(\Gamma + \Gamma^T) G + Q(\Lambda + \Lambda^T))^{-1} G^T P(\Gamma + \Gamma^T)(R - \bar{Y}) \tag{23}$$

3.3. Application of the Fractional Order EPSAC to the Nonlinear MIMO System with Distributed Structure

The fractional order MPC introduced above is for the linear case. In order to apply the FOMPC to the boiler-turbine system, the nonlinear FOMPC is studied. According to the Equation (8) and (9), the principle of superposition is applied for linear system. To get over the superposition, the optimal future input $\delta u(t+k|t)$ should be removed iteratively smaller tends to zero [36]. The procedure for the nonlinear MPC is summarized as follows:

- Choose an initial base input sequence $u_{base}(t+k|t), k = 0 \ldots N_u - 1$, this part should be as close as possible to the optimal input $u(t+k|t)$ to make the $\delta u(t+k|t)$ close to zero, which means that the term $y_{opt}(t+k|t)$ equals to zero;
- After chooseing the base future input, the $\delta u(t+k|t)$ can be calculated. The $\delta u(t+k|t)$ is not close to zero at the moment;
- Take the $u(t+k|t)$ from the second step as the new $u_{base}(t+k|t)$, and calculate $\delta u(t+k|t)$ again.
- Repeat step 2 and 3 until the $\delta u(t+k|t)$ is as close as possible to zero, then the $u_{base}(t+k|t)$ can be applied to the system at the time $t+1$.

The flow chart of nonlinear MPC is shown in Figure 2.

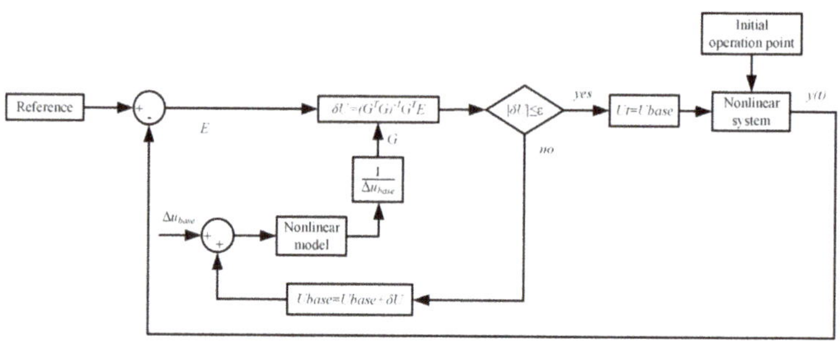

Figure 2. The flow chart of nonlinear MPC.

The boiler-turbine is a MIMO system, and there are strong interactions between variables. In order to calculate the optimal input sequence, the effect from coupling variables should be considered and the communication network should be established. In this work, the distributed structure is applied. The pseudocode is provided in Algorithm 1.

The boiler-turbine system is a nonlinear MIMO system, hence, the nonlinear MPC with distributed scheme will be applied. According to the procedure of nonlinear MPC and algorithm for the distributed MPC, the optimal future input sequence should fulfill the follow conditions:

$$\begin{cases} |||\delta \mathbf{U}_i^{iter+1} - \delta \mathbf{U}_i^{iter}|| \leqslant \varepsilon_i) \\ \delta \mathbf{U}_i^{iter+1} \approx 0. \end{cases} \tag{24}$$

Algorithm 1 Algorithm for the distributed MPC

1: The loop i receives an optimal local control action $\delta\mathbf{U}_i$ for the first time, which will be marked as $iter = 0$, and the local control action $\delta\mathbf{U}_i$ can be marked as $\delta\mathbf{U}_i^{iter}$, where $\delta\mathbf{U}_i$ indicates the vector of the optimizing future control actions with length of N_{ci};

2: The information of coupling variables $\delta\mathbf{U}_j^{iter}$ $(j \in N_i, N_i = \{j \in N : \mathbf{G}_{ij} \neq 0\})$ will be sent to the loop i, and the $\delta\mathbf{U}_i^{iter+1}$ will be recalculated with the information of $\delta\mathbf{U}_j^{iter}$ from other loops;

3: The termination condition can be designed as: $(||\delta\mathbf{U}_i^{iter+1} - \delta\mathbf{U}_i^{iter}|| \leqslant \varepsilon_i) \vee (iter + 1 > \overline{iter})$. where ε is a positive constant and $\overline{iter}$ indicates the upper bound of the number of iteration times. If the termination condition is reached, the $\delta\mathbf{U}_i^{iter+1}$ will be adopted to the system. Otherwise, $iter = iter + 1$, and return to the Step 2.

4: The final optimal control effort can be obtained as $\mathbf{U}_t = \mathbf{U}_{base} + \delta\mathbf{U}^{iter}$, which will be applied to the system;

5: $t = t + 1$, return to Step 1.

4. Simulation of the Fractional Order MPC on Boiler-Turbine System

This section shows the simulation results of the fractional order MPC. Firstly, different fractional order terms are applied to different loops. Then, the best fractional order terms are applied to the drum steam pressure loop, power for turbine loop and water level loop, and the results are compared with the integer order MPC. Finally, the results are discussed.

The parameters configuration are listed in Table 2.

Table 2. Parameters for the MPC.

Parameters	N_c	T_s	N_p	N_1	N_s
Values	$N_{c1} = 1, N_{c2} = 1,$ $N_{c3} = 1$ samples	5s	$N_{p1} = 15, N_{p2} = 15,$ $N_{p3} = 15$ samples	1	100

In Table 2, N_{ci} and N_{pi} $(i = 1, 2, 3)$ are control horizon and prediction horizon, respectively; N_s is the number of simulation steps. The termination conditions for the nonlinear iteration are set as: $\delta\mathbf{U}_i^{iter} \leqslant 0.05$; or the iteration times $iter_{nmpc} > 5$. The termination conditions for the distributed MPC are set as: $||\delta\mathbf{U}_i^{iter+1} - \delta\mathbf{U}_i^{iter}|| \leqslant 0.005$; or $iter > 5$.

4.1. The Influence of Fractional Order Terms to the Different Loops

In order to test the effect of different fractional order on the control performance, different fractional order terms are introduced to the cost function for each loop. The details for fractional order terms are listed in Table 3. In this work, the γ for the reference tracking and λ for the control effort are chosen the same for simplification.

Table 3. Fractional order terms for each loop.

Loops	Fractional Order Terms
Drum steam pressure loop	[0.5, 1, 1.5, 3]
Power loop	[0.8, 1, 1.2, 1.5]
Drum water level loop	[0.5, 1, 1.7, 2, 2.5]

According to the results shown in Figures 3–5, it can be seen that the effectiveness of different fractional order terms varies a lot. For the drum steam pressure control, the best fractional order term is 1.5; for the required power control, it is 1; and for the drum water level control, the fractional order term of 1.7 is the best.

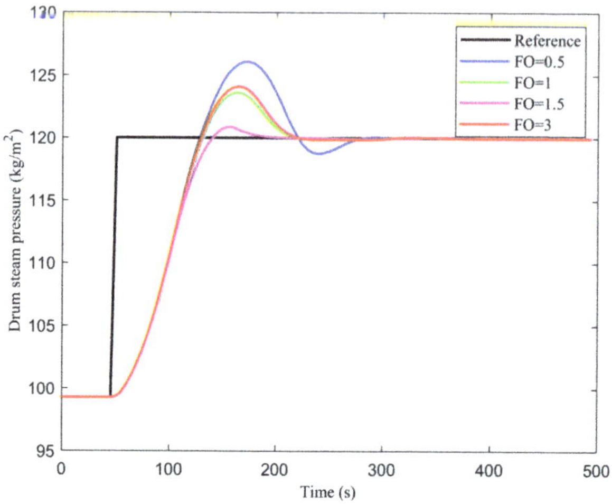

Figure 3. The drum steam pressure with different fractional orders.

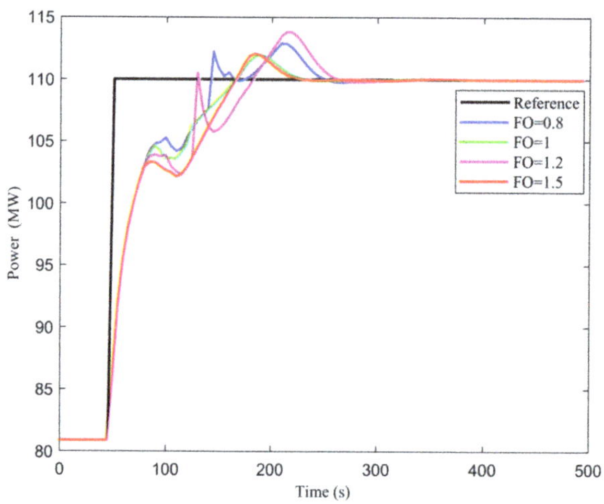

Figure 4. The required power for turbine with different fractional orders.

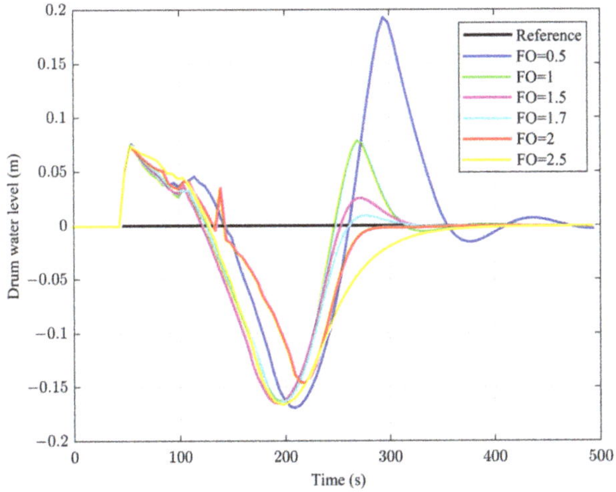

Figure 5. The drum water level with different fractional orders.

4.2. The Influence of Fractional Order Terms to the Different Loops

This section shows the comparison experiment between FOMPC and traditional MPC. The fractional order terms applied in the boiler-turbine system are chosen as 1.5, 1, 1.7 for the three loops, respectively.

In order to evaluate the effectiveness of the proposed method, the following performance indexes are compared, including Integrated Absolute Relative Error ($IARE$), Integral Secondary control output (ISU), Ratio of Integrated Absolute Relative Error ($RIARE$), Ratio of Integral Secondary control output ($RISU$) and combined index (J).

$$IARE_i = \sum_{k=0}^{N_s-1} |r_i(k) - y_i(k)|/r_i(k) \quad (i = 1,2,3) \tag{25}$$

$$ISU_i = \sum_{k=0}^{N_s-1} (u_i(k) - u_{ssi}(k))^2 \quad (i = 1,2,3) \tag{26}$$

$$RIARE_i(C_2, C_1) = \frac{IARE_i(C_2)}{IARE_i(C_1)} \quad (i = 1,2,3) \tag{27}$$

$$RISU_i(C_2, C_1) = \frac{ISU_i(C_2)}{ISU_i(C_1)} \quad (i = 1,2,3) \tag{28}$$

$$J(C_2, C_1) = \frac{1}{3}\sum_{i=1}^{3} \frac{w_1 RIARE_i(C_2, C_1) + w_2 RISU_i(C_2, C_1)}{w_1 + w_2} \tag{29}$$

where u_{ssi} is the steady state value of ith input; C_1,C_2 are the two compared controllers; the weighting factors w_1 and w_2 in equation (29) are chosen as $w_1 = w_2 = 0.5$.

The combined index J for the reference tracking case is 0.6564. From Figure 6, Tables 4 and 5, it shows that the IARE and ISU of fractional order MPC are both smaller than that of traditional MPC, which means the fractional order MPC can obtain better performance with less control effort changes. By choosing suitable fractional order terms, the fractional order MPC shows superiority compared with integer order MPC.

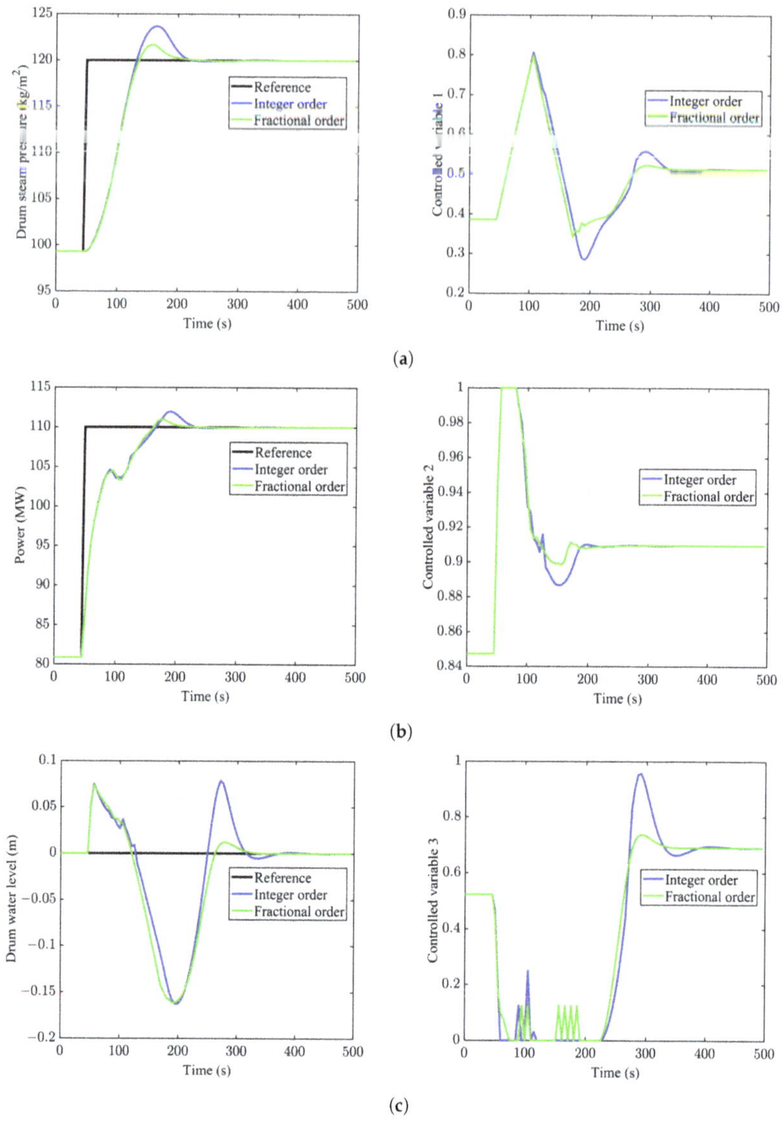

Figure 6. Outputs and inputs of the boiler-turbine system with FOMPC and MPC in the reference tracking experiment (the outputs are listed on the left hand, and the inputs are listed on the right hand). (**a**) control loop for drum steam pressure, (**b**) control loop for required power, (**c**) control loop for drum water level.

Table 4. Performance indexes for $IARE$ and ISU in the power reference tracking experiment.

Index		Drum Steam Pressure	Power	Drum Water Level
$IARE$	MPC	2.0072	1.5837	3.5389
	FOMPC	1.8342	1.4955	3.4534
ISU	MPC	1.1961	0.1068	18.8118
	FOMPC	0.9444	0.1042	17.1870

Table 5. Performance indexes for $RIARE$ and $RISU$ in the power reference tracking experiment (MPC is the C_2 and FOMPC the C_1 according to Equations (27) and (28)).

Index	Drum Steam Pressure	Power	Drum Water Level
$RIARE$	1.0943	1.0590	1.0248
$RISU$	1.2664	1.0251	1.0945

5. Conclusions

This paper proposed a fractional order model predictive controller for the boiler-turbine system. Due to the nonlinearity and multiple variables of the boiler-turbine, the nonlinear MPC with distributed scheme is designed, and the termination conditions are given. The integer order cost function is replaced with the fractional order cost function, which simplified the configration of the weighting factor matrices in the cost function. The number of weighting factors required to be tuned decreases from $N_p + N_c$ to two. According to the simulation for power tracking, it is proved that the fractional order MPC improves the control performance compared with the traditional MPC method. In this work, better control performance is obtained with fractional order MPC; however, how the fractional order effects the control performance is not clear, which can be researched further in the future.

Author Contributions: Methodology, S.Z.; software, S.W.; validation, S.W., R.C. and W.R.; formal analysis, S.W., R.C. and B.L.; writing—original draft preparation, S.Z.; writing—review and editing, S.Z., S.W., R.C., W.R. and B.L.; supervision, S.Z.; project administration, S.Z. and B.L.; funding acquisition, S.Z. and B.L. All authors have read and agreed to the published version of the manuscript.

Funding: This research was funded by Natural Science Foundation of Heilongjiang under grant KY10400210217, Foundation of Fundamental Strengthening Program for Technical Field under grant 2021-JCJQ-JJ-0026, Fundamental Research Funds for the Central Universities undergrant 3072020CFT1501 and Foundation of High-level scientific research guidance project of Harbin Engineering University under grant 3072022QBZ0406.

Institutional Review Board Statement: Not applicable.

Informed Consent Statement: Not applicable.

Data Availability Statement: Not applicable.

Conflicts of Interest: The authors declare no conflict of interest.

References

1. Yang, F.; Tang, C.; Antonietti, M. Natural and artificial humic substances to manage minerals, ions, water, and soil microorganisms. *Chem. Soc. Rev.* **2021**, *50*, 6221–6239. [CrossRef] [PubMed]
2. Zeńczak, W.; Gromadzińska, A.K. Preliminary Analysis of the Use of Solid Biofuels in a Ship's Power System. *Pol. Marit. Res.* **2020**, *27*, 67–79. [CrossRef]
3. Başhan, V.; Demirel, H. Application of Fuzzy Dematel Technique to Assess Most Common Critical Operational Faults of Marine Boilers. *Politek. Derg.* **2019**, *22*, 545–555. [CrossRef]
4. Luschtinetz, T.; Zeńczak, W.; Łuszczyński, D. The Selected Results of the Experimental Research of Solid Fuel Pneumatic Transportation to Ship's Boiler. *Manag. Syst. Prod. Eng.* **2019**, *27*, 144–148. [CrossRef]

5. Taler, J.; Zima, W.; Ocłoń, P.; Grądziel, S.; Taler, D.; Cebula, A.; Jaremkiewicz, M.; Korzeń, A.; Cisek, P.; Kaczmarski, K.; et al. Mathematical model of a supercritical power boiler for simulating rapid changes in boiler thermal loading. *Energy* **2019**, *175*, 580–592. [CrossRef]
6. Lu, X.; Liu, Z.; Ma, L.; Wang, L.; Zhou, K.; Feng, N. A robust optimization approach for optimal load dispatch of community energy hub. *Appl. Energy* **2020**, *259*, 114195. [CrossRef]
7. Piraisoodi, T.; Iruthayarajan, M.W.; Kadhar, K.M.A. An Optimized Nonlinear Controller Design for Boiler–Turbine System Using Evolutionary Algorithms. *IETE J. Res.* **2018**, *64*, 451–462. [CrossRef]
8. Wang, D.; Zhou, Y.; Li, X. A dynamic model used for controller design for fast cut back of coal fired boiler turbine plant. *Energy* **2018**, *144*, 526–534. [CrossRef]
9. Su, Z.g.; Zhao, G.; Yang, J.; Li, Y.g. Disturbance Rejection of Nonlinear Boiler–Turbine Unit Using High-Order Sliding Mode Observer. *IEEE Trans. Syst. Man Cybern. Syst.* **2018**, *50*, 5432–5443. [CrossRef]
10. Liu, X.; Cui, J. Economic model predictive control of boiler-turbine system. *J. Process. Control.* **2018**, *66*, 59–67. [CrossRef]
11. Tian, Z.; Yuan, J.; Xu, L.; Zhang, X.; Wang, J. Model-based adaptive sliding mode control of the subcritical boiler-turbine system with uncertainties. *ISA Trans.* **2018**, *79*, 161–171. [CrossRef] [PubMed]
12. Wei, Q.; Liu, Y.; Lu, J.; Ling, J.; Luan, Z.; Chen, M. A New Integral Critic Learning for Optimal Tracking Control with Applications to Boiler-Turbine Systems. *Optim. Control Appl. Methods* **2021**. [CrossRef]
13. Köhler, J.; Soloperto, R.; Müller, M.A.; Allgöwer, F. A computationally efficient robust model predictive control framework for uncertain nonlinear systems. *IEEE Trans. Autom. Control* **2020**, *66*, 794–801. [CrossRef]
14. Reynolds, J.; Rezgui, Y.; Kwan, A.; Piriou, S. A zone-level, building energy optimisation combining an artificial neural network, a genetic algorithm, and model predictive control. *Energy* **2018**, *151*, 729–739. [CrossRef]
15. Chi, C.; Cajo, R.; Zhao, S.; Liu, G.P.; Ionescu, C.M. Fractional Order Distributed Model Predictive Control of Fast and Strong Interacting Systems. *Fractal Fract.* **2022**, *6*, 179. [CrossRef]
16. Ghita, M.; Cajo Diaz, R.A.; Birs, I.R.; Copot, D.; Ionescu, C.M. Ergonomic and economic office light level control. *Energies* **2022**, *15*, 734. [CrossRef]
17. Song, D.; Chang, Q.; Zheng, S.; Yang, S.; Yang, J.; Joo, Y.H. Adaptive model predictive control for yaw system of variable-speed wind turbines. *J. Mod. Power Syst. Clean Energy* **2020**, *9*, 219–224. [CrossRef]
18. Sorcia-Vázquez, F.; Garcia-Beltran, C.; Valencia-Palomo, G.; Brizuela-Mendoza, J.; Rumbo-Morales, J. Decentralized robust tube-based model predictive control: Application to a four-tank-system. *Rev. Mex. Ingeniería Química* **2020**, *19*, 1135–1151. [CrossRef]
19. Rumbo Morales, J.Y.; López López, G.; Alvarado, V.M.; Torres Cantero, C.A.; Azcaray Rivera, H.R. Optimal Predictive Control for a Pressure Oscillation Adsorption Process for Producing Bioethanol. *Comput. Sist.* **2019**, *23*, 1593–1617.
20. Morales, J.Y.R.; López, G.L.; Martínez, V.M.A.; Vázquez, F.d.J.S.; Mendoza, J.A.B.; García, M.M. Parametric study and control of a pressure swing adsorption process to separate the water-ethanol mixture under disturbances. *Sep. Purif. Technol.* **2020**, *236*, 116214. [CrossRef]
21. Zhang, X.; Zhang, L.; Zhang, Y. Model predictive current control for PMSM drives with parameter robustness improvement. *IEEE Trans. Power Electron.* **2018**, *34*, 1645–1657. [CrossRef]
22. Shen, C.; Shi, Y.; Buckham, B. Trajectory tracking control of an autonomous underwater vehicle using Lyapunov-based model predictive control. *IEEE Trans. Ind. Electron.* **2017**, *65*, 5796–5805. [CrossRef]
23. Zhu, H.; Zhao, G.; Sun, L.; Lee, K.Y. Nonlinear predictive control for a boiler–turbine unit based on a local model network and immune genetic algorithm. *Sustainability* **2019**, *11*, 5102. [CrossRef]
24. Zhang, Y.; Decardi-Nelson, B.; Liu, J.; Shen, J.; Liu, J. Zone economic model predictive control of a coal-fired boiler-turbine generating system. *Chem. Eng. Res. Des.* **2020**, *153*, 246–256. [CrossRef]
25. Liu, X.; Cui, J. Fuzzy economic model predictive control for thermal power plant. *IET Control Theory Appl.* **2019**, *13*, 1113–1120. [CrossRef]
26. Wang, G.; Wu, J.; Ma, X. A nonlinear state-feedback state-feedforward tracking control strategy for a boiler-turbine unit. *Asian J. Control* **2020**, *22*, 2004–2016. [CrossRef]
27. Kong, L.; Yuan, J. Disturbance-observer-based fuzzy model predictive control for nonlinear processes with disturbances and input constraints. *ISA Trans.* **2019**, *90*, 74–88. [CrossRef]
28. Cui, J.; Chai, T.; Liu, X. Deep-neural-network-based economic model predictive control for ultrasupercritical power plant. *IEEE Trans. Ind. Inform.* **2020**, *16*, 5905–5913. [CrossRef]
29. Sanchez, R.O.; Rumbo Morales, J.Y.; Ortiz Torres, G.; Pérez Vidal, A.F.; Valdez Resendiz, J.E.; Sorcia Vázquez, F.d.J.; Nava, N.V. Discrete State-Feedback Control Design with D-Stability and Genetic Algorithm for LED Driver Using a Buck Converter. *Int. Trans. Electr. Energy Syst.* **2022**, *2022*, 8165149. [CrossRef]
30. Nedić, A.; Liu, J. Distributed optimization for control. *Annu. Rev. Control Robot. Auton. Syst.* **2018**, *1*, 77–103. [CrossRef]
31. Bell, R.; Åström, K.J. *Dynamic Models for Boiler-Turbine-Alternator Units: Data Logs and Parameter Estimation for a 160 MW Unit*; Technical Reports; Lund Institute of Technology: Lund, Sweden, 1987; TRFT-3192.
32. De Keyser, R. Model based predictive control for linear systems. In *UNESCO Encyclopaedia of Life Support Systems, Robotics and Automation*; Article Contribution 6.43.16.1; Eolss Publishers Co., Ltd.: Oxford, UK, 2003; Volume XI.

33. De Keyser, R.; Ionescu, C.M. The disturbance model in model based predictive control. In Proceedings of the Proceedings of 2003 IEEE Conference on Control Applications, Istanbul, Turkey, 25 June 2003; IEEE: Piscataway, NJ, USA, 2003; Volume 1, pp. 446–451.
34. Fernandez, E.; Ipanaque, W.; Cajo, R.; De Keyser, R. Classical and advanced control methods applied to an anaerobic digestion reactor model. In Proceedings of the 2019 IEEE CHILEAN Conference on Electrical, Electronics Engineering, Information and Communication Technologies (CHILECON), Valparaiso, Chile, 13–27 November 2019; IEEE: Piscataway, NJ, USA, 2019; pp. 1–7.
35. Romero, M.; de Madrid, A.P.; Vinagre, B.M. Arbitrary real-order cost functions for signals and systems. *Signal Process.* **2011**, *91*, 372–378. [CrossRef]
36. Zhao, S.; Cajo, R.; De Keyser, R.; Liu, S.; Ionescu, C.M. Nonlinear predictive control applied to steam/water loop in large scale ships. *IFAC-PapersOnLine* **2019**, *52*, 868–873. [CrossRef]

Journal of
Marine Science and Engineering

MDPI

Article

Classification of Electrical Power Disturbances on Hybrid-Electric Ferries Using Wavelet Transform and Neural Network

Aleksandar Cuculić *, Luka Draščić, Ivan Panić and Jasmin Ćelić

Faculty of Maritime Studies, University of Rijeka, Studentska 2, 51000 Rijeka, Croatia
* Correspondence: aleksandar.cuculic@pfri.uniri.hr

Abstract: Electrical power systems on hybrid-electric ferries are characterized by the intensive use of power electronics and a complex usage profile with the often-limited power of battery storage. It is extremely important to detect faults in a timely manner, which can lead to system malfunctions that can directly affect the safety and economic performance of the vessel. In this paper, a power disturbance classification method for hybrid-electric ferries is developed based on a wavelet transform and a neural network classifier. For each of the observed power disturbance categories, 200 signals were artificially generated. A discrete wavelet transform was applied to these signals, allowing different time-frequency resolutions to be used for different frequencies. Three statistical parameters are calculated for each coefficient: Standard deviation, entropy and asymmetry of the signal, providing a total of 18 variables for a signal. A neural network with 18 input neurons, 3 hidden neurons, and 6 output neurons was used to detect the aforementioned perturbations. The classification models with different wavelets were analyzed based on accuracy, confusion matrices, and other parameters. The analysis showed that the proposed model can be successfully used for the detection and classification of disturbances in the considered vessels, which allows the implementation of better and more efficient algorithms for energy management.

Keywords: hybrid-electric ferry; maritime transport; marine electrical systems; electrical power disturbances; wavelet transform; neural network

Citation: Cuculić, A.; Draščić, L.; Panić, I.; Ćelić, J. Classification of Electrical Power Disturbances on Hybrid-Electric Ferries Using Wavelet Transform and Neural Network. *J. Mar. Sci. Eng.* **2022**, *10*, 1190. https://doi.org/10.3390/jmse10091190

Academic Editors: Marco Altosole, Maria Acanfora, Flavio Balsamo and Bowen Xing

Received: 20 July 2022
Accepted: 22 August 2022
Published: 25 August 2022

1. Introduction

Ferry transport of goods and passengers plays an important socio-economic role in most coastal countries. This is particularly true in Europe, where an estimated 900 ferries are currently active, accounting for about 70% of global ferry traffic [1]. Despite its undoubted importance, the increase in ferry traffic brings with it a number of problems, of which environmental pollution is perhaps the most important.

The majority of ferries in operation today use conventional marine diesel engines for propulsion and electrical power generation, which pose a significant challenge in meeting the requirements of future environmental standards [2]. For this reason, there is a trend towards hybridization and electrification of the existing ferry fleet, especially in EU countries, which is further encouraged by generous government subsidies for the use of alternative and environmentally friendly energy sources [3].

Key technologies for the development of the current generation of hybrid-electric ferries and associated land-based infrastructure are electric propulsion and energy storage systems (ES) [4–6]. At the beginning of the application, in the first decade of the 21st century, all-electric and hybrid ferries had a relatively small capacity of passengers and cars and operated on short routes. These first ships served primarily as test platforms for evaluating new technologies and gaining in-depth knowledge of the advantages of use and possible disadvantages compared to conventional propulsion systems. As operational experience has shown that such solutions offer significant potential to reduce fuel consumption, pollutant emissions and operating costs [7], ferry operators have started to introduce all-electric and hybrid vessels with higher capacity and range in their fleets. The largest hybrid-electric plug-in ferry currently in operation has a capacity of 2000 passengers, 419 cars and 642 trailers, with an installed 5000 kWh Li-Ion battery ES [8].

Conventional ferries with diesel-mechanical propulsion (which still represent the majority in the global ferry fleet) have simple power grids fed by relatively low-power diesel generators, requiring simple rule-based power management systems and electrical protection systems. On the other hand, due to the peculiarities of the power plant on such ships and the increased requirements for installed power, it is necessary to develop new design solutions based on the integration of several different power sources that often requires the use of a combination of DC and AC networks within the same system. One of the challenges is the need to connect the ship's own power plant to the shore-side power grid for cold ironing and ES charging.

It can be said that electrical networks on hybrid-electric ships have become very similar to terrestrial micro grids mainly due to topology, islanding and increased use of power electronics and energy storage devices, but in some segments, they have to meet much stricter requirements. This refers primarily to very high reliability requirements, the deployment restraints in relatively limited ship space and fast dynamic load changes, especially in consideration of large pumps and electric propulsion [9].

An increased number of power electronics converters for energy storage grid connection and power flow control can be the source of power quality disturbances (PQ). A review of the literature found that these disturbances are mainly manifested in voltage and frequency fluctuations, transients, sags, swells, harmonic distortions, power factor variations, etc. [10–13]. The mentioned types of interference and their characteristics are clearly defined by industry standards, which apply regardless of whether it is a ship- or land-based system [14,15]. The maximum permissible voltage and frequency deviations of the ship's network, as well as the duration of transients, are specified in the regulations of the leading classification societies, and the electrical protection devices are adjusted accordingly [16–18]. On most ships with classical propulsion and a standard rule-based power management system (PMS), the protection is activated only when a certain parameter exceeds the set point, but most phenomena, especially transients of short duration, harmonics and voltage and frequency changes, whose values are below the limits at which the protection operates, are not detected. It is precisely such unrecognized phenomena that very often lead to disruptions in the ship's energy supply system, often with serious consequences. This problem, which points to the necessity of continuous monitoring detecting the aforementioned phenomena, is discussed in great detail in the paper [19].

On hybrid-electric ships with fast dynamic load changes as well as island grids with limited power, such undetected changes can have a very strong impact on the power quality [20], and may cause various side effects, such as failure of digital and communication devices, unwanted tripping of circuit breakers and protection devices, overheating of electric motors, etc. Micro grid topology on hybrid and electric ferries, as well as the number and type of power sources, depend on the planned route, speed, number of voyages, and characteristics of land infrastructure at ports. This ultimately leads to very different power grid solutions, which often also react differently to the above phenomena. It presents a major challenge to the PMS, which must respond to any changes in power system parameters that may jeopardize the safety of the ship and the availability of power through the timely coordination of multiple power sources.

Research and practice have shown that conventional rule-based PMSs are not up to these challenges and that optimization-based PMSs and machine learning are better suited for hybrid and electric ships [21–25]. In order for such a PMS to properly perform its function, it is important that, among other things, it receives timely and accurate information about disturbances and changes in the parameters of the power system. In addition, the detected disturbances must be classified so that the PMS can associate them with a possible system failure or human error and take the necessary action.

The first step towards overcoming the aforementioned challenges is to identify the faults present in the system which often involve a very large number of complex or non-linear perturbations that must be generalized, quickly identified and classified. In addition, the method used should be generally applicable to all types of the aforementioned faults, detect them all quickly and accurately, and be relatively easy to implement in the ship's PMS. The information by which the computer can distinguish different disturbances is obtained by signal analysis. Nowadays, it is possible to analyze the disturbances using different transforms, of which the best known are the Fourier transform, the short-time Fourier transform and the Wavelet transform, which is used in this work.

By reviewing recent scientific literature [26–30] and similar examples from other industrial branches, it has been shown that it can be beneficial to use neural networks for classification of such signals. With proper preparation and processing of input data, the neural network can classify faults quickly and with great accuracy, which is a great help in achieving the ultimate goal of timely detection of faults and improvement of power quality and power supply reliability on board hybrid-electric ferries. Although this work deals with hybrid electric ferries, because these vessels present special challenges for power and energy management, mentioned in the introduction, the proposed method can be useful in identifying power quality problems on all other types of vessels, regardless of the type and propulsion systems, and can contribute to the safety of the vessel, increase the efficiency of the power system, and develop better and more efficient PMS systems.

This paper is organized as follows. Electrical power disturbances and their effects in ship micro grids are analyzed in Section 2. The motivation and reasons for using the wavelet transform are presented in Section 3. Analysis of electrical power disturbances using discrete wavelet transform is described in Section 4. In Section 5, neural network classifier is presented. Neural network input data are explained in Section 6. Description of the power disturbance classification process is described in Section 7. Classification model performance analysis was carried out in Section 8. Finally, conclusions are outlined in Section 9.

2. Power Disturbances and Their Effects in Ship Micro Grids

Power disturbances can have harmful consequences for power grid components, but also for the devices of the end users. What makes this phenomenon even more problematic is the fact that the impact on equipment is usually not visible until a fault occurs. Even if equipment failure does not occur, poor power quality increases losses and reduces equipment lifetime [31].

Electromagnetic interference can be divided into low-frequency interference with a range up to 9 kHz and high-frequency interference with a range above 9 kHz. Here, each frequency range is subdivided into the conducted range and the radiated range, depending on the propagation mode. In addition to frequency range, disturbances can be divided by state (stable and unstable), by duration (very short to 3 periods of fundamental frequency, short to 3 min, long to 3 h, and very long over 3 h), and by waveform [31]. Some of the most common low-frequency disturbances that occur in marine electrical systems are: Voltage surges, voltage dips, higher order harmonic disturbances, oscillatory transients, and voltage notches.

Voltage surges are a phenomenon in the electrical system when the RMS value of the voltage increases by 10% to 80% of the nominal value. They can generate additional thermal load on the equipment and wiring, stressing and accelerating the wear of the insulation material. Typically, voltage spikes last between 10 ms and 1 min [15]. They can occur during a ground fault of a single-phase conductor in an insulated neutral system, which is the predominant topology on ships. In this case, the voltages of the healthy phases toward ground rise from the phase value to the line value, i.e., there is an overvoltage on the right phases that continues as long as the ground fault is not removed. Internal voltage surges can also occur when the system is suddenly unloaded due to intentional or unintentional disconnection of a large load or when capacitor banks are switched on [32].

Voltage dips are defined as the reduction of the RMS value of the grid voltage to between 10% and 90% of the nominal value. Most voltage dips do not fall below 50% of the nominal value and usually last between 10 ms and 1 min [31]. They are usually caused by switching on large loads with high inrush current, such as large induction motors or propulsion transformers. An internal voltage drop in the ship's electrical network can also occur in the event of a phase short circuit. Voltage drops can cause computers and programmable logic controllers (PLCs) to reset or shut down, relays and contactors to trip unintentionally [33], and frequency inverter operation problems due to unstable firing circuits that generate control pulses for semiconductor valves. In electric motors, the voltage drop causes a reduction in torque, which can result in the motor not starting when loaded.

Both voltage surges and voltage dips can be represented by Equation (1):

$$v(t) = A \cdot (1 + \alpha \cdot (h(t - t_1) - h(t - t_2))) \cdot \sin(\omega t - \varphi), \quad (1)$$

where $v(t)$ is instantaneous voltage, A voltage magnitude, α coefficient that determines the amplitude of the interference, $h(t)$ Heaviside step function, t time, t_1 moment of the beginning of interference, t_2 moment of the end of interference, ω angular frequency and φ phase angle.

Harmonics are considered sinusoidal voltages or currents whose frequency differs from the fundamental frequency of the network and can be divided into three groups: integer harmonics, interharmonics, and subharmonics. The main cause of harmonic voltage distortion in the electrical network of ships is power electronic converters used to control propulsion and general service motors. The presence of harmonics in the grid voltage causes a number of problems, mainly in electric motors and transformers. Increased iron losses occur due to hysteresis losses proportional to frequency and eddy currents proportional to the square of the frequency. Harmonic currents in electric motors cause torsional vibrations that can damage the bearings and shaft, especially if there is resonance between the torsional vibrations and the shaft.

Oscillatory transients are defined as momentary deviations in voltage or current from steady state. There is no clear boundary between voltage fluctuations and oscillatory transients, but any event lasting less than 10 ms can be considered a transient. [34]. They can be divided into low frequency with a frequency of less than 5 kHz and a duration of 0.3 ms to 50 ms, medium frequency with frequencies of 5 kHz to 500 kHz and a duration of several tens of microseconds, and high frequency with a transition frequency of more than 500 kHz and a duration of several microseconds [35]. Oscillatory transients can be represented by Equation (2):

$$v(t) = A \cdot \left[\sin(\omega t - \varphi) + \alpha e^{\frac{-(t - t_1)}{\tau}} \cdot \sin(\omega_n (t - t_1) - \vartheta) \cdot (h(t - t_2)\, h(t - t_1)) \right], \quad (2)$$

where τ is time constant and ϑ disturbance phase angle. The rest of the parameters are same as in (1).

Microcomputers and PLCs are particularly sensitive to oscillatory transients, which may significantly reduce their service life [36]. If the oscillatory transient voltage is applied to the input of the voltage source frequency converter, a current flow, which charges the capacitor used to stabilize the voltage in the DC circuit. If this current is not limited by serial chokes or a transformer, the capacitor is suddenly charged to a value higher than the predicted one, which creates a state of overvoltage. In this case, a surge protection is activated which disconnects the frequency converter and the electric motor from the mains [37].

Voltage notches are periodic short-term disturbances of power quality caused by operation of power electronics devices during current commutation. This type of interference is located between harmonics and transients. The reason is that, on the one hand, notches occur during normal operation and can be isolated as part of the harmonic spectrum of the voltage signal, but on the other hand their frequency is high and cannot be analyzed with standard equipment for harmonic distortion analysis [35]. Voltage notches can be represented by Equation (3):

$$v(t) = A[\sin(\omega t - \varphi) - sgn(\sin(\omega t - \varphi)) \sum_{n=0}^{i} k(v(t - (t_c + s{\cdot}n)) - v(t - (t_d + s{\cdot}n)))] \quad (3)$$

where k is coefficient that determines the depth of the notch, n certain period in which the disturbance occurs, i the total number of periods in which the disturbance occurs, t_c and t_d the moment of the start or end of the disturbance, respectively.

Voltage notches in the mains voltage cause problems with synchronization clocks and counters that use the natural voltage zero crossing for counting. If a voltage notch that reaches zero occurs, then the counter will increase the value and thus count more than the actual value [38]. Also, in the event that the additional mains voltage passes through zero, the circuit breakers may break prematurely. If voltage notches occur together with harmonic distortion in the voltage and frequency control circuits for generators, voltage and frequency instability in the network can occur [39].

A summary of power disturbances and their causes and effects is given in Table 1.

Table 1. Summary of power disturbances, their causes and effects.

Type of Disturbance and Typical Waveform	Causes of Disturbance	Negative Effects
Voltage surges	Ground fault of a single-phase conductor in an insulated neutral system. Sudden disconnection of large load. Switching on capacitor bank.	Thermal loading on the equipment and wiring. Stressing the and accelerating the wear of the insulation material.
 Voltage dips	Switching on large loads with high inrush current. Interphase short circuit.	Reset or shutting down PLCs and computers. Trip of relays and contactors. Unstable firing circuits that generate control pulses for semiconductor valves. Torque reduction in electric motors.
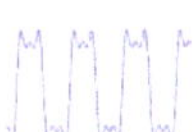 Higher order harmonic disturbances	Power electronic converters used to control propulsion and general service motors.	A number of problems, mainly in electric motors and transformers. Iron losses due to hysteresis losses and eddy currents proportional to frequency. Torsional vibrations that can damage the bearings and shaft.
 Oscillatory transients	Switching on/off inductive loads. Loose connections. Power electronics switching. Sudden circuit breaker trips. Electrostatic discharge. Arc faults.	Reducing service life of PLCs, microcomputers and other sensitive equipment. Overvoltage at DC link capacitors in voltage source converters.
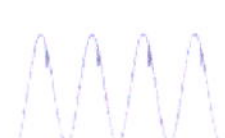 Voltage notches	Current commutation in power electronics devices	Problems with synchronization clocks and counters. Circuit breakers may break prematurely.May cause voltage and frequency instability.

3. Motivation for Using Wavelet Transform

Displaying signals in the time domain does not provide enough information to effectively identify different types of disturbances in the power system. Therefore, it is necessary to apply a mathematical transformation to the base signal to obtain additional information about it and to obtain relevant and sufficiently accurate parameters required for neural network training.

The Fourier transform, which is most commonly used to analyze periodic signals, is a reversible transform, meaning that it can be switched between the time and frequency domains at will, but only one of the two domains can be represented at a time. This means that information from the frequency domain is not available in the time domain and vice versa. For stationary signals whose frequency content does not change over time, i.e., which have the same frequency components throughout, this is not a problem, since it is not necessary to know at what time certain frequency components of the signal occur. On the other hand, the disturbances mentioned in the previous chapter are typical representatives of non-stationary signals whose time-frequency characteristics and duration depend on a series of random events such as disturbances and operating conditions.

The above-mentioned drawbacks of the Fourier transform for the analysis of non-stationary signals can be solved by applying the short-time Fourier transform (STFT), in which the original signal is divided into equal parts at which the signal can be considered stationary. Using a suitable window function, a three-dimensional diagram is obtained in which the vertical axis represents amplitude and the horizontal axis represents time and frequency. From such a diagram, one can see at what time there are frequency components belonging to that part of the signal, i.e., one obtains the time-frequency representation of the signal [40].

The disadvantage of STFT is that one does not know which individual frequency components are present at any given time. Instead, it is only possible to know the time intervals in which certain frequency bands are present. This problem is caused by the window function having a finite width that covers only a portion of the signal, resulting in poor frequency resolution. When the window is infinite, the same result is obtained as with the Fourier transform, i.e., excellent resolution in the frequency domain and no resolution in the time domain. The narrower the window, the better the time resolution and the better the approximation to a stationary signal, but at the same time the frequency resolution is lower and vice versa. Therefore, the main problem of the short-time Fourier transform is the correct choice of the width of the window function that can be used to analyze different signals.

The wavelet transform allows the analysis of signals with multiple resolutions by using different resolutions at different frequencies. Multiresolution analysis is particularly suitable for signals where high-frequency components are short-lived while low-frequency components are long-lived, and these are precisely the interference signals considered in this work. Short-lived high-frequency components require very accurate temporal localization achieved by a narrow wavelet, resulting in poor frequency resolution, and vice versa. Low frequency components often determine most of the signal characteristics, and these characteristics are best quantified when the frequency resolution is as good as possible [41,42]. In view of the above, a multilevel discrete wavelet transform (DWT) is chosen for the power disturbance signal analysis.

4. Power Disturbance Analysis Using Multilevel DWT

The DWT is the most common method for implementing the wavelet transform in computers. It allows signal analysis (decomposition) and synthesis (reconstruction), and filters are used for this purpose. Multilevel decomposition and signal reconstruction is most often used in practice. The reason is that it allows higher frequency resolution, which means that the presence of individual frequency bands in time can be determined with greater accuracy.

The DWT uses filters with different cutoff frequencies to analyze signals at different scales. At each level of signal analysis, two half-band filters with impulse response are used, one of which is a low-pass filter and the other a high-pass filter. Each filter consists of a number of coefficients that differ for the low-pass filter, the high-pass filter, and for signal decomposition and reconstruction. The number of coefficients depends on the type of wave. In signal processing, only orthonormal waves such as Daubechies, Coiflet, and Symlet waves are used because they have both the DWT and inverse DWT capability.

Daubechies waves used in this study are referred to as dbN, where N is the order of the waves, i.e., the number of vanishing moments. The length of the filter or the number of filter coefficients is 2N. The largest wave order is 45, but waves up to order 10 are most commonly used [43].

As the order of the waves increases, the time required to perform the transformation increases. On the other hand, such waves are smoother and better localized in the time domain, which is why the oscillations in the original signal can be better represented.

The operation of DWT can be explained with reference to Figure 1. The discrete input signal x[n] consists of n samples and contains the maximum frequency f. HP and LP denote high and low bandwidth filters, the symbol $\downarrow 2$ denotes the subsampling method with 2, while cD and cA denote the detail coefficients and the approximation coefficients, respectively. The signal x[n] must be sampled at twice the frequency of the signal bandwidth. Since the bandwidth is B = f - 0, the sampling frequency is 2f.

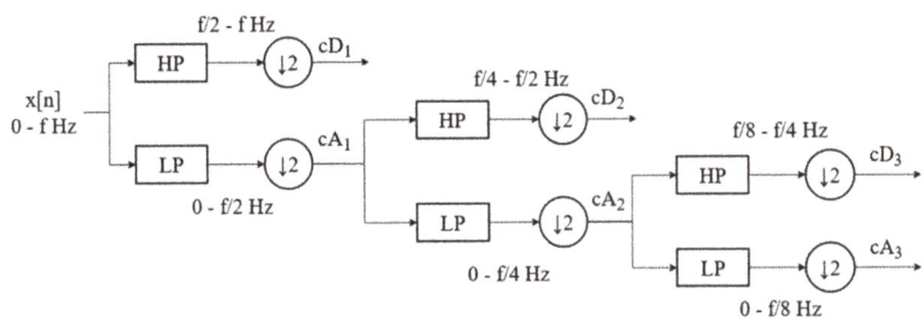

Figure 1. Example of signal decomposition (3 levels).

The interfering signal x[n] is first passed through a LP which transmits frequencies from 0 Hz to half the maximum frequency of the signal. Since the signal after filtering contains half the bandwidth than before filtering, the signal contains twice as many samples as required. A subsampling of 2 is performed (every other sample of the filtered signal is discarded) to remove this redundancy of information on the filtered signal and obtain cA.

Then, the same input signal x[n] is passed through a HP that sweeps frequencies from f/2 to f. As with LP, the bandwidth is reduced by half, and subsampling must be performed. During subsampling, the content of the frequency components in the range f/2 to f is shifted to the new range 0 to f/2. Since there are no frequency components in this frequency range because they were previously removed by a HP filter, there is no loss of information and it is possible to reconstruct the signal if necessary. In this way, the cD is obtained, which are twice the number of samples of the input signal x [n]. The process that constitutes one stage of the decomposition can be described mathematically by Equations (4) and (5),

$$cA[k] = \sum_{n} x[n] \cdot LP[2k - n] \tag{4}$$

$$cD[k] = \sum_{n} x[n] \cdot HP[2k - n] \tag{5}$$

where *k* is the number of approximation samples (details), *n* is the number of samples, and LP and HP are the low-pass and high-pass filter functions, respectively.

The approximation coefficients represent the general trend of the original signal, while the details contain the high frequency components of the signal. The approximation is a low-resolution representation of the original signal, and the details represent the difference between two successive approximations [44].

The values of the low-pass and high-pass filters with length L are not chosen arbitrarily, but the rule described in Equation (6) applies.

$$HP[L-1-n] = (-1)^n \cdot LP[n] \tag{6}$$

The coefficients of the high-pass filter are actually the coefficients of the low-pass filter in reverse order, and every other coefficient has the opposite sign. Such filters are called quadrature mirror filters and are often used in signal processing.The symbolic representation of the frequency responses of the filters obtained by decomposition into three stages is shown in Figure 2.

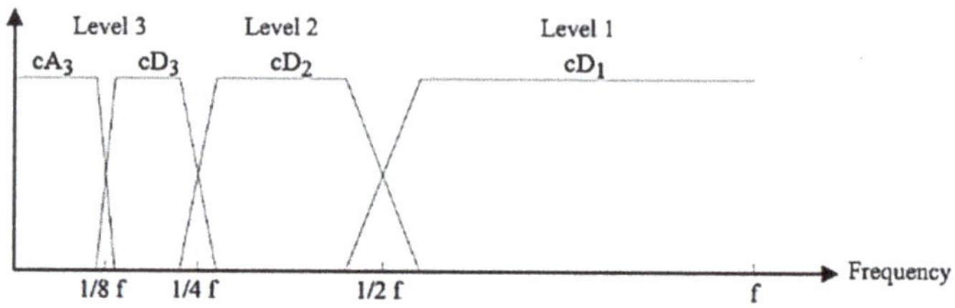

Figure 2. Symbolic representation of filter frequency responses for three level decomposition.

As the degree of decomposition increases, narrower frequency bands are obtained. The narrowest and lowest frequency band is always that of the approximation coefficients of the last stage. The bandwidth can be calculated according to Equations (7) and (8) [45],

$$B_{cA} = \left[0,\ 2^{-p-1} \cdot f_s\right] \tag{7}$$

$$B_{cDp} = \left[2^{-p-1} \cdot f_s,\ 2^{-p} \cdot f_s\right] \tag{8}$$

where p is the decomposition level, B_{cA} is the bandwidth of the approximation coefficients, B_{cDp} is the bandwidth of the detail coefficients at the level p of the samples (details), and f_s is the sampling frequency.

The signal reconstruction process is shown in Figure 3. The oversampling procedure with 2 (↑ 2) is performed over the coefficients cA and cD. This means that another sample with the value 0 or an interpolated value of adjacent samples is added between each of the two samples of the mentioned coefficient. Such signals are passed through a high-pass filter HP′ and a low-pass filter LP′ for signal synthesis [40]. These filters are responsible for returning the coefficients to the original frequency domain [46]. The filters for signal synthesis are identical to those for signal analysis, except that their coefficients are listed in reverse order. After the signal has passed through the filters, they add up to obtain the output signal x[n].

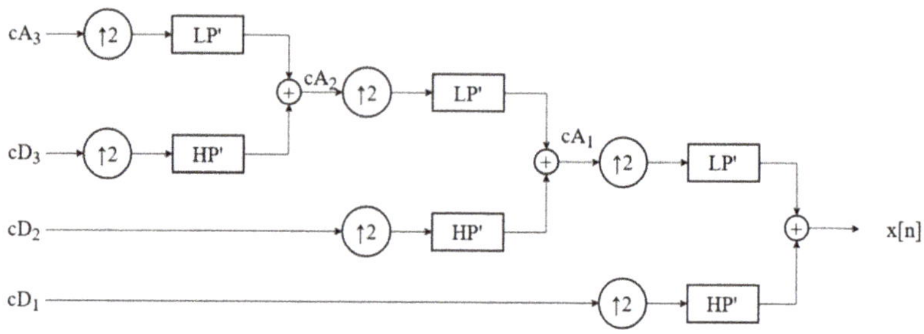

Figure 3. Example of signal reconstruction (3 levels).

The signal reconstruction can be described by Equation (9) [40].

$$x[n] = \sum_{k=-\infty}^{\infty} \left(cA[k] \cdot LP'[2k-n]\right) + \left(cD[k] \cdot HP'[2k-n]\right) \tag{9}$$

As an example, the DWT decomposition of the oscillatory transient is performed with a db6 wavelet with composition level p = 5. The example noise signal lasts 1 s and is sampled at fs = 8 kHz, which means that the maximum frequency that can be detected is 4 kHz. The transform coefficients of DWT have the frequency bands calculated using Equations (7) and (8) and listed in Table 2.

Table 2. Frequency range of DWT coefficients for db6 wavelet and f_s = 8 kHz.

DWT Coefficient	cA	cD5	cD4	cD3	cD2	cD1
Frequency range (Hz)	0–125	125–250	250–500	500–1000	1000–2000	2000–4000

The DWT decomposition of the oscillatory transient is shown in Figure 4, which shows a total of seven plots. The first diagram shows the original signal, the second the approximation coefficient cA, and from the third to the last diagram the cD1-cD5 are plotted.

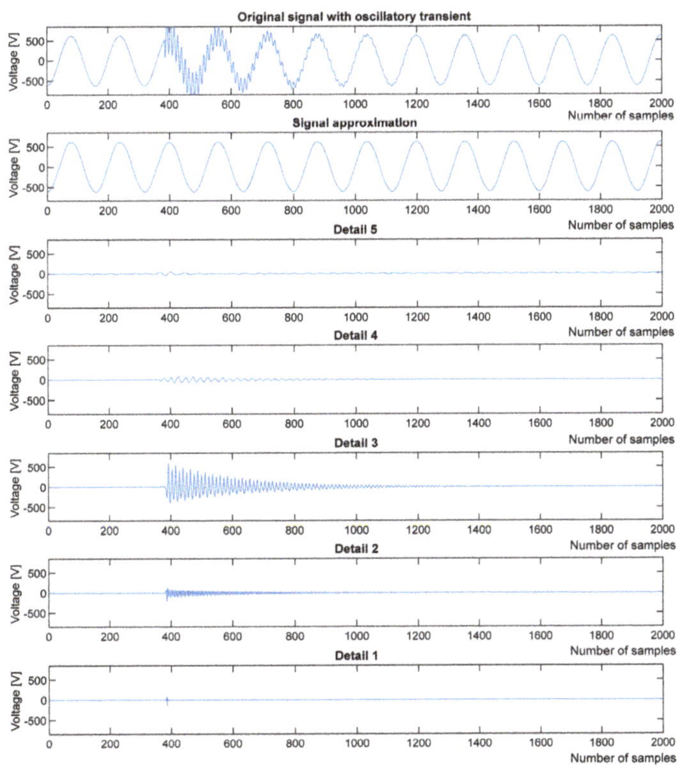

Figure 4. DWT decomposition on the example of oscillatory transient.

It should be noted that for clarity, Figure 4 shows only the first 2000 samples of the signal, but each signal displayed consists of 8000 samples. The original signal has a fundamental frequency of 50 Hz, while a frequency of 733 Hz occurs during the oscillating transient. The cA contains a frequency component of 50 Hz, which makes the signal shown in this diagram look like a pure sinusoidal voltage.

The graphs of cD5 and cD4 are zero because there is no frequency component of the original signal in these frequency bands. The frequency component of the 733 Hz signal is included in the cD3 graph, where it can be seen that it begins at the 380th sample, after which the amplitude decreases as the oscillatory transient phase in the original signal decreases until it disappears. In the cD2 diagram, the same phenomenon can be observed as in the cD3 diagram, but with a much smaller amplitude. Ideally, this detail should have a value of zero, but due to the imperfection of the wave filters, the frequency bands overlap, as symbolically shown in Figure 2. Finally, in the cD1 diagram, all values are close to zero because the original signal has no frequency components in this frequency band.

Similarly, the DWT decomposition of voltage rise, fall and dip also shows the fundamental frequency component in the approximation diagram as in the previous cases. A sudden change of the voltage value in the original signal at the beginning and at the end of the transient causes peaks in the detail plots at the same time.

5. Neural Network Classifier

The proposed model for classifying electromagnetic disturbances in the energy system of the considered ships is based on the use of a shallow feed-forward neural network.

The neurons receive input signals multiplied by the associated weights, add the obtained products, add a sensitivity threshold to this sum, and pass it through the activation function. The output of the jth neuron y_j can be expressed by Equation (10) as follows:

$$y_j = \psi\left(\sum_{i=0}^{n} w_{ij}x_i + \theta\right) \tag{10}$$

where ψ is transfer or activation function of the j-th neuron, x_i input signal, w_{ij} weight coefficients at the input of j-th neuron and ι sensitivity threshold.

Input layer receives the input signals and forwards them to the hidden layer. It does not perform any processing on the input signals, nor are weights or sensitivity thresholds assigned to them. The number of neurons in the input layer is equal to the number of input variables.

The central layer is called the hidden layer and contains neurons that perform data processing. The number of neurons in the hidden layer is determined by the trial-and-error method, starting with the smallest number and observing the resulting error. Then, a minimal number of neurons is chosen at which a satisfactorily small error is obtained. Too few hidden neurons will result in large learning errors and poor generalization due to undertraining, while too many hidden neurons will result in a small learning error but learning will be unnecessarily slow. The activation function in the hidden layer must be nonlinear to approximate the nonlinear and linear relationships between input and output variables [47]. In this classification model, the hyperbolic tangent function is used. It provides values in the bounded range from -1 to 1. The outputs are oriented to zero, so that it can be achieved that the mean of all outputs in a layer is zero, which facilitates and accelerates learning in the next layer of neurons.

The outputs of the hidden layer are directed to the last, the output layer, which is the output of the network. The number of neurons in the output layer corresponds to the number of categories used in classification, i.e., each neuron represents a category. When the input vector of the corresponding category is input to the network, the corresponding neuron in the output layer should provide output 1 and the other neurons should provide output 0 [47]. To achieve this, a SoftMax function is used in the output layer to generate a probability vector for each category.

After the neural network is formed, the weight coefficients and thresholds must be determined. This procedure is important because the time required for the neural network to learn "well" depends on these parameters, i.e., it directly affects the convergence rate of the objective function towards a minimum [48]. If the network is known a priori, this knowledge can be used to set weights to specific values. In most cases, such knowledge is not available, so the weights are initialized with random values that are uniformly distributed in a certain interval. One of the methods used to achieve this is the Nguyen–Widrow method, which is standard in the MATLAB environment. This method generates random values in the interval $[-1, 1]$. This distributes the neuron inputs (sum of weights) approximately uniformly in the active region of the neuron, avoiding saturation of the neuron and slowdown in learning the network. The interval $[-2, 2]$ is considered as the active region for the hyperbolic tangent function. It should also be mentioned that this method can only be used for those transfer functions that have a limited active range, as is the case for the sigmoidal function or the hyperbolic tangent function [49].

When using a neural network, it goes through three phases of work: Learning, Validation, and Testing. Each phase requires separate data sets, which are obtained by dividing the total number of samples into three subsets in certain proportions. Often 70% of the samples are reserved for learning, 15% for validation, and 15% for testing the network, but other ratios are possible. The partitioning is usually done by random sampling.

Training of the neural network is an iterative process to adjust the network parameters according to a given algorithm. The learning process aims to determine such values of the network parameters for which the error is minimal for the whole set of learning patterns. Simultaneously with the learning process, the validation of the network is performed on a set of samples for validation. During the validation process, only the input variables of the sample without results are passed to the network to check whether the network has the property of generalization, and this can be done only on a set of samples that has not been used for learning.

The learning and validation process end when one of the conditions for the completion of the learning process is met. The most common conditions are: a sufficiently small squared error has been achieved in all samples, a sufficiently small gradient has been achieved, the maximum number of learning epochs has been achieved, the set learning time has elapsed, the allowed number of consecutive epochs has been exceeded, etc. The network is then trained and can be used to evaluate the category of new data.

6. Preprocessing of Data

Signals obtained via DWT cannot be fed directly into the neural network input layer because they contain a different number of samples. Even if this problem could be solved by interpolating the values or by other means, the problem of too many input signals remains. Therefore, it is necessary to perform statistical data processing. Each signal obtained by DWT is represented using standard deviation, asymmetry and entropy.

The standard deviation of DWT coefficients is expressed by Equation (11):

$$\sigma_i = \sqrt{\frac{1}{N} \cdot \sum_{j=1}^{N} \left(D_{ij} - \mu_i\right)^2} \tag{11}$$

where σ_i is standard deviation of the i-th level of DWT coefficient, N number of samples the i-th level of DWT coefficient, D_{ij} the j-th sample i-th level of DWT coefficient and μ_i arithmetic mean i-th level of DWT coefficient.

The arithmetic mean of the samples in the individual DWT coefficient can be calculated according to the Equation (12) and the asymmetry according to the Equation (13).

$$\mu_i = \frac{\sum_{j=1}^{N} D_{ij}}{N} \tag{12}$$

$$Sk_i = \frac{1}{N} \cdot \sum_{j=1}^{N} \left(\frac{D_{ij} - \mu_i}{\sigma_i} \right)^3 \quad (13)$$

where Sk_i is asymmetry of the i-th level of DWT coefficient.

The entropy of the i-th level of DWT coefficient H_i can be calculated according to the Equation (14).

$$H_i = -\sum_{j=1}^{N} D_{ij}^2 log\left(D_{ij}^2\right) \quad (14)$$

Some of the calculated parameters may have values very close to zero, while others may have very large values. If such data were input directly into the neural network, the error function would converge very slowly toward the minimum, i.e., the neural network would take a very long time to learn. To avoid this, it is necessary to scale the input data to the same scale. The input data should be reduced to the interval where the activation functions have the largest derivative to increase the learning speed [46]. Since the hyperbolic tangent function or the sigmoidal function is widely used, the input data is often scaled to the interval [−1, 1] and in some cases to the interval [0, 1]. There are two common methods for this purpose: Normalization and Standardization. The *patternnet* tool in MATLAB automatically scales the input data to the interval [−1, 1] using the *mapminmax* function for normalization and *mapstd* for standardization. Tests have shown that standardization of this particular input data results in less variation in classification accuracy than normalization, so this method is used in this work.

7. Power Disturbance Classification

The low frequency power disturbances used to test the neural network classifier were artificially generated using the MATLAB script pqmodel.m developed by R. Igual et al. [15]. With this script it is possible to generate all the mentioned low frequency interferences as well as many others. This part of the program is executed only the first time, because after that the generated signals are stored and recalled at each restart, so that the classification is done with the same signals.

To create power disturbance signals, six parameters must be entered into the script, namely: number of signals per disturbance type n_s, signal sampling frequency f_s, fundamental frequency of the electrical signal f, number of fundamental frequency periods in a signal n, signal amplitude A and disturbance category number. The current disturbance parameters used in this work are listed in Table 3.

Table 3. Power disturbance parameters.

Parameter	Value
Number of signals per type of interference n_s	200
Sampling frequency f_s (Hz)	8000
Fundamental frequency of electric signal f (Hz)	60
Number of fundamental frequency periods in one signal n	60
Signal amplitude A	$440 \cdot \sqrt{2}$
Power disturbance category number	1, 2, 3, 4, 5, 6

The numbers of the current disturbance categories have the following meanings: 1. pure sinusoidal voltage, 2. voltage sag, 3. voltage swell, 4. oscillatory transient, 5. harmonically distorted voltage, and 6. voltage notches. Figure 5 shows examples of the training data for each of the disturbance categories (three randomly selected signals from each category are shown).

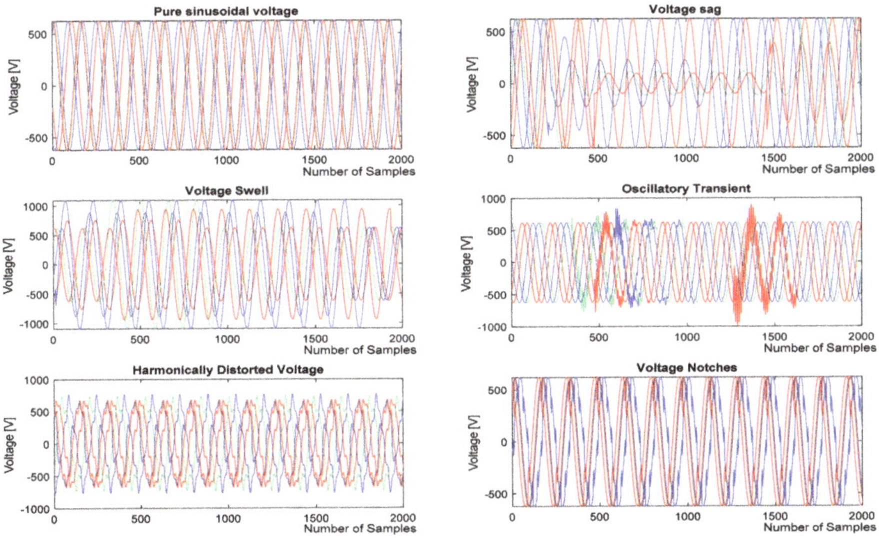

Figure 5. Samples of figures of the training data for each disturbance category.

With a total of six disturbance categories to classify and 200 signals in each category, the total data set consists of 1200 signals. When signals are generated, they are in a variable representing a three-dimensional matrix with the following dimensions: Number of signals per disturbance type × Number of points of the discrete signal × Total number of disturbance types.

In the script, the time of the beginning of the disturbance is chosen randomly. It is possible to define the limits within which the duration of the disturbance is randomly chosen in terms of the number of fundamental signal periods. Thus, for voltage sags and swells, a minimum duration of one fundamental signal period and a maximum duration of 25 periods, or half the duration of the entire signal, was chosen. In addition, it is possible to specify the limits of voltage dip or overshoot. In case of voltage drop, the minimum amplitude is 40% and the maximum amplitude is 70% of the nominal amplitude. In case of overvoltage, the minimum amplitude increase is 40% and the maximum is 70% of the nominal amplitude. It is also possible to change the proportion of harmonics and the limits of the proportion of each harmonic component in relation to the nominal value of the voltage. In the case of harmonic disturbances, third-, fifth- and seventh-order harmonics are generated, and the proportion can be arbitrarily selected between 5% and 15% of the nominal value. For the oscillatory transients, a minimum frequency of 300 Hz and a maximum frequency of 900 Hz were chosen, and the duration from half of the fundamental period to one third of the total number of fundamental periods in a signal. Finally, for the voltage dips, the minimum and maximum dip depths are set to 10% and 40% of the nominal voltage, while the number of voltage dips in a period is randomly chosen between 1, 2, 4 and 6.

In the next part of the program, the DWT is performed over the input signals. The signal decomposition was performed in five stages. Since there are no subharmonics or interharmonics in the observed disturbances, increasing the decomposition level would not do much good since only the lowest frequency band (0, 125) Hz would be further decomposed.

The decomposition into five levels results in one approximation coefficient and five detail coefficients, which are forwarded to the third part of the program, where the statistical parameters are calculated. The obtained statistical parameters form a vector of features or variables that are input into the neural network. For each signal, a total of 18 variables are entered into the network. In addition, a matrix of the actual signal categories is formed, which is also input into the neural network.

A neural network with 18 input neurons, 3 hidden neurons and 6 output neurons is used (Figure 6). This corresponds to 54 weight and 3 threshold values between the input and hidden layer as well as 18 weight and 6 threshold values between the hidden and output layer.

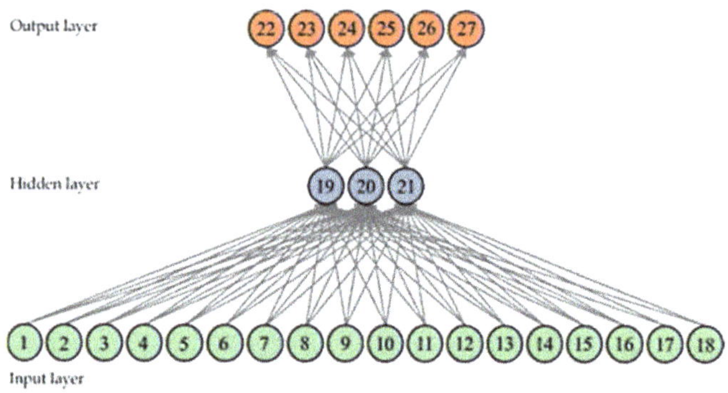

Figure 6. Neural network structure.

The hyperbolic tangent function is used in hidden neurons and the softmax function in output neurons. The Algorithms section lists in order: the method of data division, the training algorithm used, performance, and the method of computation. The settings listed correspond to those specified in the program. The Progress section lists in order: the number of epochs, the time needed for learning, the value of the achieved error function, the value of the achieved gradient and the number of epochs in which successive validation checks take place. After the neural network has completed the phase of learning, validation and testing, the classification results are obtained, i.e., the parameters for evaluating the performance of the classification model.

Neural network weight and threshold values obtained for db6 wavelet transformed signal are presented in Table 4.

Table 4. Weights and threshold values for db6 wavelet neural network classifier.

Weights and Thresholds between Input and Hidden Layer					
$w_{1,19}$	−0.298519580949239	$w_{1,20}$	0.573554788819570	$w_{1,21}$	0.286054507247907
$w_{2,19}$	0.156077024142990	$w_{2,20}$	−0.137243161824565	$w_{2,21}$	−0.064347075470886
$w_{3,19}$	−0.052284659140329	$w_{3,20}$	0.021402634499325	$w_{3,21}$	−0.190678878045148
$w_{4,19}$	−0.011912221291351	$w_{4,20}$	0.127952132194397	$w_{4,21}$	0.120522335003772
$w_{5,19}$	−0.021996393296008	$w_{5,20}$	−0.275408989161263	$w_{5,21}$	−0.074618183723548
$w_{6,19}$	0.063521112301365	$w_{6,20}$	−0.018403108555472	$w_{6,21}$	0.313055123012981
$w_{7,19}$	−0.392382737896585	$w_{7,20}$	0.002803919613340	$w_{7,21}$	0.240030895482742
$w_{8,19}$	0.176613943184293	$w_{8,20}$	0.103925620290560	$w_{8,21}$	−0.269552733626524
$w_{9,19}$	−0.011425555997867	$w_{9,20}$	0.032749310865204	$w_{9,21}$	0.020971568691560
$w_{10,19}$	−0.193378589493338	$w_{10,20}$	−0.000671270540329	$w_{10,21}$	0.044525515844779
$w_{11,19}$	0.363169912275016	$w_{11,20}$	0.267588491178083	$w_{11,21}$	−0.221081241558572
$w_{12,19}$	−0.122999692890526	$w_{12,20}$	−0.015892009080086	$w_{12,21}$	0.175425105871244
$w_{13,19}$	−0.403948257488173	$w_{13,20}$	−0.112707134890068	$w_{13,21}$	−0.105929848861931

Table 4. *Cont.*

Weights and Thresholds between Input and Hidden Layer					
$w_{14,19}$	0.393404059820330	$w_{14,20}$	0.124871188772520	$w_{14,21}$	−0.058015684884989
$w_{15,19}$	0.029308887371915	$w_{15,20}$	0.010596256476507	$w_{15,21}$	0.085798887395684
$w_{16,19}$	0.024963910356675	$w_{16,20}$	0.547350170841016	$w_{16,21}$	−0.538364141596885
$w_{17,19}$	−0.110016756323621	$w_{17,20}$	−0.493349919024467	$w_{17,21}$	0.449691963917287
$w_{18,19}$	0.002318311565705	$w_{18,20}$	−0.001846135528970	$w_{18,21}$	0.043504571855962
θ_{19}	2.284738654822070	θ_{20}	0.014763925589916	θ_{21}	1.303520578192060
Weights and Thresholds between Hidden and Output Layer					
$w_{19,22}$	−0.704382524220114	$w_{20,22}$	1.250463967932700	$w_{21,22}$	0.463049317886266
$w_{19,23}$	−0.868528085817271	$w_{20,23}$	−0.334297999377398	$w_{21,23}$	−0.318359623499615
$w_{19,24}$	−0.291087228904490	$w_{20,24}$	0.117701354689571	$w_{21,24}$	0.296370784297989
$w_{19,25}$	0.338912896407098	$w_{20,25}$	1.092644168370200	$w_{21,25}$	−1.169136285597290
$w_{19,26}$	0.115646559642188	$w_{20,26}$	−0.679149729403622	$w_{21,26}$	−0.723165104726665
$w_{19,27}$	−0.113249804098582	$w_{20,27}$	0.404736715990662	$w_{21,27}$	0.176437998074745
θ_{22}	0.256026164	θ_{23}	0.719171291	θ_{24}	−0.578629812
θ_{25}	0.189845986	θ_{26}	0.030325979	θ_{27}	−0.830338618

Since the training, testing, and validation datasets are randomly partitioned from the overall dataset, the weights and thresholds of a given neural network classifier may vary. However, the Hinton diagram can be used to visualize the value of weights and thresholds within each layer of the neural network, where a particular rectangle correlates with the influence of the particular weight or threshold [50,51]. The overall configuration of weights and thresholds for the db6 wavelet transform neural network classifier is shown with the Hinton plot in Figure 7.

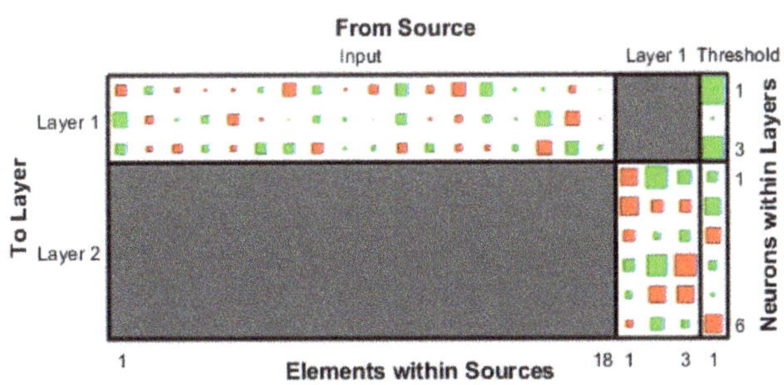

Figure 7. Hinton plot of weights and thresholds for db6 wavelet transform-based neural network.

8. Performance Analysis of Classification Models

Performance analysis is performed for three classification models that use different wavelets, while all other parameters remain unchanged. The root wavelets to be used are db1, db4 and db6 wavelets. The classification models are analyzed using a confusion matrix for the test data set, which consists of 180 randomly selected signals used to test the neural network. A total of 30 signals are sinusoidal (category 1), 27 with voltage sag (category 2), 32 with voltage swell (category 3), 24 with oscillatory transients (category 4), 31 with harmonic distortions (category 5), and 36 with voltage notches (category 6). Since each signal category contains approximately the same number of signals and the signal categories are equally important, the accuracy of classification can be considered as a relevant parameter.

The classification model performance results for all three wavelets used are shown in Figure 8. The execution time refers to the execution of the entire program. The true-positive rate (TPR), precision, and F1 value are calculated separately for each signal category. These parameters are not critical for selecting a classification model but may help in selecting models where the other parameters are approximately the same, especially for large datasets.

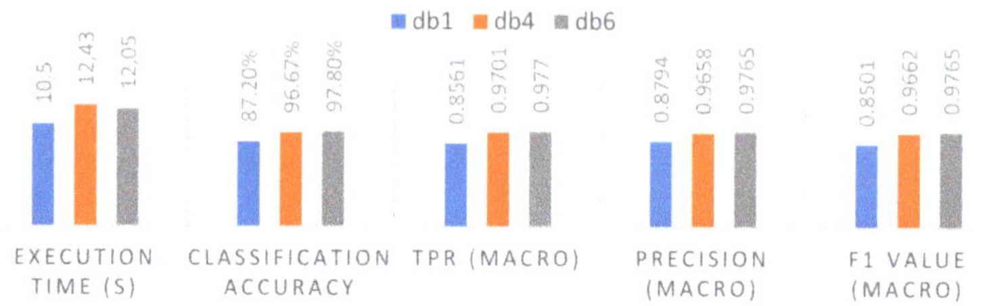

Figure 8. Classification model performance results.

The confusion matrix provides more detailed information about the classification model. The neural network interface provides four confusion matrices: the first for the learning dataset, the second for the validation dataset, the third for the testing dataset, and the fourth is the overall confusion matrix. Only the third matrix, which is based on a test data set, is of interest for performance analysis. The confusion matrix for the classification model with db1 wavelet is shown in Figure 9.

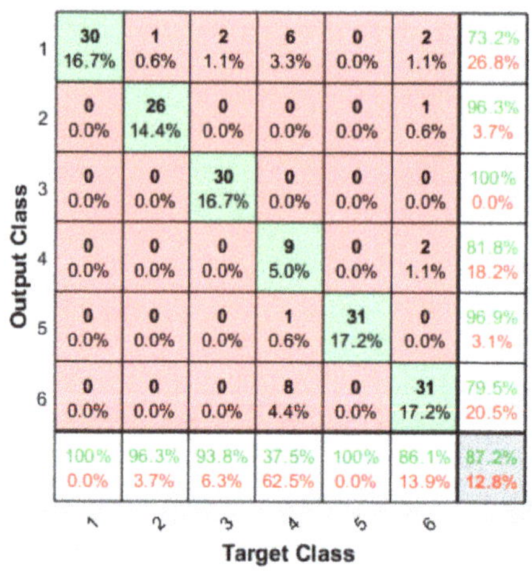

Figure 9. Confusion matrix for classification model with db1 wavelet.

Figure 9 shows that the classifier poorly detects category 4 signals or oscillatory transients. This conclusion comes from examining the TPR in the fourth column, which is only 37.5%. Out of a total of 24 signals of this type, only nine signals are correctly classified, while six signals are classified as sinusoidal voltage, one signal as a signal with harmonic distortion, and eight signals as voltage notches. This means that this model is very poor at distinguishing oscillatory transitions from sinusoidal voltage and voltage notches. Examination of column 6 shows that the TPR is 86.1% and out of a total of 36 signals with voltage dips, 31 are correctly classified, two as sinusoidal voltage, one as voltage dips and two signals as oscillatory transients. Other signal categories are well distinguished by the classifier. Looking at the first row of the confusion matrix, it is found that the classifier classified a total of 41 signals as sinusoidal voltage, which is why the lowest accuracy in this row is 73.2%. This data shows that the classifier often predicts the category of sinusoidal voltage for different signals and was correct only 73.2% of the time. Low prediction accuracy also exists in row 6, where it is 79.5%. The confusion matrix shows that this model detects voltage sags, voltage boosts and harmonic distortions very well, since the values of the TPR parameters and the precision are high for the mentioned signal categories. It can be concluded that, despite the high precision, the classification model with db1 or Haar wavelet is not satisfactory since it has a very poor knowledge of oscillatory transients and therefore distinguishes them poorly from sinusoidal voltages and voltage notches.

The confusion matrix for the classification model with a db4 wavelet is shown in Figure 10. Looking at this confusion matrix and the one shown in Figure 9, a significant improvement can be seen in the classification of signals with oscillatory transients. This classification model correctly classified all signals with oscillatory transients, however, it classified four signals with voltage notches as oscillatory transients, which still indicates that the model does not perfectly distinguish between these two categories of disturbances. It is also noted that the model achieves higher accuracy for the sinusoidal signal category, as fewer signals are misclassified as sinusoidal voltage. Looking at the 6th column of the matrix, it is noticeable that out of a total of 36 signals with voltage notches, 32 were correctly classified, while the remaining 4 were classified as oscillatory transients. Thus, this is the only relevant direction for further improvement of the model.

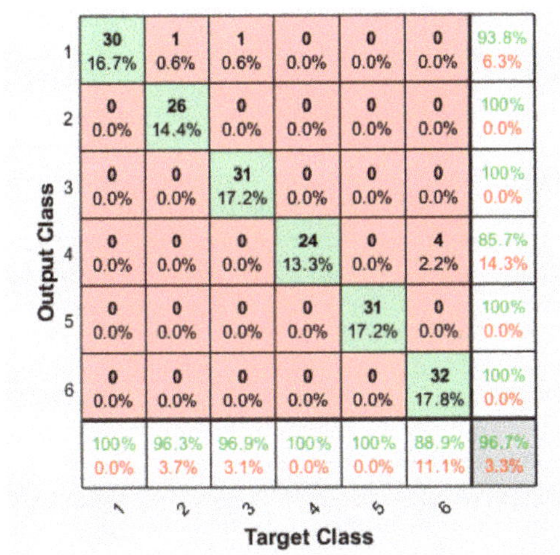

Figure 10. Confusion matrix for classification model with db4 wavelet.

The confusion matrix for the classification model with db6 wavelet is shown in Figure 11. It can be seen that this classification model distinguishes the signal categories very clearly. Examination of the confusion matrix shows that this model is an improvement over the db4 wavelet model because it detects voltage notches better and thus classifies with fewer errors. It classifies other signal categories as well as the db4 model, and the entire program executes 0.4 s faster, making it the best classification model of all the models examined in this dataset. In case of an increase in the size of the data set, equally good results or even an improvement of the results can be expected, since the network then has more learning patterns.

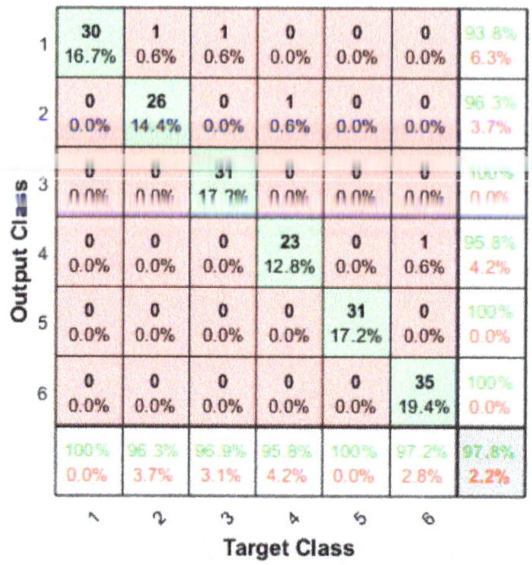

Figure 11. Confusion matrix for classification model with db6 wavelet.

9. Conclusions

Any electrical power system is subject to disturbances, most of which are caused by the operation of power electronics equipment, outages, the connection and disconnection of loads, or human error. In complex marine microgrids, which consist of several different power sources and energy storage devices, require connection to shore infrastructure during port calls, and are characterized by frequent rapid dynamic load changes, timely detection and classification of disturbances is a prerequisite for their elimination or mitigation. In this work, voltage dips, overvoltage, oscillatory transients, harmonic distortions and voltage notches were observed. The aforementioned disturbances must be analyzed to obtain additional information by which the neural network will be able to distinguish the disturbances. The Fourier transform is not satisfactory because it does not provide information about when the frequency component occurs, and in the short-time Fourier transform there is only time-frequency resolution, which makes it difficult to analyze fast and slow changes in the signal simultaneously. Therefore, a discrete wavelet transform was chosen for the analysis, which analyzes low-frequency components with high time resolution and high-frequency components with high frequency resolution. The decomposition of each signal was performed in five stages and separately with db1, db4 and db6 filters. For each obtained coefficient (frequency band), three statistical parameters are determined: standard deviation, entropy and signal asymmetry. This results in a total of 18 variables representing a signal, which are introduced into the neural network. For interference detection, a probabilistic feed-forward neural network with 18 input neurons, 3 hidden neurons and 6 output neurons was used.

Classification models with different filters were tested on a separate dataset of 180 interferences with an approximately uniform distribution of samples across interference categories. The models were analyzed based on program execution time, accuracy, precision, TPR parameters, and F1 value. The confusion matrices of each classification model were also analyzed. The analysis showed that the model with the db1 valve had the shortest program execution time and satisfactory values for all parameters. However, the analysis of the confusion matrix shows that it is very poor at distinguishing oscillatory transients from sinusoidal voltages and voltage notches. Therefore, this model is still not satisfactory. The model with the db4 valve distinguishes the mentioned disturbances better and gives significantly better results for all parameters, except for the program execution time, which increases by 18.4% compared to the model with the db1 valve. The last tested model with a db6 valve gives the best results in terms of accuracy and other performance parameters, and also has a slightly shorter execution time than the model with a db4 valve. Thus, it is the best model for the given data set. The proposed model can be successfully applied to the detection and classification of faults in the considered vessels, which can contribute to the safety and reliability of the power supply and serve as a basis for the development of advanced machine learning-based power management systems.

Author Contributions: Methodology, A.C., L.D., I.P. and J.Ć.; Software, A.C., L.D., I.P. and J.Ć.; Resources, A.C., L.D., I.P. and J.Ć.; Writing—original draft preparation, A.C. and L.D.; Writing—review and editing, A.C., L.D., I.P. and J.Ć.; Supervision, A.C. and L.D. All authors have read and agreed to the published version of the manuscript.

Funding: This research received no external funding.

Institutional Review Board Statement: Not applicable.

Informed Consent Statement: Not applicable.

Data Availability Statement: Not applicable.

Conflicts of Interest: The authors declare no conflict of interest.

References

1. Anwar, S.; Zia, M.Y.I.; Rashid, M.; Rubens, G.; Enevoldsen, P. Towards Ferry Electrification in the Maritime Sector. *Energies* **2020**, *13*, 6506. [CrossRef]
2. Martino, A.; Brambilla, M. *Research for TRAN Committee-The EU Maritime Transport System: Focus on Ferries*; European Parliament, Policy Department for Structural and Cohesion Policies: Brussels, Belgium, 2016.
3. Gagatsi, E.; Estrup, T.; Halatsis, A. Exploring the Potentials of Electrical Waterborne Transport in Europe: The E-Ferry Concept. *Transp. Res. Procedia* **2016**, *14*, 1571–1580. [CrossRef]
4. Nguyen, H.P.; Hoang, A.T.; Nizetic, S.; Nguyen, X.P.; Le, A.T.; Luong, C.N.; Chu, V.D.; Pham, V.V. The Electric Propulsion System as a Green Solution for Management Strategy of CO_2 Emission in Ocean Shipping: A Comprehensive Review. *Int. Trans. Electr. Energy Syst.* **2020**, *31*, e12580.
5. Kumar, D.; Zare, F. A Comprehensive Review of Maritime Microgrids: System Architectures, Energy Efficiency, Power Quality, and Regulations. *IEEE Access* **2019**, *7*, 67249–67277. [CrossRef]
6. Mutarraf, M.U.; Terriche, Y.; Niazi, K.A.K.; Vasquez, J.C.; Guerrero, J.M. Energy Storage Systems for Shipboard Microgrids—A Review. *Energies* **2018**, *11*, 3492. [CrossRef]
7. All-Electric Ferry Cuts Emission by 95% and Costs by 80%, Brings in 53 Additional Orders—Electrek. Available online: https://electrek.co/2018/02/03/all-electric-ferry-cuts-emission-cost/ (accessed on 7 April 2021).
8. Plug-in Hybrid Ship—the Largest in the World! | Color Line. Available online: https://www.colorline.com/about-us/worlds-largest-plug-in-hybrid-ship (accessed on 8 April 2021).
9. Zhao-Xia, X.; Tianli, Z.; Huaimin, L.; Guerrero, J.; Su, C.-L.; Vasquez, J.C. Coordinated Control of a Hybrid-Electric-Ferry Shipboard Microgrid. *IEEE Trans. Transp. Electrif.* **2019**, *5*, 828–839. [CrossRef]
10. Torreglosa, J.P.; González-Rivera, E.; García-Triviño, P.; Vera, D. Performance Analysis of a Hybrid Electric Ship by Real-Time Verification. *Energies* **2022**, *15*, 2116. [CrossRef]
11. Shagar, V.; Gamini, S.; Enshaei, H. Effect of Load Changes on Hybrid Electric Ship Power Systems. In Proceedings of the 2016 IEEE 2nd Annual Southern Power Electronics Conference (SPEC), Auckland, New Zealand, 5–8 December 2016; pp. 1–5.
12. Han, Y.; Feng, Y.; Yang, P.; Xu, L.; Xu, Y.; Blaabjerg, F. Cause, Classification of Voltage Sag, and Voltage Sag Emulators and Applications: A Comprehensive Overview. *IEEE Access* **2020**, *8*, 1922–1934. [CrossRef]
13. Tarasiuk, T.; Jayasinghe, S.G.; Gorniak, M.; Pilat, A.; Shagar, V.; Liu, W.; Guerrero, J.M. Review of Power Quality Issues in Maritime Microgrids. *IEEE Access* **2021**, *9*, 81798–81817. [CrossRef]
14. Patel, N.; Gandhi, K.; Mahida, D.; Chudasama, P. A Review on Power Quality Issues and Standards. *Int. Res. J. Eng. Technol.* **2017**, *4*, 247–250.
15. Igual, R.; Medrano, C.; Arcega, F.J.; Mantescu, G. Integral Mathematical Model of Power Quality Disturbances. In Proceedings of the IEEE 2018 18th International Conference on Harmonics and Quality of Power (ICHQP), Ljubljana, Slovenia, 13–16 May 2018; pp. 1–6.
16. Det Norske Veritas GL. *Rules for Classification: Ships; Part 4, Chapter 8 Electrical Installations*; DNV GL: Høvik, Norway, 2019.
17. International Association of Classification Societies (IACS). *Requirements Concerning Electrical and Electronic Installations*; IACS: Hamburg, Germany, 2019.
18. *Standard 60092-101*; Electrical Installations in Ships—Part 101: Definitions and General Requirements. IEC: Geneva, Switzerland, 2018.

19. Mindykowski, J.; Tarasiuk, T.; Gnaciński, P. Review of Legal Aspects of Electrical Power Quality in Ship Systems in the Wake of the Novelisation and Implementation of IACS Rules and Requirement. *Energies* **2021**, *14*, 3151. [CrossRef]
20. Shagar, V.; Jayasinghe, S.G.; Enshaei, H. Effect of Load Changes on Hybrid Shipboard Power Systems and Energy Storage as a Potential Solution: A Review. *Inventions* **2017**, *2*, 21. [CrossRef]
21. Gan, M.; Hou, H.; Wu, X.; Liu, B.; Yang, Y.; Xie, C. Machine Learning Algorithm Selection for Real-Time Energy Management of Hybrid Energy Ship. *Energy Rep.* **2022**, *8*, 1096–1102. [CrossRef]
22. Xie, P.; Guerrero, J.M.; Tan, S.; Bazmohammadi, N.; Vasquez, J.C.; Mehrzadi, M.; Al-Turki, Y. Optimization-Based Power and Energy Management System in Shipboard Microgrid: A Review. *IEEE Syst. J.* **2022**, *16*, 578–590. [CrossRef]
23. Chua, L.; Tjahjowidodo, T.; Seet, G.; Chan, R. Implementation of Optimization-Based Power Management for All-Electric Hybrid Vessels. *IEEE Access* **2018**, *6*, 74339–74354. [CrossRef]
24. Fang, S.; Wang, H. *Optimization-Based Energy Management for Multi-Energy Maritime Grids*; Springer Series on Naval Architecture, Marine Engineering, Shipbuilding and Shipping; Springer: Singapore, 2021; Volume 11, ISBN 978-981-336-733-3.
25. Haseltalab, A.; Negenborn, R.R. Model Predictive Maneuvering Control and Energy Management for All-Electric Autonomous Ships. *Appl. Energy* **2019**, *251*, 113308. [CrossRef]
26. Miraftabzadeh, S.M.; Longo, M.; Foiadelli, F.; Pasetti, M.; Igual, R. Advances in the Application of Machine Learning Techniques for Power System Analytics: A Survey. *Energies* **2021**, *14*, 4776. [CrossRef]
27. Li, X.; Zhou, K.; Xue, F.; Chen, Z.; Ge, Z.; Chen, X.; Song, K. A Wavelet Transform-Assisted Convolutional Neural Network Multi-Model Framework for Monitoring Large-Scale Fluorochemical Engineering Processes. *Processes* **2020**, *8*, 1480. [CrossRef]
28. Lee, C.-Y.; Cheng, Y.-H. Motor Fault Detection Using Wavelet Transform and Improved PSO-BP Neural Network. *Processes* **2020**, *8*, 1322. [CrossRef]
29. Yu, Q.; Hu, Y.; Yang, Y. Identification Method for Series Arc Faults Based on Wavelet Transform and Deep Neural Network. *Energies* **2019**, *13*, 142. [CrossRef]
30. Aker, E.; Othman, M.L.; Veerasamy, V.; Aris, I.B.; Wahab, N.I.A.; Hizam, H. Fault Detection and Classification of Shunt Compensated Transmission Line Using Discrete Wavelet Transform and Naive Bayes Classifier. *Energies* **2020**, *13*, 243. [CrossRef]
31. Fuchs, E.; Masoum, M.A. *Power Quality in Power Systems and Electrical Machines*; Academic Press: Cambridge, MA, USA, 2011.
32. Abedini, M.; Davarpanah, M.; Sepehr, A.; Ajaei, F.B. Shunt Capacitor Bank: Transient Issues and Analytical Solutions. *Int. J. Electr. Power & Energy Syst.* **2020**, *120*, 106025.
33. Karady, G.G.; Saksena, S.; Shi, B.; Senroy, N. Effects of Voltage Sags on Loads in a Distribution System. Final Project Report. Power System Engineering Research Center. Arizona State University, Department of Electrical Engineering 2005. Available online: https://pserc.wisc.edu/wp-content/uploads/sites/755/2018/08/T-16_Final-Report_Oct-2005.pdf (accessed on 1 July 2022).
34. Bollen, M.; Gu, I. *Signal Processing of Power Quality Disturbances*; John Wiley & Sons, Inc.: Hoboken, NJ, USA, 2006.
35. Smith, V.; Ilango, V.; Perera, S.; Gosbell, V.; Robinson, D. Power Quality and Reliability Center, University of Wollongong, Wollongong, Australia. Available online: http://eisweb.adeis.uow.edu.au/apqrc/content/technotes/APQRC%20TN008-0504%20Transient%20Overvoltages%20-%20Classification,%20Causes%20and%20Propogation.pdf (accessed on 1 July 2022).
36. Bickel, J. An Overview of Transients in Power Systems. Power Applications, Schneider Electric, Vols. *IEEE Standard 1159-2019 2019, 1–12*. Available online: https://download.schneider-electric.com/files?p_enDocType=White+Paper&p_File_Name=998-20579579_GMA.pdf&p_Doc_Ref=998-20579579_GMA (accessed on 5 July 2022).
37. Plant Engineering | Preventing Overvoltage Tripping of Adjustable-Speed Drives. Available online: https://www.plantengineering.com/articles/preventing-overvoltage-tripping-of-adjustable-speed-drives/ (accessed on 2 July 2022).
38. ENA Customer Guide to Electricity Supply—PDF Free Download. Available online: https://docplayer.net/1485020-Ena-customer-guide-to-electricity-supply.html (accessed on 13 January 2022).
39. ABS Guidance Notes on Control of Harmonics in Electrical Power Systems. Available online: https://dokumen.tips/documents/abs-guidance-notes-on-control-of-harmonics-in-electrical-power-systems.html (accessed on 2 July 2022).
40. The Wavelet Tutorial. Available online: https://users.rowan.edu/~{}polikar/WTtutorial.html (accessed on 9 July 2022).
41. Continuous Wavelet Transform (CWT). Available online: https://www.weisang.com/en/documentation/timefreqspectrumalgorithmscwt_en/ (accessed on 9 July 2022).
42. Tary, J.B.; Herrera, R.H.; van der Baan, M. Analysis of Time-Varying Signals Using Continuous Wavelet and Synchrosqueezed Transforms. *Philos. Trans. R. Soc. A Math. Phys. Eng. Sci.* **2018**, *376*, 20170254. [CrossRef] [PubMed]
43. Wavelet Families—MATLAB & Simulink. Available online: https://www.mathworks.com/help/wavelet/ug/wavelet-families-additional-discussion.html (accessed on 9 July 2022).
44. Kim, C.H.; Aggarwal, R. Wavelet Transforms in Power Systems. II. Examples of Application to Actual Power System Transients. *Power Eng. J.* **2001**, *15*, 193–202. [CrossRef]
45. Merry, R.J.E.; Steinbuch, M. *Wavelet Theory and Applications, Literature Study*; Eindhoven University of Technology, Department of Mechanical Engineering, Control Systems Technology Group: Eindhoven, The Netherlands, 2005.
46. Schneiders, M.G.E. Wavelets in Control Engineering. Master's Thesis, Eindhoven University of Technology, Eindhoven, The Netherlands, August 2001.
47. Classify Patterns with a Shallow Neural Network—MATLAB & Simulink. Available online: https://www.mathworks.com/help/deeplearning/gs/classify-patterns-with-a-neural-network.html (accessed on 10 July 2022).

48. Poklepović, T. Identifikacija Strukture Neuronske Mreže u Mjerenju Očekivane Inflacije. Ph.D. Thesis, Faculty of Economics University of Split, Split, Croatia, 2017.
49. Nguyen-Widrow Layer Initialization Function—MATLAB Initnw. Available online: https://www.mathworks.com/help/deeplearning/ref/initnw.html (accessed on 10 July 2022).
50. Saikia, P.; Vij, P.; Baruah, R.D. Unsupervised Pre-Training on Improving the Performance of Neural Network in Regression. In Proceedings of the 2018 International Joint Conference on Neural Networks (IJCNN), Rio de Janeiro, Brazil, 8–13 July 2018; pp. 1–6.
51. Nowlan, S.J.; Hinton, G.E. Simplifying Neural Networks by Soft Weight Sharing. In *The Mathematics of Generalization*; CRC Press: Boca Raton, FL, USA, 2018; pp. 373–394.

Journal of
Marine Science and Engineering

MDPI

Article

Improved Robust High-Degree Cubature Kalman Filter Based on Novel Cubature Formula and Maximum Correntropy Criterion with Application to Surface Target Tracking

Tianjing Wang, Lanyong Zhang * and Sheng Liu

College of Intelligent Systems Science and Engineering, Harbin Engineering University, Harbin 150001, China
* Correspondence: zhanglanyong@hrbeu.edu.cn

Abstract: Robust nonlinear filtering is an important method for tracking maneuvering targets in non-Gaussian noise environments. Although there are many robust filters for nonlinear systems, few of them have ideal performance for mixed Gaussian noise and non-Gaussian noise (such as scattering noise) in practical applications. Therefore, a novel cubature formula and maximum correntropy criterion (MCC)-based robust cubature Kalman filter is proposed. First, the fully symmetric cubature criterion and high-order divided difference are used to construct a new fifth-degree cubature formula using fewer symmetric cubature points. Then, a new cost function is obtained by combining the weighted least-squares method and the MCC loss criterion to deal with the abnormal values of non-Gaussian noise, which enhances the robustness; and statistical linearization methods are used to calculate the approximate result of the measurement process. Thus, the final fifth-degree divided difference–maximum correntropy cubature Kalman filter (DD-MCCKF) framework is constructed. A typical surface-maneuvering target-tracking simulation example is used to verify the tracking accuracy and robustness of the proposed filter. Experimental results indicate that the proposed filter has a higher tracking accuracy and better numerical stability than other common nonlinear filters in non-Gaussian noise environments with fewer cubature points used.

Keywords: maximum correntropy criterion; fully symmetric cubature criterion; weighted least-squares method; cubature Kalman filter; surface target tracking

Citation: Wang, T.; Zhang, L.; Liu, S. Improved Robust High-Degree Cubature Kalman Filter Based on Novel Cubature Formula and Maximum Correntropy Criterion with Application to Surface Target Tracking. *J. Mar. Sci. Eng.* **2022**, *10*, 1070. https://doi.org/10.3390/jmse10081070

Academic Editor: Sergei Chernyi

Received: 24 July 2022
Accepted: 2 August 2022
Published: 4 August 2022

Publisher's Note: MDPI stays neutral with regard to jurisdictional claims in published maps and institutional affiliations.

1. Introduction

Accurate and robust state estimation is important for the stable target tracking of conventional ships and surface unmanned ships. It is one of the main target-tracking processes for realizing sensor data fusion and anti-interference performance via a filtering algorithm. For linear Gaussian state space models, the Kalman filter (KF) is a powerful optimal estimation algorithm based on minimum mean square error. It is the most widely used adaptive filter because of its analytical optimality, algorithm stability, and simplicity. However, most commonly used target-tracking models are nonlinear, and this limits the role of the traditional KF, which only applies to linear models in practical applications.

Therefore, a nonlinear filtering algorithm in the Gaussian filter framework is required for target tracking. The extended Kalman filter (EKF) [1–3] is a common filtering method that linearizes the nonlinear model by using the multivariate Taylor formula of the nonlinear function to perform local linear approximation for obtaining a linear model, which degrades the model to the general KF model. However, for functions with strong local nonlinearity, the fitting accuracy is poor, the effect of filtering is not ideal, and the calculation of the Jacobian matrices of complex multivariate functions is difficult. As a better alternative to the EKF, the unscented Kalman filter (UKF) [3–5] was proposed to deal with highly nonlinear-state estimation problems. While optimizing the model performance, the Jacobian matrix in the EKF need not be calculated, solving the problem of complex calculations. It is a widely used nonlinear filtering method, but has a disadvantage in that the weight may have a negative value in the process of untraced transformation, and the positive definiteness of the covariance matrix is difficult to maintain with an increase in the system dimension, eventually leading to filtering divergence [6]. Thus, filtering stability is poor for systems with strong nonlinearity.

To avoid the Jacobian matrix that must be calculated in the EKF, Norgarrd et al. proposed the central difference Kalman Filter (CDKF) [7]. The CDKF uses the Stirling polynomial interpolation formula to approximate the nonlinear system and inserts it into the nonlinear Bayesian filter framework for obtaining a new nonlinear KF. However, the accuracy and computational performance of the filter must be improved for a highly nonlinear system. The Gauss–Hermite quadrature filter (GHQF) [8] is a more accurate nonlinear filter method that uses the Gauss–Hermite numerical integration formula to estimate the parameters of the nonlinear KF and embed it into the framework of the KF to form a new nonlinear filter. Although the numerical integration formula of GHQF improves the parameter estimation accuracy of filtering obviously compared with CDKF, the number of points required in the integral formula grows exponentially as the system dimension increases, which increases the computational burden for parameter estimation, leading to the problem of "dimension explosion".

Arasaratnam et al. [6] proposed a cubature Kalman filter (CKF) based on the third-degree spherical–radial criterion, which uses the spherical–radial cubature criterion to solve the probability density integrals in the framework of the nonlinear KF, providing a systematic solution to the problem of high-dimensional nonlinear filtering. Furthermore, the CKF based on the square-root criterion was derived to improve the numerical stability in the calculation process [9]. The growth rate of the cubature points used in the spherical–radial numerical integration formula with an increase in the dimension is significantly lower than that for the GHQF, avoiding the problem of "dimension explosion" and bringing considerable advantages with regard to computational complexity and stability. However, the filter constructed by the third-degree spherical–radial criterion is not as accurate as the GHQF. Bin Jia et al. [10] proposed a high-degree spherical–radial criterion that can calculate an arbitrary order accuracy according to the third-degree spherical–radial criterion, and on this basis, they proposed the high-degree cubature Kalman filter (HDCKF). Because high-degree cubature formulas are used, the nonlinear KF has higher precision, and better numerical stability is achieved. According to the spherical–radial criterion, Dong Meng et al. [11] proposed a high-degree CKF calculation formula for the seventh-degree spherical–radial criterion. Table 1 shows the performance comparison of some commonly used filters.

Table 1. Performance comparison of some commonly used filters.

Filters	Linearization Error	Suitable for Nonlinear Systems	Dimension Error Problem	Dimension Explosion Problem	Jacobian Matrix Calculation
EKF	Yes	Yes	No	No	Yes
UKF	No	Yes	Yes	No	No
CKF	No	Yes	No	No	No
GHQF	No	Yes	No	Yes	No

Xinchun Zhang et al. [12] used the fully symmetric cubature criterion of J. McNamee [13] to approximate the probability density function integral in the nonlinear KF and then combined it with the KF framework to obtain an embedded CKF (ECKF). Compared with the previous five-degree filter constructed according to the spherical–radial cubature criterion, the embedded cubature filter has fewer cubature points and can reduce the number of computations while maintaining the filtering accuracy. Additionally, because the coordinates of the cubature points do not increase with the system dimension n, compared with spherical–radial quadrature filter, the stability of the nonlinear filter is enhanced. However, even the embedded KF with fewer cubature points and a lower structural complexity than the HDCKF has problems related to computational stability. It is necessary to develop a quadrature formula that can improve the stability of CKF iteration while maintaining the filtering accuracy and controlling the number of computations.

Conventional CKFs and high-order spherical–radial filtering algorithms have high performance under Gaussian noise conditions, but their accuracies decrease or even diverge under non-Gaussian noise or mixed Gaussian noise conditions because they are based on second-order information estimation via the KF framework. Unfortunately, in most practical applications, because the system is affected by the surrounding environment, e.g., unmanned equipment maneuvering extensively in a short time, process noise and measurement noise typically do not obey the simple Gaussian distribution, which degrades the performance of the conventional target-tracking algorithm of the cubature-based filter.

To solve this problem, scholars proposed some robust filters based on the conventional KF framework, which can adapt to the noise of non-Gaussian systems. For a general linear system containing non-Gaussian noise, Izanloo R et al. [14] developed a new optimization objective function based on the maximum correntropy criterion (MCC) and combined it with the weighted least squares (WLS) method. The fixed-point iteration method was used to obtain the optimal solution of the state estimation equation, which was inserted into the standard flow of the conventional KF to obtain the MCC-KF. The MCC-KF has the same structure as the KF and uses higher-order (>2) statistics to obtain state estimation parameters. Compared with the UKF and the Gaussian sum filter (GSF) [15], the MCC-KF has a smaller estimation error and does not require the use of multiple filters or sigma points; additionally, it has a lower computational complexity and less computations than the UKF and GSF. Guoqing Wang et al. [16] proposed the maximum correlation entropy unscented Kalman filter (MC-UKF) and unscented information filter (MC-UIF) based on the MCC combined with the framework of the UKF and information filtering to solve the filtering problem of nonlinear systems in non-Gaussian noise environments. Compared with the existing UKF algorithm, similar or better estimation results are obtained. When the core bandwidth is infinite, the proposed MC-UKF and MC-UIF converge to the UKF and UIF, respectively. Qingwen Meng et al. [17] proposed a robust KF based on the third-degree spherical–radial CKF and the smallest Cauchy kernel loss (CKL) function. Under the filtering framework of the third-degree CKF, a new optimization objective function was obtained by combining the WLS method with the smallest CKL function. The simulation results of typical nonlinear systems verify that the MCK-CKF has strong robustness and a high filtering efficiency against non-Gaussian noise. He et al. [18] proposed an adaptive and robust CKF based on the MCC of the variable decibel Bayesian (VB) method to solve the problems of unknown measurement noise covariance and outliers in a visual and dual inertial measurement unit integrated-attitude system.

To overcome the shortcomings of robust KFs based on the MCCKF, MC-UKF, and GSF algorithms with regard to the filtering accuracy and numerical stability, a new robust nonlinear KF, based on a novel cubature formula and MCC is proposed in this study. In contrast to the general spherical–radial criterion-based CKF, a new numerical integral quadrature formula was first constructed using a fully symmetric quadrature criterion and high-order divided difference formula to approximate the probability density of the Gaussian weighted integral form in the CKF state and measurement update. A cubature formula with good comprehensive performance is obtained, which considers the number of cubature points, numerical stability, and calculation accuracy. Then, a new optimization objective function and parameter estimation equation are defined by combining the WLS method and MCC. The solving process is combined with the filtering process of the constructed high-degree CKF framework to obtain a nonlinear KF, i.e., the fifth-degree divided difference-maximum correntropy cubature Kalman filter (DD-MCCKF). Finally, typical surface-target-tracking simulation examples were used to verify the performance of the filter. The experimental results indicate that the fifth-degree DD-MCCKF has high filtering accuracy and stability as compared to third-degree MCCKF, fifth-degree MCCKF, embedded MCCKF, and MC-UKF when there are two different types of non-Gaussian mixture noise.

2. Construction of New High-Degree Cubature Formula

2.1. Nonlinear Filtering Problem and Gaussian Weighted Integral (GWI)

Consider the following nonlinear systems that can be described by discrete nonlinear state–space models:

$$\begin{cases} \boldsymbol{x}_{k+1} = \boldsymbol{f}(\boldsymbol{x}_k, \boldsymbol{u}_k) + \boldsymbol{w}_k \\ z_k = \boldsymbol{h}(\boldsymbol{x}_k, \boldsymbol{u}_k) + \boldsymbol{v}_k \end{cases}, \quad (1)$$

where $\boldsymbol{f}(\boldsymbol{x},\boldsymbol{u})$ and $\boldsymbol{h}(\boldsymbol{x},\boldsymbol{u})$ are arbitrary nonlinear functions and $\boldsymbol{w}_k$ and $\boldsymbol{v}_k$ are the mutually independent system process noise and measurement noise with covariance matrices $\boldsymbol{Q}_k$ and $\boldsymbol{R}_k$, respectively. Further, $\boldsymbol{u}_k$ represents the control input, and $\boldsymbol{x}_k$ and z_k represent the system state and measurement, respectively, at time k.

The state posterior distribution $p(\boldsymbol{x}_k|\boldsymbol{Z}_k)$ of the above discrete system at time k can be estimated using the measurement set $\boldsymbol{Z}_k = \{z_1, z_2, \ldots, z_k\}$ formulated in Equation (1), according to the Bayesian estimation theory. Using the Chapman–Kolmogorov equation, the posterior density can be estimated and updated as follows:

$$p(\boldsymbol{x}_k|\boldsymbol{Z}_{k-1}) = \int p(\boldsymbol{x}_k|\boldsymbol{x}_{k-1})p(\boldsymbol{x}_{k-1}|\boldsymbol{Z}_{k-1})d\boldsymbol{x}_{k-1} \quad (2)$$

$$p(\boldsymbol{x}_k|\boldsymbol{Z}_k) = \frac{p(z_k|\boldsymbol{Z}_k)p(\boldsymbol{x}_{k-1}|\boldsymbol{Z}_{k-1})}{\int p(z_k|\boldsymbol{Z}_k)p(\boldsymbol{x}_k|\boldsymbol{Z}_{k-1})d\boldsymbol{x}_k} \quad (3)$$

For nonlinear systems, the posterior density cannot be directly calculated because the high-dimensional integral in the equation does not have an exact analytical solution. Therefore, approximate or suboptimal Bayesian algorithms must be used for nonlinear systems. There are some limitations to using the existing methods to filter nonlinear non-Gaussian systems.

Because Equations (2) and (3) cannot be calculated accurately, and in consideration of the accuracy and computational complexity, CKF is typically used as a Gaussian approximation filtering algorithm. Before using it, the following key assumptions of the one-step posterior predictive PDF of the state $\boldsymbol{x}_k$ and measurement z_k conditioned by $\boldsymbol{Z}_k$ must first be made:

$$p(\boldsymbol{x}_k|\boldsymbol{Z}_{k-1}) = N\left(\boldsymbol{x}_k; \hat{\boldsymbol{x}}_{k|k-1}, \boldsymbol{P}_{k|k-1}\right) \quad (4)$$

$$p(z_k|\boldsymbol{Z}_{k-1}) = N\left(z_k; \hat{z}_{k|k-1}, \boldsymbol{P}^{zz}_{k|k-1}\right) \quad (5)$$

By Equations (4) and (5) and the Bayesian rule, the posterior PDF of the state is also Gaussian, that is, $p(\boldsymbol{x}_k|\boldsymbol{Z}_k) = N\left(\boldsymbol{x}_k; \hat{\boldsymbol{x}}_{k|k}, \boldsymbol{P}_{k|k}\right)$. In this manner, we transform the general nonlinear filtering problem into a Kalman filtering problem under a Gaussian framework.

CKF is a suboptimal filtering algorithm that combines precision and computational performance. The difficulty in the CKF filtering process lies mainly in calculating the following Gaussian weighted integral (GWI):

$$G(\boldsymbol{f}) = \int_{R^n} \boldsymbol{f}(\boldsymbol{x})N(\boldsymbol{x};\boldsymbol{\mu},\boldsymbol{\Sigma})d\boldsymbol{x}\,. \quad (6)$$

where $\boldsymbol{f}(\boldsymbol{x})$ is a multivariate function, $\boldsymbol{x} = (x_1, x_2, \ldots, x_n)$, which does not yield exact results that conform to analytical expressions when $\boldsymbol{f}(\boldsymbol{x})$ is nonlinear and must be calculated by numerical integral approximation methods.

As we know the expression of normal distribution function $N(\boldsymbol{x};\boldsymbol{\mu},\boldsymbol{P})$ from Equation (7), this integral can be simplified by linear transformation of the integral variable.

$$N(\boldsymbol{x};\boldsymbol{\mu},\boldsymbol{\Sigma}) = \frac{1}{(2\pi)^{n/2}|\Sigma|^{1/2}}\exp\left(-\frac{(\boldsymbol{x}-\boldsymbol{\mu})^T\Sigma^{-1}(\boldsymbol{x}-\boldsymbol{\mu})}{2}\right). \quad (7)$$

Then, let $x = \sqrt{2\Sigma}v + \mu$. The specific integral form of the Gaussian weighted integral can be simplified as follows:

$$\begin{aligned} G(f) &= \int_{R^n} \frac{f(x)}{(2\pi)^{n/2}|\Sigma|^{1/2}} \exp\left(-\frac{(x-\mu)^T\Sigma^{-1}(x-\mu)}{2}\right)dx \\ &= \frac{1}{(\pi)^{n/2}} \int_{R^n} f\left(\sqrt{2\Sigma}v + \mu\right) \exp\left(-v^T v\right) dv \end{aligned} \quad (8)$$

The integral can be approximated numerically via many proposed numerical approximation methods. A typical example is the use of unscented transform (UT) or the spherical–radial cubature criterion for approximation, which can be combined with the KF framework to obtain the UKF or CKF, respectively. The third-degree spherical–radial numerical integration formula is as follows:

$$\int_{R^n} f\left(\sqrt{2\Sigma}v + \mu\right) \exp\left(-v^T v\right) dv \approx \frac{(\pi)^{n/2}}{2n} \sum_{k=1}^{N} \left(f\left(\sqrt{2n\Sigma}e_k + \mu\right) + f\left(-\sqrt{2n\Sigma}e_k + \mu\right)\right) \quad (9)$$

where $N = n$ represents the system dimension, and e_k is the kth column in the n-order identity matrix E. The identity matrices E and $-E$ form the first set of fully symmetric cubature points in the numerical approximation formula. This formula can stably approach the original integral with a minimum number of cubature points.

From the numerical analysis point of view, the formula of untraced transformation (UT) shows that when the dimension of the system, n, exceeds three, its stability decreases linearly with the increase in dimension N, thus causing a significant disturbance in the numerical estimation of the moment integral. Because there is no square root solution in the UKF, when the pseudo-square root operation is performed on the error covariance matrix, a non-positively determined updated matrix can be obtained owing to the existence of sigma points with negative weights in the UKF. Therefore, it is impossible to express the square root UKF with a numerical advantage similar to the square root-CKF by formula. The covariance matrix calculated by the UKF is not always guaranteed to be positive definite, and the unavailability of the square root covariance causes the UKF to stop running. However, the set of cubature points in the CKF does not have these problems. The cubature point method is mathematically more accurate and principled than the sigma point method [6].

2.2. Commonly Used High-Degree Cubature Rules

The accuracy of the numerical integration formula is determined primarily by the order of the fitting polynomial. Cubature formulas of the fifth degree can obtain higher numerical approximation accuracy at the cost of using more cubature points. In the third-degree cubature formula, only the GWI corresponding to polynomial $\{1, x_1^2\}$ is accurate, and its approximation error mainly comes from the fourth order or higher polynomial integration in the expansion of function $f(x)$. In the formula of the fifth degree, the GWI corresponding to $\{1, x_1^2, x_1^4, x_1^2 x_2^2\}$ is accurate, and its approximation error mainly comes from the GWI corresponding to the sixth and higher order polynomials. To ensure numerical stability, the fifth degree can approximate GWI with higher numerical accuracy and obtain more accurate integral results in simple numerical calculation problems.

To increase the approximation accuracy of the numerical integration for the cubature formula, the third-degree cubature criterion in the CKF can be extended to higher degrees. For simplicity, we use $u(x)$ instead of $f\left(\sqrt{2\Sigma}x + \mu\right)$ in this section.

2.2.1. Basic Formulas and Theorems

We consider the fully symmetric numerical integration formula of the following form:

$$G(u) = \int_{R^n} u(x) \exp\left(x^T x\right) dx \approx \sum_{k=1}^{N} W_k \sum_{FS} u\left(r_{1k}, r_{2k}, \ldots, r_{pk}, 0\right)_k \quad (10)$$

where $\left(r_{1k}, r_{2k}, \ldots, r_{pk}, 0\right)_k$ represents the kth generator of cubature coordinate points and r_{pk} represents the pth coefficients of the points. Further, W_k is the weight of the corresponding part of each generator, and FS is a set of fully symmetric cubature points. The integral region R^n and the integrand weighted function $\exp(-x^T x)$ are completely symmetric, with $\exp(-x^T x) > 0$. In R^n, if the set of evaluation points is fully symmetric and S is the union of the fully symmetric set S_i, Equation (10) is called a fully symmetric numerical integration formula. The cubature points are generated by different generators $\left(r_{1k}, r_{2k}, \ldots, r_{pk}, \mathbf{0}\right)_k$, and each generator corresponds to exactly one weight.

In addition, another integration method, the spherical–radial quadrature method, was proposed. In Gaussian weighted integrals, the integrand weight function is of the form $\exp\left(-\boldsymbol{x}^T\boldsymbol{x}\right)$. Thus, Cartesian integration can be converted into a spherical integration via the n-dimensional spherical coordinate transformation $x = rp$ with $p = (\cos\theta_1, \sin\theta_1\cos\theta_2, \ldots, \sin\theta_1 \ldots \sin\theta_{n-2}\cos\theta_{n-1}, \sin\theta_1 \ldots \sin\theta_{n-2}\sin\theta_{n-1})$. Using the above coordinate transformation, variable substitution of the Gaussian weighted integral is performed as follows:

$$\begin{aligned} G(\boldsymbol{u}) &= \int_{R^n} \boldsymbol{u}(\boldsymbol{x}) \exp\left(-\boldsymbol{x}^T\boldsymbol{x}\right) d\boldsymbol{x} \\ &= \int_{S_n} dS \int_0^\infty \boldsymbol{u}(r\boldsymbol{p}) r^{n-1} \exp\left(-r^2\right) dr \approx \sum_{k=1}^{N_r} \sum_{l=1}^{N_p} w_{rk} w_{pl} \boldsymbol{u}(r_k \boldsymbol{s}_l) \end{aligned} \tag{11}$$

This is the spherical radial integral. The integral region of the surface integral $S_n = \left\{\boldsymbol{p} \in R^n : p_1^2 + p_2^2 + \ldots + p_n^2 = 1\right\}$ is an n-dimensional hypersphere with radius 1. This special surface integral can be approximated using the spherical isomorphic integration criterion and Stroud's integration formula. The radial integral can be calculated using the moment matching algorithm and the generalized Gauss–Laguerre quadrille criterion.

In the approximate expression of Equation (11), r_k and w_{rk} are the corresponding points of radial integrals and their weights respectively. $\boldsymbol{s}_l$ and w_{pl} are vectors corresponding to spherical integrals and their weights, respectively. According to the cubature integration rules and the numerical method of radial integration, the Gaussian weighted integral can be approximated using a fifth-degree numerical integration formula with the radial integral formula and spherical integral formula. All the fifth-degree cubature formulas based on the spherical–radial criterion can be summarized by the form of Equation (11) [19].

To evaluate the computational complexity and numerical stability of the cubature formula, the following two theorems were introduced:

Theorem 1. *The cubature formula of degree 2s–1 has the minimum number of cubature points and given as follows [20]:*

$$P_{min} = \begin{cases} \dbinom{n+s-1}{n} + \sum\limits_{k=1}^{n-1} 2^{k-n} \dbinom{k+s-1}{k}, s = 2p, p \in N^+ \\ \dbinom{n+s-1}{n} + \sum\limits_{k=1}^{n-1} \left(1 - 2^{k-n}\right) \dbinom{k+s-2}{k}, s = 2p-1, p \in N^+ \end{cases}. \tag{12}$$

Theorem 2. *When the system dimension n is so large that the signs of different weights of the cubature formula are not always positive, the stability of the cubature formula can be evaluated according to the stability coefficient discriminant, as follows:*

$$stb = \frac{\sum\limits_{u=0}^{M} \sum\limits_{k=1}^{N} \left|\boldsymbol{W}_{u,k}\right|}{\sum\limits_{u=0}^{M} \sum\limits_{k=1}^{N} \boldsymbol{W}_{u,k}} = \frac{\sum\limits_{u=0}^{M} \sum\limits_{k=1}^{N} \left|\boldsymbol{W}_{u,k}\right|}{G(1)} \geq 1. \tag{13}$$

According to Theorem 1, for all third-degree cubature formulas (including the common third-degree spherical–radial rule-based formula), the minimum number of cubature points is $2n$. Next, we consider formulas of the fifth degree.

2.2.2. Fifth-Degree Cubature Formulas

- ***Formula I***

Stroud et al. [21] provided a fully symmetric fifth-degree cubature formula based on the spherical–radial integration method, which is one of the most widely used cubature formulas in high-degree CKF robust algorithms. This formula uses Stroud's formula [22] to approximate the spherical integral, and the radial integral is approximated via the Gauss–Laguerre numerical integration method. It can be expressed as

$$G(\boldsymbol{u}) \approx \frac{2(\pi)^{n/2}}{n+2}\boldsymbol{u}(0) + \frac{(4-n)(\pi)^{n/2}}{2(n+2)^2}\sum_{FS}\boldsymbol{u}(s,0) + \frac{(\pi)^{n/2}}{(n+2)^2}\sum_{FS}\boldsymbol{u}(r,r,0)\,, \tag{14}$$

where $s = \sqrt{\frac{n}{2}+1}$ and $r = \sqrt{\frac{n}{4}+\frac{1}{2}}$. The generation method for fully symmetric cubature points is as follows:

$$\begin{aligned}
&(s,\mathbf{0}) = \sqrt{2}\begin{bmatrix} s & 0 & 0 & -s & 0 & 0 \\ \cdots & \cdots & \cdots & \cdots & \cdots & \cdots \\ 0 & 0 & s & 0 & 0 & -s \end{bmatrix} \\
&\boldsymbol{U}_1 = \sqrt{2}r\underbrace{(e_1+e_2,\ldots,e_1+e_n,e_2+e_3,\ldots,e_2+e_n,\ldots,e_{n-1}+e_n)}_{n(n-1)/2} \\
&\boldsymbol{U}_2 = \sqrt{2}r\underbrace{(e_1-e_2,\ldots,e_1-e_n,e_2-e_3,\ldots,e_2-e_n,\ldots,e_{n-1}-e_n)}_{n(n-1)/2} \\
&(r,r,\mathbf{0}) = (\boldsymbol{U}_1,-\boldsymbol{U}_1,\boldsymbol{U}_2,-\boldsymbol{U}_2)
\end{aligned} \tag{15}$$

Bin Jia et al. applied Stroud's cubature formula to the parameter estimation of a nonlinear KF and obtained the HDCKF. In the high-degree KF of [10], Equation (14) is rewritten as

$$\begin{aligned}
G(\boldsymbol{u}) \approx\ & \tfrac{2(\pi)^{n/2}}{n+2}\boldsymbol{u}(0) + \tfrac{(4-n)(\pi)^{n/2}}{2(n+2)^2}\sum_{k=1}^{n}\left(\boldsymbol{u}(\sqrt{n+2}\boldsymbol{e}_k) + \boldsymbol{u}(-\sqrt{n+2}\boldsymbol{e}_k)\right) \\
& + \tfrac{(\pi)^{n/2}}{(n+2)^2}\sum_{k=1}^{n(n-1)/2}\begin{pmatrix} \boldsymbol{u}(\sqrt{n+2}\boldsymbol{l}_k^+) + \boldsymbol{u}(-\sqrt{n+2}\boldsymbol{l}_k^+) \\ +\boldsymbol{u}(\sqrt{n+2}\boldsymbol{l}_k^-) + \boldsymbol{u}(-\sqrt{n+2}\boldsymbol{l}_k^-) \end{pmatrix}
\end{aligned}\ , \tag{16}$$

where $\boldsymbol{l}_k^+ = \boldsymbol{e}_k + \boldsymbol{e}_l, \boldsymbol{l}_k^- = \boldsymbol{e}_k - \boldsymbol{e}_l, k < l \leq n$. Equations (14) and (16) are essentially identical. The total number of cubature points is $2n^2+1$. According to Theorem 2, when n is sufficiently large, its asymptotic stability coefficient converges to 3, resulting in a cubature formula with good numerical stability.

- ***FormulaII***

Mysovskikh [23] derived the spherical integral formula according to the transformation group of the regular simplex. Lu and Darmofal [24] proposed a new fifth-degree cubature formula based on the integral formula of Mysovskikh, which is similar to the formula proposed by Stroud et al. It also decomposes the Gaussian weighted integral into the product of the spherical and radial integrals:

$$\begin{aligned}
G(\boldsymbol{u}) \approx\ & \tfrac{2(\pi)^{n/2}}{n+2}\boldsymbol{u}(\mathbf{0}) \\
& + \tfrac{n^2(7-n)(\pi)^{n/2}}{2(n+1)^2(n+2)^2}\sum_{k=1}^{n+1}\left(\boldsymbol{u}\left(\sqrt{\tfrac{n}{2}+1}\boldsymbol{a}_k\right) + \boldsymbol{u}\left(-\sqrt{\tfrac{n}{2}+1}\boldsymbol{a}_k\right)\right) \\
& + \tfrac{2(n-1)(\pi)^{n/2}}{(n+1)^2(n+2)^2}\sum_{k=1}^{n(n+1)/2}\left(\boldsymbol{u}\left(\sqrt{\tfrac{n}{2}+1}\boldsymbol{b}_k\right) + \boldsymbol{u}\left(-\sqrt{\tfrac{n}{2}+1}\boldsymbol{b}_k\right)\right).
\end{aligned} \tag{17}$$

In this formula, the values of the cubature points and parameters are given as follows:

$$\begin{aligned}
&\boldsymbol{a}_k = \left(a_{1,k}, a_{2,k}, \ldots, a_{n,k}\right)^T, k = 1,2,\ldots,n+1 \\
&a_{i,k} = \begin{cases} -\sqrt{\frac{n+1}{n(n-i+2)(n-i+1)}}, i < k \\ \sqrt{\frac{(n+1)(n-k+1)}{n(n-k+2)}}, i = k \end{cases},\ \boldsymbol{b}_k = \sqrt{\tfrac{n}{2(n-1)}}\boldsymbol{v}_k, \\
&\boldsymbol{V}_{n\times\frac{n(n+1)}{2}} = \underbrace{(\boldsymbol{a}_1+\boldsymbol{a}_2,\ldots,\boldsymbol{a}_1+\boldsymbol{a}_{n+1},\boldsymbol{a}_2+\boldsymbol{a}_3,\ldots,\boldsymbol{a}_2+\boldsymbol{a}_{n+1},\ldots,\boldsymbol{a}_n+\boldsymbol{a}_{n+1})}_{n(n+1)/2},
\end{aligned} \tag{18}$$

where $\boldsymbol{a}_k$ represents the $n+1$ vertices of n-dimensional hypersphere S_n, and $\boldsymbol{b}_k$ represents the topological mapping of the midpoints of the vertices of the simplex on hypersphere S_n. The number of cubature points required by this formula is n^2+3n+3. For low dimensional systems, this formula requires more cubature points than the cubature formula of HDCKF, and resulting in unnecessary costs. According to Theorem 2, the stability index of the formula can be calculated as stb = 3, indicating that the algorithm has good numerical stability. However, it is difficult to extend and improve the formula, because of the complex structure of the spherical simplex criterion.

- ***FormulaIII***

According to Stroud's invariance theory for cubature integrals, McNamee et al. [13] constructed a group of fully symmetric integration formulas with order $2k+1$ in the n-dimensional space as:

$$G(\boldsymbol{u}) \approx T(\boldsymbol{u}) = W_0\boldsymbol{u}(0) + W_1\sum_{FS}\boldsymbol{u}(r,0) + W_2\sum_{FS}\boldsymbol{u}(r,r,0) \tag{19}$$

This cubature formula is a special form of Equation (10) in the fifth-degree case. Zhang et al. [12] applied it to nonlinear Kalman filtering and formed an ECKF. The fully symmetric quadrature formula of the fifth degree is

$$\begin{aligned} G(\boldsymbol{u}) &\approx \frac{(n^2-7n+18)(\pi)^{n/2}}{18}\boldsymbol{u}(\mathbf{0}) \\ &+ \frac{(4-n)(\pi)^{n/2}}{18}\sum_{FS}\boldsymbol{u}\left(\sqrt{\tfrac{3}{2}},\mathbf{0}\right) + \frac{(\pi)^{n/2}}{36}\sum_{FS}\boldsymbol{u}\left(\sqrt{\tfrac{3}{2}},\sqrt{\tfrac{3}{2}},\mathbf{0}\right). \end{aligned} \tag{20}$$

In this formula, nonlinear equations are constructed and weight coefficients are obtained by solving the fully symmetric cubature criterion. The number of required cubature points is $2n^2+1$, leading to a simple structure and a small number of calculations. In contrast to Equations (16) and (17) obtained using spherical radial integration, the coefficient of cubature points $\sqrt{3/2}$ is a fixed value and does not increase with the system dimension n. The problem of cubature points exceeding the integral domain in the spherical integral formula is avoided. However, according to Theorem 2, the stability coefficient is $\frac{2n^2-8n+9}{9}$; thus, the numerical stability of this formula is poor. In addition, for systems with sufficiently large values of n, the weights are negative, which further affects the stability.

2.3. Novel High-Order Cubature Formula Based on Fully Symmetric Cubature Criterion and Divided Difference Formula

This section describes the construction of a novel fifth-degree cubature formula using a new method to maintain the filtering accuracy and numerical stability while controlling the number of cubature points required for integration.

First, the partial-derivative formula is used to modify the original formula to increase accuracy. Generally, it is difficult to directly calculate the partial derivative of a multivariable continuous function at a certain point. To avoid complex calculations, we use discrete high-order divided difference formulas to approximate the value of the partial derivatives and write them as linear combinations of the original functions. Because the weight function of the Gaussian weighted integral is a fully symmetric function, the integral of the partial derivatives of odd order is zero, and only the partial derivatives of the even order must be considered. From Equation (19), the new cubature formula modified by the high-order divided difference and even-order divided difference formula of $f(x)$ is expressed as follows:

$$\begin{aligned} \hat{G}(f) &= T(f) + \sum_{k=1}^{n} \nabla_4^{x_k} f(\boldsymbol{x})_{\boldsymbol{x}=0} + \sum_{k<l} \nabla_2^{x_k}(\nabla_2^{x_l} f(\boldsymbol{x}))_{\boldsymbol{x}=0} \\ \nabla_{2n}^{x} f(x) &= \sum_{k=0}^{2n} \frac{(2n)!f(x-(n-k)r)}{\prod\limits_{u=0}^{2n-k-1}(r(2n-k)-ur)\prod\limits_{u=2n-k+1}^{2n}(r(2n-k)-ur)}. \end{aligned} \tag{21}$$

For example, the first two high-order divided difference formulas can be expressed as follows:

$$\begin{aligned} \nabla_2^x f(x) &= \frac{f(x-r)-2f(r)+f(x+r)}{r^2}, \\ \nabla_4^x f(x) &= \frac{f(x-2r)-4f(x-r)+6f(r)-4f(x+r)+f(x+2r)}{r^4}\ldots \end{aligned}, \tag{22}$$

where r represents the selected step size for the divided differences. By combining the above formulas, Equation (21) with the fully symmetric integral formula with Equation (19), a new form of the fifth-degree cubature formula is obtained as follows:

$$\hat{G}(\boldsymbol{u}) = W_0\boldsymbol{u}(\mathbf{0}) + W_1\sum_{FS}\boldsymbol{u}(r,\mathbf{0}) + W_2\sum_{FS}\boldsymbol{u}(2r,\mathbf{0}) + W_3\sum_{FS}\boldsymbol{u}(r,r,\mathbf{0}). \tag{23}$$

The definitions of cubature generators $(r, \mathbf{0})$ and $(r, r, \mathbf{0})$ are identical to those in Equation (15). Compared with Equations (14), (17), and (20), the newly constructed cubature formula uses discrete derivatives and has a higher accuracy. However, it uses more cubature points. To determine the weight coefficient of the above equation, it is necessary to construct a set of higher-order algebraic equations. According to the fully symmetric cubature criterion, consider the Gaussian weighted integral of the following even-power monomial function:

$$I_{2s_1,2s_2,\ldots 2s_v} = \int_0^\infty (x_1)^{2s_1}(x_2)^{2s_2}\ldots(x_v)^{2s_v}\exp(-\mathbf{x}^T\mathbf{x})d\mathbf{x} \\ i \le j, 0 \le s_i \le s_j, 2\sum_{k\le v} s_k \le 5. \tag{24}$$

Note the above conditions for the s_k value; in the fifth-degree cubature formula, only the integrals $I_{0,0}$, $I_{0,2}$, $I_{0,4}$ and $I_{2,2}$ must be calculated. The following formula based on the gamma function is used to calculate the integrals:

$$\begin{aligned} I_{2a,2b} &= \int_{R^n} (x_1)^{2a}(x_2)^{2b}\exp(-\mathbf{x}^T\mathbf{x})d\mathbf{x} \\ &= \left(\int_0^\infty (x_1)^{2a}e^{-x_1^2}dx_1\right)^2\left(\int_0^\infty (x_2)^{2b}e^{-x_2^2}dx_2\right)^2\int_{R^{n-2}}\exp\left(-\left(\mathbf{x}^{(n-2)}\right)^T\mathbf{x}^{(n-2)}\right)d\mathbf{x}^{(n-2)} \\ &= \Gamma\left(a+\tfrac{1}{2}\right)\Gamma\left(b+\tfrac{1}{2}\right)(\pi)^{n/2-1}, a, b \ge 0. \end{aligned} \tag{25}$$

By combining the coefficients of the above monomial function with the newly constructed cubature formula of Equation (23), the following fully symmetric polynomial can be obtained [13]:

$$M_{u_{a,b},I_{a,b}} = 2^{r_N}\sum_{k_1=1}^{p}\cdots\sum_{k_v=1}^{p}\frac{(n-v)!r_{k_1}^{2s_1}\ldots r_{k_v}^{2s_v}}{(n-r_N)!}\left(\prod_{j=1}^{p}\left(p_j-q_j\right)!\right)^{-1}_{(k_1,\ldots,k_v)} \tag{26}$$

This polynomial is used to calculate the coefficient-matrix elements of the higher-order algebraic equations corresponding to the newly constructed cubature formula. Here, r_N represents the number of nonzero vector coefficients at the cubature points used in each part of Equation (23), and p represents the number of nonzero vector coefficients in each generator of the new formula which is different from the others. Thus, r_N and p can be easily obtained using the above formula. p_j and q_j represent the numbers of times that the coefficient r of the cubature points and the integer j appear in the counting units $k_1, \ldots, k_v$ of the sum, respectively.

According to the different cubature points in Equation (23), different integral trajectories are selected, and the parameters are calculated using Equations (25) and (26). Thus, the coefficient matrix of the higher-order algebraic equation can be obtained:

$$\left\{\begin{array}{c} r_N = 2, p = 1 \Rightarrow (r, r, \mathbf{0}) \\ r_N = 1, p = 1 \Rightarrow (ar, \mathbf{0}) \\ r_N = 0, p = 0 \Rightarrow \mathbf{0} \end{array}\right. \rightarrow \begin{array}{cccc} \mathbf{0} & (r,\mathbf{0}) & (2r,\mathbf{0}) & (r,r,\mathbf{0}) \\ \Downarrow & \Downarrow & \Downarrow & \Downarrow \end{array} \begin{pmatrix} 1 & 2n & 2n & 2n(n-1) \\ 0 & 2r^2 & 2(2r)^2 & 4(n-1)r^2 \\ 0 & 2r^2 & 2(2r)^2 & 4(n-1)r^4 \\ 0 & 0 & 0 & 4r^4 \end{pmatrix}_{M_{u,I}}. \tag{27}$$

Using Equations (23), (25), and (26), we can construct a higher-order algebraic system for solving weight coefficients:

$$\begin{pmatrix} W_0 + 2nW_1 + 2nW_2 + 2(n-1)nW_3 = I_{0,0} \\ 2W_1r^2 + 8W_2r^2 + 4(n-1)W_3r^2 = I_{0,2} \\ 2W_1r^4 + 32W_2r^4 + 4(n-1)W_3r^4 = I_{0,4} \\ 4r^4W_3 = I_{2,2} \end{pmatrix}. \tag{28}$$

To solve the above algebraic equations, the following unique solution of the weight coefficients can be obtained:

$$\begin{aligned} W_0 &= \frac{\pi^{n/2}\left(2n^2-10nr^2+n+16r^4\right)}{16r^4}, W_1 = \frac{\pi^{n/2}\left(8r^2-3n\right)}{24r^4}, \\ W_2 &= \frac{\pi^{n/2}\left(3-2r^2\right)}{96r^4}, W_3 = \frac{\pi^{n/2}}{16r^4} \end{aligned}. \tag{29}$$

To maintain the approximate accuracy and numerical stability while minimizing the number of cubature points used, we can take $r = \sqrt{3n/8}$ or $\sqrt{3/2}$ obtain a new set of weight coefficients. When $r = \sqrt{3/2}$, we obtain the cubature formula of the embedded KF in Equation (20), which has inadequate numerical stability; thus, we take $r = \sqrt{3n/8}$. The weight solution is as follows:

$$w_0 = \frac{2(n+2)\pi^{n/2}}{9n}, w_1 = 0, w_2 = -\frac{(n-4)\pi^{n/2}}{18n^2}, w_3 = \frac{4\pi^{n/2}}{9n^2} \quad (30)$$

Therefore, Equation (23) can be written as

$$\begin{aligned} \hat{G}(\boldsymbol{u}) &= \tfrac{2(n+2)\pi^{n/2}}{9n}\boldsymbol{u}(\mathbf{0}) - \tfrac{(n-4)\pi^{n/2}}{18n^2}\sum_{FS}\boldsymbol{u}\left(\sqrt{\tfrac{3n}{2}},\mathbf{0}\right) \\ &+ \tfrac{4\pi^{n/2}}{9n^2}\sum_{FS}\boldsymbol{u}\left(\sqrt{\tfrac{3n}{8}},\sqrt{\tfrac{3n}{8}},\mathbf{0}\right). \end{aligned} \quad (31)$$

The total number of points in the new cubature formula is $2n^2 + 1$, and the stability coefficient is $\frac{11n-8}{9n} > 1$ (when $n > 4$). The numerical stability of this formula is superior to that of Equation (20) for the embedded KF, and the number of cubature points used is equal to that for Equation (16). The new formula has a simpler structure and a smaller cubature-points coefficient r than Equations (14) and (17), which improves the numerical stability. To ameliorate the non-local sampling problem of point coordinates caused by the increase in system dimensions, we adjust the coordinate parameters of the cubature points in the above cubature formula:

$$\begin{aligned} \hat{G}(\boldsymbol{u}) &= \tfrac{2(n+2)\pi^{n/2}}{9n}\boldsymbol{u}(0) - \tfrac{(n-4)\pi^{n/2}}{18n^2}\sum_{FS}\boldsymbol{u}\left(\sqrt{\tfrac{3(n-c)}{2}},0\right) \\ &+ \tfrac{4\pi^{n/2}}{9n^2}\sum_{FS}\boldsymbol{u}\left(\sqrt{\tfrac{3(n-c)}{8}},\sqrt{\tfrac{3(n-c)}{8}},0\right), 0 < c < 1. \end{aligned} \quad (32)$$

Fine-tuning the parameters can not only maintain the accuracy of the cubature formula but also reduce the influence of nonlocal sampling problems.

3. Robust KF Based on New Cubature Formula and MCC

3.1. Cross-Correntropy Formula

As a statistical measure of the similarity between two random variables, the cross-correntropy formula has been widely used in non-Gaussian noise-signal processing. Cross-correntropy is a generalized variable that characterizes the correlation between a pair of scalar random variables and can measure not only second-order information but also high-order moments of the joint probability density. The cross-correntropy between two scalar random variables can be expressed by the mathematical expectation of the positive-definite kernel function $K_\sigma(x,y)$ [14]:

$$H_\sigma(X,Y) = E(K_\sigma(X,Y)) = \iint_{R^2} K_\sigma(x,y)f_{XY}(x,y)d\tau. \quad (33)$$

Here, $f_{XY}(x,y)$ is the joint probability density function between two random variables. Under normal circumstances, the joint distribution function between random variables cannot be accurately obtained; thus, it can only be estimated using a discrete approximation formula and a limited number of sample data points:

$$H_{\sigma E} = \frac{1}{n}\sum_{k=1}^{n} K_\sigma(x_k, y_k). \quad (34)$$

In the above expression, $\{x_k, y_k\}_1^n$ is a sample extracted from the joint distribution function $F_{XY}(x,y)$. We selected the Gaussian correlation entropy function for the estimation, which is a positive definite kernel function that satisfies Mercer's theory:

$$K_\sigma(x,y) = G_\sigma(e) = \exp\left(-\frac{e^2}{2\sigma^2}\right). \quad (35)$$

where $e = x - y$ and σ represents the bandwidths of the kernel function. The Gaussian function has a series of advantageous properties. For example, it is positive and bounded, and the maximum point is zero. The MCC is obtained using the correlation entropy estimation formula based on the Gaussian function. Expanding the kernel function in Equation (34) using the Taylor series:

$$H_{\sigma E} = \sum_{k=0}^{\infty} \frac{1}{k!}\left(-\frac{1}{2\sigma^2}\right)^k E\left[(X-Y)^{2k}\right] \tag{36}$$

According to the series expression above, the correlation entropy can extract the information of all even moments of $X - Y$ under the appropriate kernel bandwidth. This helps us to use the higher-order moment information of the signal better.

3.2. Parameter Estimation Method Based on Correntropy Criterion

According to the derivation process of the original Kalman filter, the state update equation of the conventional Kalman filter can be obtained under the assumption that the innovation conforms to a Gaussian distribution. When interference values and outliers exist in the noise of the state or observation information, the hypothesis that innovation conforms to the Gaussian distribution in Equations (4) and (5) is no longer satisfied, resulting in a reduced filtering effect. Therefore, a robust method should be introduced to optimize the original Kalman filter framework.

Izanloo R et al. [14] combined the weighted matrix with the cost function of the c-filter and obtained the cost function of the WLS method and the MCC for a linear system, so that the least-variance estimation and high-order moment information were combined in the filtering process, and the estimation process was embedded into the conventional KF framework. A KF with robustness was obtained. According to this cost function, a new generalized cost function is designed, and its parameter estimation process is extended to the filtering process of nonlinear systems. Using the nonlinear filter framework presented in Section 2.1, together with the WLS method and MCC, we give the generalized cost function as follows:

$$J = G_\sigma\left(\|x_k - \hat{x}_{k|k-1}\|^2_{P^{-1}_{k|k-1}}\right) + \gamma G_\sigma\left(\|z_k - \overline{H}_k x_k - \hat{z}_{k|k-1} + \overline{H}_k \hat{x}_{k|k-1}\|^2_{\overline{R}_k^{-1}}\right), \tag{37}$$

where $\|x\|^2_R = x^T R x$, $x_k, \hat{x}_{k|k-1}, z_k \hat{z}_{k|k-1}$ and $P_{k|k-1}$ are the prediction, the update state vector, the observation vector and the state covariance matrix in the KF framework, respectively. γ is an undetermined constant. We use statistical linearization [25] to solve nonlinear filtering problems. $\overline{R}_k = P_{zz} - \overline{H}_k P_{k|k-1} \overline{H}_k^T$ is the noise covariance matrix of the statistically linearized observation vector, and $\overline{H}_k = P^T_{xz} P^{-1}_{k|k-1}$ is the coefficient matrix of the statistically linearized observation vector.

To minimize the above objective function J, we first calculate its derivative with respect to x_k:

$$\begin{aligned}\frac{\partial J}{\partial x_k} = \gamma &\frac{\overline{H}_k^T \overline{R}_k^{-1} G_\sigma\left(\|z_k - \overline{H}_k x_k - \hat{z}_{k|k-1} + \overline{H}_k \hat{x}_{k|k-1}\|^2_{\overline{R}_k^{-1}}\right)\left(z_k - \overline{H}_k x_k - \hat{z}_{k|k-1} + \overline{H}_k \hat{x}_{k|k-1}\right)}{\sigma^2} \\ &- G_\sigma\left(\|x_k - \hat{x}_{k|k-1}\|^2_{P^{-1}_{k|k-1}}\right)\frac{P^{-1}_{k|k-1}\left(x_k - \hat{x}_{k|k-1}\right)}{\sigma^2}\end{aligned}. \tag{38}$$

By setting this to 0, the following matrix equation is obtained:

$$\begin{aligned}&\gamma G_k \overline{H}_k^T \overline{R}_k^{-1}\left(z_k - \overline{H}_k x_k - \hat{z}_{k|k-1} + \overline{H}_k \hat{x}_{k|k-1}\right) = P^{-1}_{k|k-1}\left(x_k - \hat{x}_{k|k-1}\right) \\ &G_k = \frac{G_\sigma\left(\|z_k - \overline{H}_k x_k - \hat{z}_{k|k-1} + \overline{H}_k \hat{x}_{k|k-1}\|^2_{\overline{R}_k^{-1}}\right)}{G_\sigma\left(\|x_k - \hat{x}_{k|k-1}\|^2_{P^{-1}_{k|k-1}}\right)}\end{aligned} \tag{39}$$

According to the equation, we take the constant parameter $\gamma = 1$, and simplify the above equation to obtain a simple equation involving x_k:

$$\begin{aligned}&\left(P^{-1}_{k|k-1} + G_k \overline{H}_k^T \overline{R}_k^{-1} \overline{H}_k\right) x_k = G_k \overline{H}_k^T \overline{R}_k^{-1}\left(z_k - \hat{z}_{k|k-1}\right) \\ &+ \left(P^{-1}_{k|k-1} + G_k \overline{H}_k^T \overline{R}_k^{-1} \overline{H}_k\right)\hat{x}_{k|k-1} \\ &x_k = \left(P^{-1}_{k|k-1} + G_k \overline{H}_k^T \overline{R}_k^{-1} \overline{H}_k\right)^{-1} G_k \overline{H}_k^T \overline{R}_k^{-1}\left(z_k - \hat{z}_{k|k-1}\right) + \hat{x}_{k|k-1}\end{aligned} \tag{40}$$

Then, we obtain the new state-estimation expression for a robust KF:

$$\begin{aligned}\hat{x}_{k|k} &= \hat{x}_{k|k-1} + \hat{K}_k^G\left(z_k - \hat{z}_{k|k-1}\right);\\ \hat{K}_k^G &= \left(P_{k|k-1}^{-1} + G_k\overline{H}_k^T\overline{R}_k^{-1}\overline{H}_k\right)^{-1}G_k\overline{H}_k^T\overline{R}_k^{-1}\end{aligned} \tag{41}$$

In the KF framework, we replace $\overline{H}_k x_k$ with $\overline{H}_k\hat{x}_{k|k-1}$ and substitute the corresponding state estimation vector of x_k.

$$G_k = \frac{G_\sigma\left(\|z_k - \overline{H}_k x_k - \hat{z}_{k|k-1} + \overline{H}_k\hat{x}_{k|k-1}\|^2_{\overline{R}_k^{-1}}\right)}{G_\sigma\left(\|x_k - \hat{x}_{k|k-1}\|^2_{P_{k|k-1}^{-1}}\right)} \Rightarrow \frac{G_\sigma\left(\|z_k - \hat{z}_{k|k-1}\|^2_{\overline{R}_k^{-1}}\right)}{G_\sigma\left(\|\hat{x}_{k|k-1} - \hat{x}_{k-1|k-1}\|^{2-1}_{P_{k|k-1}}\right)} \tag{42}$$

Then, we obtain the following state-estimation expression for a robust KF:

$$\underbrace{\begin{cases} K_k = P_{k|k-1}^{xz}(P_{k|k-1}^{zz})^{-1}\\ \hat{x}_{k|k} = \hat{x}_{k|k-1} + K_k\left(z_k - \hat{z}_{k|k-1}\right)\\ P_{k|k} = P_{k|k-1} - K_k P_{k|k-1}^{P_z^z} K_k^T \end{cases}}_{\text{traditional}(a)} \Rightarrow \underbrace{\begin{cases} \overline{H}_k = \left(P_{k|k-1}^{xz}\right)^T P_{k|k-1}^{-1}\\ \overline{R}_k = P_{k|k-1}^{zz} - \overline{H}_k P_{k|k-1}\overline{H}_k^T\\ G_k = \dfrac{G_\sigma\left(\|z_k - \hat{z}_{k|k-1}\|^2_{\overline{R}_k^{-1}}\right)}{G_\sigma\left(\|\hat{x}_{k|k-1} - \hat{x}_{k-1|k-1}\|^2_{P_{k-1}^{-1}}\right)}\\ \hat{K}_k^G = \left(P_{k|k-1}^{-1} + G_k\overline{H}_k^T\overline{R}_k^{-1}\overline{H}_k\right)^{-1}G_k\overline{H}_k^T\overline{R}_k^{-1}\\ \hat{x}_{k|k} = \hat{x}_{k|k-1} + \hat{K}_k^G\left(z_k - \hat{z}_{k|k-1}\right)\\ P_{k|k} = \left(E - \hat{K}_k^G\overline{H}_k\right)P_{k|k-1}\left(E - \hat{K}_k^G\overline{H}_k\right)^T\\ +\hat{K}_k^G\overline{R}_k\left(\hat{K}_k^G\right)^T \end{cases}}_{\text{robust}(b)} \tag{43}$$

The above formula is used to replace the gain in the original Kalman filtering process and to update the state vector and covariance estimation. The algorithm flow of robust Kalman filtering can be obtained using the new cubature formula derived in Section 2.3.

3.3. Robust KF Based on New Fifth-Degree Cubature Formula and MCC

In the following algorithm, we use diagonalization transformation to solve the square root of the matrix. With regard to numerical stability, the diagonalization transformation is better than the Cholesky decomposition method used in the general CKF algorithm. Cholesky decomposition requires the matrix to be positive-definite, which may lead to process instability or even to algorithm divergence.

We now present the time update and measurement update processes of the proposed robust KF algorithm.

3.3.1. Initialization of Cubature Points and Parameters

The state vector and covariance matrix are initialized as follows:

$$\hat{x}_{0|0} = E(x_0), P_{0|0} = E\left[\left(x_0 - \hat{x}_{0|0}\right)\left(x_0 - \hat{x}_{0|0}\right)^T\right]. \tag{44}$$

The trajectory nodes of the integration required by the fifth-degree algorithm are generated:

$$\begin{aligned} U_1 &= \sqrt{\tfrac{3(n-c)}{4}}\underbrace{(e_1 + e_2, \ldots, e_1 + e_n, e_2 + e_3, \ldots, e_2 + e_n, \ldots, e_{n-1} + e_n)}_{n(n-1)/2}\\ U_2 &= \sqrt{\tfrac{3(n-c)}{4}}\underbrace{(e_1 - e_2, \ldots, e_1 - e_n, e_2 - e_3, \ldots, e_2 - e_n, \ldots, e_{n-1} - e_n)}_{n(n-1)/2} \end{aligned} \tag{45}$$

Then, we initialize the different weights in the cubature formula and their corresponding cubature points:

$$\begin{cases} N_f = 2n^2 + 1\\ \chi_{(1)} = 0\\ \chi_{(2,\ldots,2n+1)} = \sqrt{3(n-c)}(E, -E)\\ \chi_{(2n+2,\ldots,2n^2+1)} = (U_1, -U_1, U_2, -U_2) \end{cases}, \begin{cases} w_{(1)} = \frac{2(n+2)}{9n}\\ w_{(2,\ldots,2n+1)} = -\frac{(n-4)}{18n^2}\\ w_{(2n+2,\ldots,2n^2+1)} = \frac{4}{9n^2} \end{cases}. \tag{46}$$

3.3.2. Time Update

The diagonalization transformation given above is used to calculate the square-root matrix of the covariance matrix [26]:

$$P_{k|k} = U_I \Lambda_{k|k} U_I^T, U_{k|k} = U_I \sqrt{\Lambda_{k|k}} U_I^T$$

$$\Lambda_{k|k} = \begin{pmatrix} \sqrt{\lambda_1} & 0 & \cdots & 0 \\ 0 & \sqrt{\lambda_2} & \cdots & 0 \\ \cdots & \cdots & \cdots & \cdots \\ 0 & 0 & \cdots & \sqrt{\lambda_n} \end{pmatrix}_{P_{k|k}} . \tag{47}$$

Then, the new cubature points are calculated by substituting the integral variable into Equation (8) and its corresponding state function value:

$$X^l_{k-1|k-1} = U_{k-1|k-1}\chi_{k-1} + \hat{x}_{k-1|k-1}, X^{(l)}_{k-1|k-1} = f\left(X^l_{k-1|k-1}\right) \tag{48}$$

Next, the state-vector prediction and state error covariance matrix prediction are calculated:

$$\begin{cases} \hat{x}_{k|k-1} = \sum_{l=1}^{N_f} w_l f\left(X^l_{k-1|k-1}\right) = \sum_{l=1}^{N_f} w_l X^{(l)}_{k-1|k-1} \\ P_{k|k-1} = \sum_{l=1}^{N_f} w_l \left(X^{(l)}_{k-1|k-1} - \hat{x}_{k|k-1}\right)\left(X^{(l)}_{k-1|k-1} - \hat{x}_{k|k-1}\right)^T + Q_{k-1} \end{cases} . \tag{49}$$

3.3.3. Measurement Update

Similar to Equation (46), diagonalization transformation is performed on the state prediction covariance matrix to obtain the square-root matrix:

$$P_{k|k-1} = U_J \Lambda_{k|k-1} U_J^T, U_{k|k-1} = U_J \sqrt{\Lambda_{k|k}} U_J^T . \tag{50}$$

The new cubature points are calculated after the Gaussian integral substitution, along with the corresponding measurement-function value:

$$X^l_{k|k-1} = U_{k|k-1}\chi_k + \hat{x}_{k|k-1}, Z^l_{k|k-1} = h\left(X^l_{k|k-1}\right) . \tag{51}$$

The measurement vector estimation, autocorrelation covariance matrix, and cross-covariance matrix are evaluated:

$$\begin{cases} \hat{z}_{k|k-1} = \sum_{l=1}^{N_f} w_l h\left(X^l_{k|k-1}\right) = \sum_{l=1}^{N_f} w_l Z^l_{k|k-1} \\ P^{zz}_{k|k-1} = \sum_{l=1}^{N_f} w_l \left(Z^l_{k|k-1} - \hat{z}_{k|k-1}\right)\left(Z^l_{k|k-1} - \hat{z}_{k|k-1}\right)^T + R_k \\ P^{xz}_{k|k-1} = \sum_{l=1}^{N_f} w_l \left(X^l_{k|k-1} - \hat{x}_{k|k-1}\right)\left(Z^l_{k|k-1} - \hat{z}_{k|k-1}\right)^T . \end{cases} \tag{52}$$

We calculate the statistical linearization update matrix and covariance matrix of the measurement function:

$$\overline{H}_k = \left(P^{xz}_{k|k-1}\right)^T P^{-1}_{k|k-1}, \overline{R}_k = P^{zz}_{k|k-1} - \overline{H}_k P_{k|k-1} \overline{H}_k^T . \tag{53}$$

We then compute the correlation entropy coefficient based on the Gaussian kernel function and weighted norm:

$$G_k = \frac{G_\sigma\left(\|z_k - \hat{z}_{k|k-1}\|^2_{\overline{R}_k^{-1}}\right)}{G_\sigma\left(\|\hat{x}_{k|k-1} - \hat{x}_{k-1|k-1}\|^2_{P^{-1}_{k|k-1}}\right)} . \tag{54}$$

Finally, we calculate the new gain matrix of the robust KF; then, the updated state vector estimation and updated covariance matrix are computed according to the new gain matrix:

$$\begin{cases} \hat{K}^G_k = \left(P^{-1}_{k|k-1} + G_k \overline{H}_k^T \overline{R}_k^{-1} \overline{H}_k\right)^{-1} G_k \overline{H}_k^T \overline{R}_k^{-1} \\ \hat{x}_{k|k} = \hat{x}_{k|k-1} + \hat{K}^G_k\left(z_k - \hat{z}_{k|k-1}\right) \\ P_{k|k} = \left(E - \hat{K}^G_k \overline{H}_k\right) P_{k|k-1} \left(E - \hat{K}^G_k \overline{H}_k\right)^T + \hat{K}^G_k \overline{R}_k \left(\hat{K}^G_k\right)^T \end{cases} . \tag{55}$$

4. Simulation Experiment of Target Tracking Based on Proposed Algorithm

4.1. Filtering Simulation Experiment and Numerical Analysis of Cubature Formulas

We evaluated the filtering performance of the CKF using the cubature formula proposed in Section 2 through simulation experiments.

Consider the following nonlinear system equation with trigonometric functions, exponential functions, and power functions [20]:

$$\begin{cases} Process\ equation: \\ x_k = \begin{pmatrix} x_k^{(1)} \\ x_k^{(2)} \\ x_k^{(3)} \end{pmatrix} = \begin{pmatrix} 3\sin^2\left(5x_{k-1}^{(2)}\right) \\ x_{k-1}^{(1)} + e^{-0.05x_{k-1}^{(3)}} + 10 \\ 0.2x_{k-1}^{(1)}\left(x_{k-1}^{(2)} + x_{k-1}^{(3)}\right) \end{pmatrix} + w_{k-1} \\ Measurement\ equation: \\ z_k = \cos\left(x_k^{(1)}\right) + x_k^{(2)}x_k^{(3)} + v_k \end{cases} \tag{56}$$

The process noise is $w_k \sim N(0, Q)$, $Q = 0.1E_3$, where E_n is the n-dimensional identity matrix. The measurement noise is $v_k \sim N(0, R)$, $R = 1.0$. In the simulation experiment, the initial vector was set as $x_0 = (1, 1, 1)^T$, and the initial covariance matrix was set as $P_{0|0} = \text{diag}(0.1, 0.1, 0.1)$. In addition to the proposed 5th-degree DDCKF (Equation (32)), the following HDCKFs were used for comparison:

- Equation (11), 3rd-degree CKF;
- Equation (14), 5th-degree simplified CKF (SCKF);
- Equation (17), 5th-degree CKF; and
- Equation (20), 5th-degree ECKF.

The initial conditions of all the filters were identical at the beginning of each simulation, and the simulation time was 40 s. The root-mean-square error (*RMSE*) was used as an evaluation index for the filtering performance. The *RMSE* and average *RMSE* (*ARMSE*) were defined as follows:

$$\begin{aligned} RMSE_{(x,x_l)} &= \sqrt{\frac{1}{N}\sum_{k=1}^{N}\sum_{l=1}^{3}\left(x_k^{(l)} - \hat{x}_{k|k}^{(l)}\right)^2} \\ ARMSE_{(x)} &= \frac{1}{A}\sum_{k=1}^{A} RMSE_{(x,x_k)}(k) \end{aligned} \tag{57}$$

where N represents the number of Monte Carlo simulation runs, and A represents the timestep (in s) of the simulation experiment. We set N = 100 and A = 40. Additionally, we set c = 0 in Equations (45) and (46). The simulation results are shown in Figure 1.

From the *RMSE* results of the simulation experiment in Figure 1, it can be concluded that the filtering accuracy of the 3rd-degree CKF was the lowest when the system was highly nonlinear, because it only used $2N$ cubature points and the low-order cubature formula with low accuracy to approximate the probability density; thus, it could not obtain a good approximation. Among the 5th-degree formulas, the SCKF had a slightly higher average filtering accuracy than the ECKF, and the proposed DDCKF had the highest accuracy. In such a low-dimension nonlinear system, the cubature of the 5th-degree CKF is $n^2 + 3n + 3$; thus, the number of cubature points required by the 5th-degree CKF exceeded those for the other filters, leading to a larger number of computations. According to the experimental results, compared with the other 5th-degree algorithms, the DDCKF has better filtering performance for systems with a high degree of nonlinearity. The *ARMSE* values of the algorithms are presented in Table 2.

Table 2. Performance indices of the different filters used in Experiment 4.1.

Cubature Filters	Number of Points	*ARMSE* Pos
3rd -degree CKF	$2n = 6$	2.2091
5th-degree SCKF	$2n^2 + 1 = 19$	1.3428
5th-degree CKF	$n^2 + 3n + 3 = 21$	1.5959
5th-degree ECKF	$2n^2 + 1 = 19$	1.5284
(Proposed) 5th-degree DDCKF	$2n^2 + 1 = 19$	**1.0193**

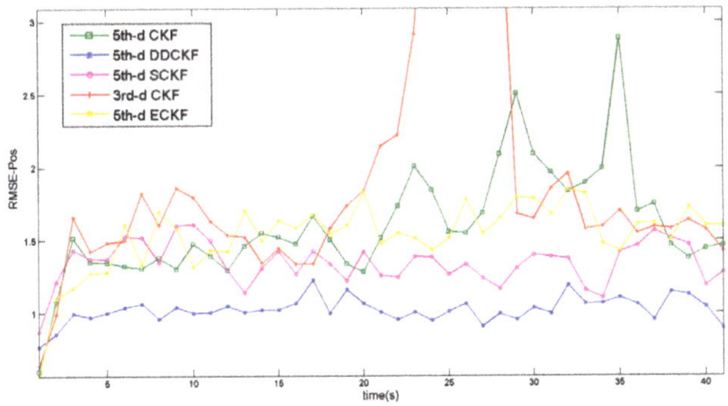

Figure 1. RMSE for different filters in Experiment 4.1.1.

According to the *ARMSE* results, the 3rd-degree CKF requires the least number of cubature points, but it has the lowest filtering accuracy. The proposed DDCKF algorithm achieved the highest filtering accuracy when the number of cubature points was maintained.

4.2. Simulation Experiment of Robust Target Tracking Based on DD-MCCKF and Surface Target-Tracking Models

4.2.1. Simulation Experiment of Robust Filtering Based on the Constant Velocity (CV) Tracking Model of the Surface Target

In this experiment, the following target-tracking model is used to verify the filtering performance of the proposed robust CKF algorithm for the model with high maneuvering speed and non-Gaussian noise. Consider the following constant velocity (CV) surface target-tracking model:

$$\begin{cases} State\ equation: \\ x_k = \begin{pmatrix} x_k \\ \dot{x}_k \\ y_k \\ \dot{y}_k \end{pmatrix} = \begin{pmatrix} 1 & T & 0 & 0 \\ 0 & 1 & 0 & 0 \\ 0 & 0 & 1 & T \\ 0 & 0 & 0 & 1 \end{pmatrix} \begin{pmatrix} x_{k-1} \\ \dot{x}_{k-1} \\ y_{k-1} \\ \dot{y}_{k-1} \end{pmatrix} + \begin{pmatrix} \frac{T^2}{2} & 0 \\ T & 0 \\ 0 & T \\ 0 & \frac{T^2}{2} \end{pmatrix} w_{k-1} \\ Measurement\ equation: \\ z_k = \begin{pmatrix} \sqrt{x_k^2 + y_k^2} \\ arctan\left(\frac{y_k - y_l}{x_k - x_l}\right) \end{pmatrix} + v_k \end{cases} . \quad (58)$$

The CV model is a type of coordinate-uncoupled model. These models assume that the target maneuvering process in three orthogonal directions is uncoupled in three-dimensional space. Target maneuvering is caused by acceleration changes caused by external forces. Therefore, the difficulty of maneuvering modeling lies in the description of the target acceleration. For high-speed surface targets, the CV model is often used to describe the movement of such targets.

In this model, w_k and v_k are the mutually independent system process noise and measurement noise with covariance matrices $\boldsymbol{Q}_k$ and $\boldsymbol{R}_k$, and the sampling period is T.

The position *RMSE*, velocity *RMSE*, and *ARMSE* were defined as the filtering-accuracy evaluation criteria:

$$\begin{aligned} RMSE_{CVpos} &= \sqrt{\frac{1}{N}\sum_{k=1}^{N}\left[\left(x_k - \hat{x}_{k|k}\right)^2 + \left(y_k - \hat{y}_{k|k}\right)^2\right]} \\ RMSE_{CVvel} &= \sqrt{\frac{1}{N}\sum_{k=1}^{N}\left[\left(\dot{x}_k - \hat{\dot{x}}_{k|k}\right)^2 + \left(\dot{y}_k - \hat{\dot{y}}_{k|k}\right)^2\right]} \\ ARMSE_{any} &= \frac{1}{A}\sum_{k=1}^{A} RMSE_{any}(k), \end{aligned} \quad (59)$$

where N represents the number of Monte Carlo simulation runs, and A represents the timestep of the simulation experiment. The initial values of the state variable and the error covariance matrix are $\boldsymbol{x}_0$ and $\boldsymbol{P}_{0|0}$, respectively. In this experiment, the parameters were initialized as follows:

$$\begin{aligned} &Q_1 = q_1^2\begin{bmatrix} M_1 & 0 \\ 0 & M_1 \end{bmatrix}, Q_2 = q_2^2\begin{bmatrix} M_2 & 0 \\ 0 & M_2 \end{bmatrix} \\ &q_1 = 0.2, q_2 = 0.3, M_k = \begin{bmatrix} T_k^3/3 & T_k^2/2 \\ T_k^2/2 & T_k \end{bmatrix}, T_1 = 1, T_2 = \tfrac{1}{2} \\ &\boldsymbol{R}_1 = 0.1\text{diag}\left\{(20\text{ m})^2, \left(\tfrac{6\pi}{180}\right)^2\text{rad}\right\}, \boldsymbol{R}_2 = 0.1\text{diag}\left\{(30\text{ m})^2, \left(\tfrac{8\pi}{180}\right)^2\text{rad}\right\} \end{aligned} \tag{60}$$

To set different types of the non-Gaussian state noise and measurement noise environment based on $\begin{cases} w_k \sim (1-\eta)N(\boldsymbol{0}, Q_1) + \eta N(\boldsymbol{0}, Q_2) \\ r_k \sim (1-\eta)N(\boldsymbol{0}, \boldsymbol{R}_1) + \eta N(\boldsymbol{0}, \boldsymbol{R}_2) \end{cases}$, we considered the following conditions:

Condition 1. *Under mixture noise, $\eta = \frac{1}{2}$:*

$$w_{k1} \sim \frac{1}{2}N_1(\boldsymbol{0}, Q_1) + \frac{1}{2}N_2(\boldsymbol{0}, Q_2)\,,\; r_{k1} \sim \frac{1}{2}N_1(\boldsymbol{0}, \boldsymbol{R}_1) + \frac{1}{2}N_2(\boldsymbol{0}, \boldsymbol{R}_2)\,. \tag{61}$$

Condition 2. *Under mixture noise, $\eta = \frac{2}{3}$:*

$$w_{k2} \sim \frac{1}{3}N(\boldsymbol{0}, Q_1) + \frac{2}{3}N(\boldsymbol{0}, Q_2)\,,\; r_{k2} \sim \frac{1}{3}N(\boldsymbol{0}, \boldsymbol{R}_1) + \frac{2}{3}N(\boldsymbol{0}, \boldsymbol{R}_2)\,. \tag{62}$$

Experiment 1. *Comparison of the filtering performance between robust and regular 3rd-degree CKF in the non-Gaussian noise Environment (61).*

First, to verify the strong tracking performance of the algorithm proposed in this study and compare it with the traditional nonlinear Kalman filter in a non-Gaussian noise environment, a target tracking experiment based on a 3rd-degree CKF was carried out for the above CV tracking model, in which the status update process of CKF(regular) was implemented using Formula (43a). The status updating process of CKF(robust) was realized using Formula (43b).

We set the initial value of the state variable and the error covariance matrix as $\boldsymbol{x}_0 = (100\text{ m}, 30\text{ m/s}, 100\text{ m}, 20\text{ m/s})^T$ and $\boldsymbol{P}_{0|0} = \text{diag}\left(10\text{ m}^2, 1\text{ m}^2/\text{s}^2, 10\text{ m}, 1\text{ m}^2/\text{s}^2\right)^T$, respectively. In this experiment, we set N = 200 times and A = 150 s. Additionally, we set c in Equations (45) and (46) as 1/3. Other parameters and initial values in the experiment were evaluated according to Equation (60).

The simulation results are shown in Figures 2 and 3:

Therefore, from the above experiments, the state-updating process proposed in this study is more robust than the traditional nonlinear Kalman filter in a strong non-Gaussian noise environment. This experiment proved that the proposed method is effective and feasible.

Experiment 2. *Comparison of the filtering performance of five types of robust nonlinear Kalman filters in a non-Gaussian noise environment, Equations (61) and (62), respectively:*

In the experiment in this section, the robust filter MCUKF proposed in [16], MCSCKF proposed in [27], and E-MCCKF and S-MCCKF, which are formed by Equations (14) and (20), and the MCC method proposed in this study were compared with the robust algorithm DD-MCCKF proposed in this study in a non-Gaussian state and measured noise Environment (61) and (62).

We set the initial value of the state variable and the error covariance matrix as $\boldsymbol{x}_0 = (100\text{ m}, 30\text{ m/s}, 100\text{ m}, 20\text{ m/s})^T$ and $\boldsymbol{P}_{0|0} = \text{diag}\left(10\text{ m}^2, 1\text{ m}^2/\text{s}^2, 10\text{ m}, 1\text{ m}^2/\text{s}^2\right)^T$, respectively. The number of Monte Carlo simulation $N = 1 \times 10^5$ times and A = 120 s. Additionally, we set c in Equations (45) and (46) as 1/3, and the other parameters and initial values in the experiment were evaluated according to Equation (60).

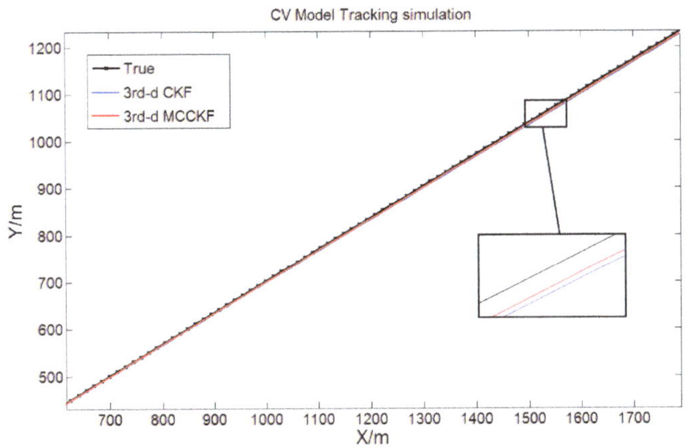

Figure 2. Tracking trajectory of the 3rd-drgree CKF and the 3rd-degree MCCKF under non-Gaussian noise conditions in Equation (61).

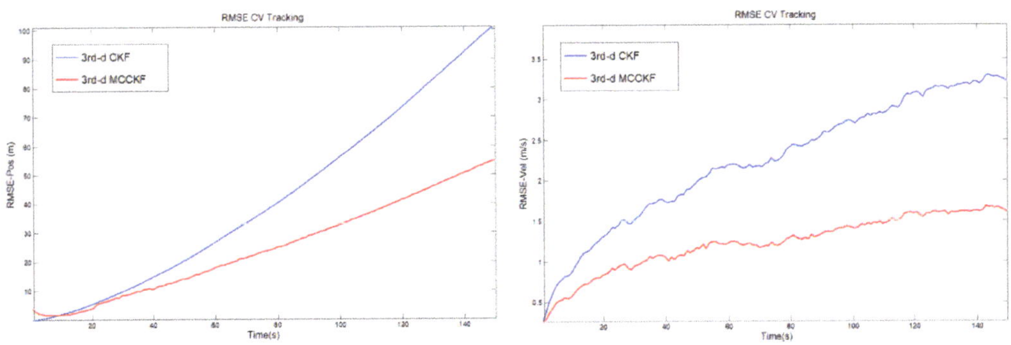

(a) (b)

Figure 3. *RMSE* of the position and velocity for different filters under non-Gaussian noise conditions in Equation (61) ((**a**) is the position RMSE and (**b**) is the velocity RMSE).

The simulation results are shown in Figures 4–6:

The running time of the algorithm was 120 s. The results show that, similar to related properties introduced in Section 2.1, the comprehensive numerical performance of UKF was not as good as that of 3rd-degree CKF, resulting in the highest mean square error and the lowest filtering accuracy of MCUKF. In addition, when $n = 4$, the number of cubature points used by 5th-degree SCKF and ECKF was the same as those used by the DDKCF proposed in this study. The formula structure was similar. Thus, the calculation amount was very similar. Because DD-CKF maintains the numerical stability when the system dimension is large while calculating with fewer cubature points, and Formula (32) carries out a certain displacement of the coefficient of cubature points to reduce the non-sampling error, the accuracy of DD-MCCKF was slightly higher than that of other algorithms in tracking experiments.

As shown in Table 3, the proposed DD-MCCKF algorithm had the highest estimation accuracy under the premise of using fewer cubature points in a non-Gaussian noise environment.

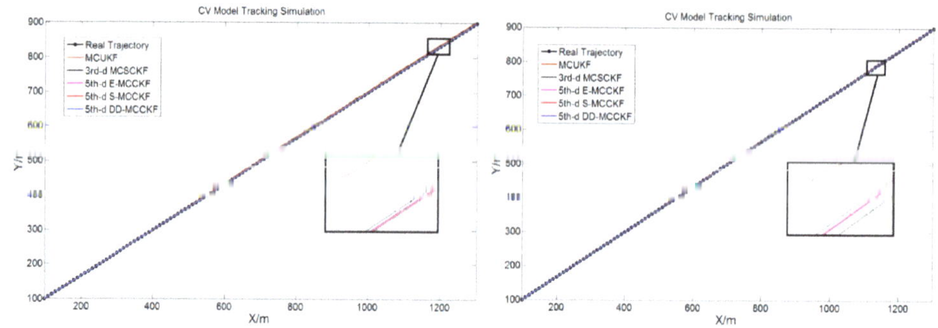

(**a**) (**b**)

Figure 4. Tracking trajectories of a surface target (unmanned surface vessel) under non-Gaussian noise conditions ((**a**) is the result under Equation (61) and (**b**) is the result under Equation (62)).

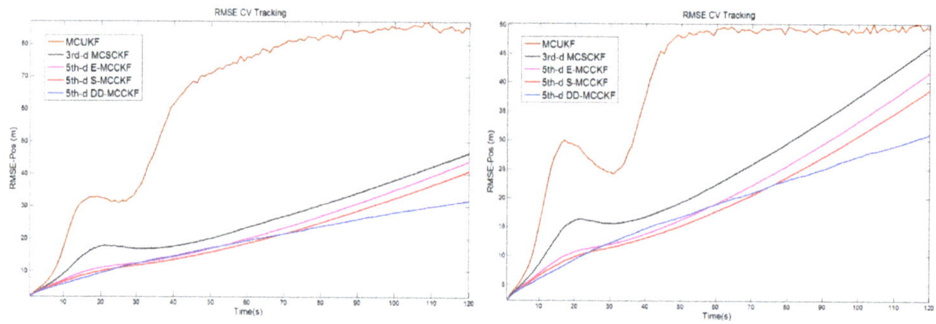

(**a**) (**b**)

Figure 5. Position RMSEs of robust target-tracking algorithms under non-Gaussian noise conditions ((**a**) is the result under Equation (61) and (**b**) is the result under Equation (62)).

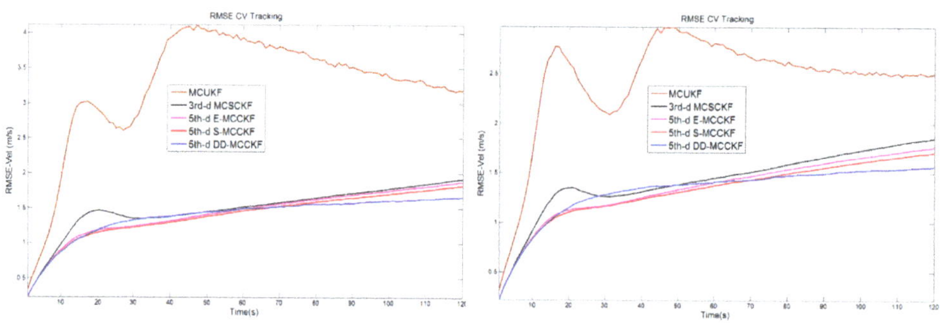

(**a**) (**b**)

Figure 6. Velocity RMSEs of robust target-tracking algorithms under non-Gaussian noise conditions ((**a**) is the result under Equation (61) and (**b**) is the result under Equation (62)).

Table 3 shows the number of points and ARMSE of each algorithm in Experiment 2 above.

Table 3. Number of points used by different filtering algorithms in Experiment 1, along with the *ARMSE* for the position and velocity.

Robust Filters	Number of Points	*ARMSE* Pos (C1, C2)		*ARMSE* Vel (C1, C2)	
MCUKF	$2n = 8$	63.26	39.98	3.261	2.444
3rd-degree MCSCKF	$2n = 8$	25.28	24.47	1.496	1.428
5th-degree E-MCCKF	$2n^2 - 2n + 1 = 25$	21.92	20.91	1.432	1.354
5th-degree S-MCCKF	$2n^2 - 2n + 1 = 25$	20.37	19.43	1.401	1.323
(Proposed) 5th-degree DD-MCCKF	$2n^2 - 2n + 1 = 25$	**18.96**	**18.49**	**1.392**	**1.320**

4.2.2. Simulation Experiment of Robust Filtering Based on Cooperative Turning Tracking Model of Surface Targets

The cooperative turning (*CT*) model is a coordinate coupling model. In most cases, the coordinate coupling model refers to a turning motion model. Because this type of model is closely related to the coordinates, it can be divided into two types: two-dimensional and three-dimensional turning models. The two-dimensional turning model is also called the planar turning model, that is, the *CT* model.

The *CT* model is one of the most important maneuvering models in surface target tracking. It is a commonly used model to describe maneuvering target in USV tracking. The state equation and measurement equation of the *CT* model are as follows:

$$\begin{cases} \boldsymbol{x}_k = \begin{pmatrix} x_k \\ \dot{x}_k \\ y_k \\ \dot{y}_k \\ \omega_k \end{pmatrix} = \begin{pmatrix} 1 & \frac{\sin(\omega_{k-1}T)}{\omega_{k-1}} & 0 & \frac{\cos(\omega_{k-1}T)-1}{\omega_{k-1}} & 0 \\ 0 & \cos(\omega_{k-1}T) & 0 & -\sin(\omega_{k-1}T) & 0 \\ 0 & \frac{1-\cos(\omega_{k-1}T)}{\omega_{k-1}} & 1 & \frac{\sin(\omega_{k-1}T)}{\omega_{k-1}} & 0 \\ 0 & \sin(\omega_{k-1}T) & 0 & \cos(\omega_{k-1}T) & 0 \\ 0 & 0 & 0 & 0 & 1 \end{pmatrix} \begin{pmatrix} x_{k-1} \\ \dot{x}_{k-1} \\ y_{k-1} \\ \dot{y}_{k-1} \\ \omega_{k-1} \end{pmatrix} + \boldsymbol{w}_{k-1} \\ z_k = \begin{pmatrix} \sqrt{x_k^2 + y_k^2} \\ arctan\left(\frac{y_k - y_l}{x_k - x_l}\right) \end{pmatrix} + \boldsymbol{v}_k \end{cases}, \quad (63)$$

where $\boldsymbol{x}_k$ is the system state variable, and x_k, y_k and $\dot{x}_k\dot{y}_k$ represent the position and velocity of the target in the x and y directions, respectively. T represents the sampling period, and ω_k represents the steering angular velocity. $\boldsymbol{w}_k$ represents the process noise, which has covariance matrix $\boldsymbol{Q}_k$, and $\boldsymbol{v}_k$ represents the measurement noise, which has covariance matrix $\boldsymbol{R}_k$. The initial value of the state variable is $\boldsymbol{x}_0$, and the correlation covariance matrix is $\boldsymbol{P}_{0|0}$. To enhance the mobility of the surface target, the parameters were initialized as follows:

$$\begin{aligned} & Q_1 = q_1^2 \begin{bmatrix} \boldsymbol{M}_1 & 0 & 0 \\ 0 & \boldsymbol{M}_1 & 0 \\ 0 & 0 & T_1/3 \end{bmatrix}, Q_2 = q_2^2 \begin{bmatrix} \boldsymbol{M}_2 & 0 & 0 \\ 0 & \boldsymbol{M}_2 & 0 \\ 0 & 0 & T_2/3 \end{bmatrix} \\ & q = 0.05, q_1 = 0.2, q_2 = 0.3, \boldsymbol{M}_k = \begin{bmatrix} T_k^3/3 & T_k^2/2 \\ T_k^2/2 & T_k \end{bmatrix}, T_1 = \frac{1}{4}, T_2 = \frac{1}{6} \\ & x_0 = \left(100\ \text{m}, 80\ \text{m/s}, 100\ \text{m}, 120\ \text{m/s}, -\frac{8\pi}{180}\ \text{rad}\right)^T \\ & \boldsymbol{P}_{0|0} = \text{diag}\left(10\ \text{m}^2, 1\ \text{m}^2/\text{s}^2, 10\ \text{m}^2, 1\ \text{m}^2/\text{s}^2, 0.1\ \text{rad}^2/\text{s}^2\right) \\ & \boldsymbol{R}_1 = q\text{diag}\left\{(25\ \text{m})^2, \left(\frac{3\pi}{180}\right)^2 \text{rad}\right\}, \boldsymbol{R}_2 = q\text{diag}\left\{(30\ \text{m})^2, \left(\frac{9\pi}{180}\right)^2 \text{rad}\right\} \end{aligned} \quad (64)$$

To set up the different types of non-Gaussian state noise and measurement noise environment based on $\begin{cases} w_k \sim (1-\eta)N(\boldsymbol{0}, \boldsymbol{Q}_1) + \eta N(\boldsymbol{0}, \boldsymbol{Q}_2) \\ r_k \sim (1-\eta)N(\boldsymbol{0}, \boldsymbol{R}_1) + \eta N(\boldsymbol{0}, \boldsymbol{R}_2) \end{cases}$, we considered the conditions as follows:

Condition 3. *Under mixture noise,* $\eta = \frac{1}{2}$*:*

$$w_{k3} \sim \frac{1}{2}N_1(\mathbf{0}, \mathbf{Q}_1) + \frac{1}{2}N_2(\mathbf{0}, \mathbf{Q}_2)\ ,\ \mathbf{r}_{k3} \sim \frac{1}{2}N_1(\mathbf{0}, \mathbf{R}_1) + \frac{1}{2}N_2(\mathbf{0}, \mathbf{R}_2)\ . \tag{65}$$

Condition 4. *Under mixture noise,* $\eta = \frac{2}{3}$*:*

$$w_{k4} \sim \frac{1}{3}N(\mathbf{0}, \mathbf{Q}_1) + \frac{2}{3}N(\mathbf{0}, \mathbf{Q}_2)\ ,\ \mathbf{r}_{k4} \sim \frac{1}{3}N(\mathbf{0}, \mathbf{R}_1) + \frac{2}{3}N(\mathbf{0}, \mathbf{R}_2)\ . \tag{66}$$

In the simulation experiment, we used the *RMSE* and *ARMSE* to evaluate the filtering performance. The *RMSE* and *ARMSE* of the displacement, velocity, and steering angle were defined as follows:

$$\begin{aligned}
RMSE_{CTpos} &= \sqrt{\tfrac{1}{N}\sum_{k=1}^{N}\left[\left(x_k - \hat{x}_{k|k}\right)^2 + \left(y_k - \hat{y}_{k|k}\right)^2\right]} \\
RMSE_{CTvel} &= \sqrt{\tfrac{1}{N}\sum_{k=1}^{N}\left[\left(\dot{x}_k - \hat{\dot{x}}_{k|k}\right)^2 + \left(\dot{y}_k - \hat{\dot{y}}_{k|k}\right)^2\right]} \\
RMSE_{CTomg} &= \sqrt{\tfrac{1}{N}\sum_{k=1}^{N}\left[\left(\omega_k - \hat{\omega}_{k|k}\right)^2\right]} \\
ARMSE_{any} &= \tfrac{1}{A}\sum_{l=1}^{A} RMSE_{any}(l).
\end{aligned} \tag{67}$$

Here, N represents the total number of Monte Carlo simulations, and A represents the timestep of each Monte Carlo simulation. In the simulation experiment, We set c in Equations (45) and (46) as 1/3, and we set $\mathbf{N = 1 \times 10^5}$ times and A = 140 s. Lines 1–3 of Equation (67) give the position *RMSEs* obtained from the real value and estimated value of the position vector and the velocity *RMSE* and angular-velocity *RMSE* obtained via the same method as the position *RMSE*, respectively. Finally, the calculation formula for the *ARMSE* is presented.

Because the *CT* model has a strong nonlinearity, and the collaborative *CT* model can better describe the USV high-speed maneuvering steering process on the water surface in reality than the uniform turning model, this experiment used the *CT* model to conduct the last robust filtering experiment based on the nonlinear system in a non-Gaussian environment.

As with Experiment 2 in Section 4.2.1, in the experiment of this section, the robust filter MCUKF proposed in reference [16], the MCSCKF proposed in reference [27] and E-MCCKF and S-MCCKF which are formed by Formula (14), Formula (20) and the MCC method proposed in this paper are used to compare with the robust algorithm DD-MCCKF proposed in this paper in a non-Gaussian state and measured noise environments above. The simulation results for the *CT* target-tracking model are presented in Figures 7–10.

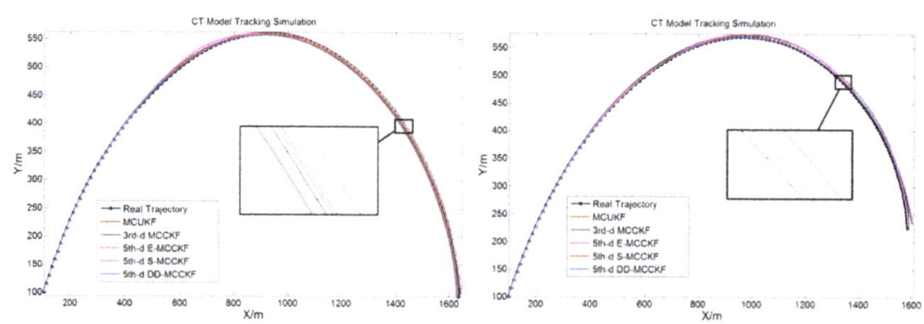

(**a**) (**b**)

Figure 7. Tracking trajectories of surface target (unmanned surface vessel) under non-Gaussian noise conditions ((**a**) is the result under Equation (65) and (**b**) is the result under Equation (66)).

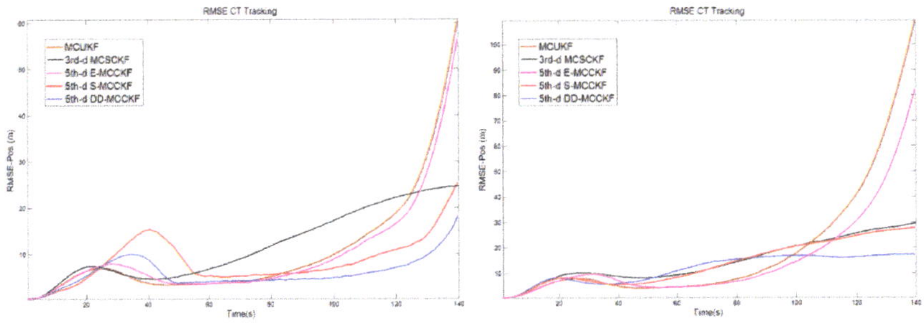

(**a**) (**b**)

Figure 8. Position RMSEs of robust target-tracking algorithms under non-Gaussian noise conditions ((**a**) is the result under Equation (65) and (**b**) is the result under Equation (66)).

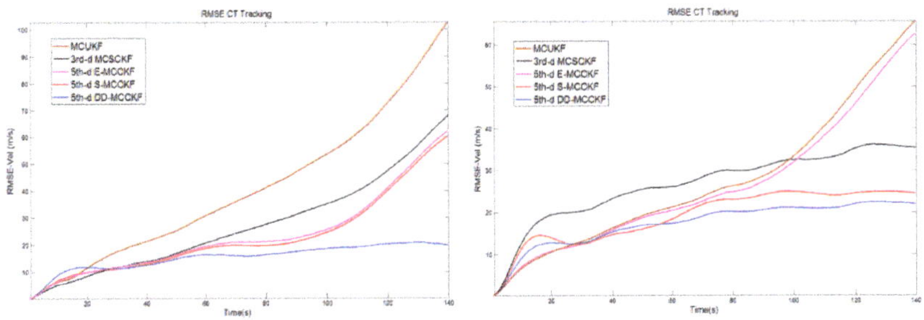

(**a**) (**b**)

Figure 9. Velocity RMSEs of robust target-tracking algorithms under non-Gaussian noise conditions ((**a**) is the result under Equation (65) and (**b**) is the result under Equation (66)).

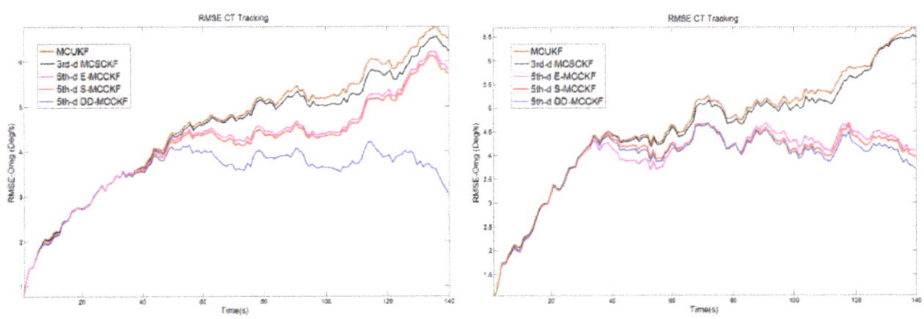

(**a**) (**b**)

Figure 10. Course angular-velocity *RMSE*s of robust target-tracking algorithms under non-Gaussian noise conditions ((**a**) is the result under Equation (65) and (**b**) is the result under Equation (66)).

The running time of the algorithm was 140 s. Among the position errors of the simulation results, the accuracy of the MCUKF was the worst owing to the numerical instability and inaccurate UT. Because of the high system dimensions, the approximation process of the ECKF was unstable, and its filtering performance was worse than that of the S-MCCKF. Especially in the position *RMSE* index, the ECKF was particularly affected by the instability of its formula value, and its filtering accuracy is even lower than that of the 3rd-degree MCCKF.

The DD-MCCKF and S-MCCKF were similar with regard to the structure of the cubature formula and the number of cubature points used. Thus, they had similar filtering accuracies. Because the DD-CKF integrates the two indexes of numerical stability and the number of cubature points used, it is better at improving accuracy and reducing the calculation amount. In addition, the smaller coefficient of the cubature points of the DD-CKF can better reduce non-local sampling problems, which helps further improve its filtering accuracy.

By fine-tuning parameter c in Equations (45) and (46), the non-local sampling problem caused by the increase in the system dimension was alleviated, and the filtering accuracy was increased. For the estimation of the velocity and course angular velocity, the UKF and 3rd-degree CKF had the lowest accuracy, whereas the 5th-degree ECKF and SCKF had similar performance, and the DD-MCCKF had the highest accuracy. Furthermore, as indicated in Table 4, the ARMSEs obtained via the proposed algorithm were lower than those for the 5th-degree ECKF and SCKF.

Table 4. Number of points and the ARMSEs of the position, velocity and course angular velocity for different robust filtering algorithms in the target tracking simulation of Experiment 4.2.2.

Robust Filters	Number of Points	ARMSE Pos-C1	ARMSE Pos-C2	ARMSE Vel-C1	ARMSE Vel-C2	ARMSE Omg-C1	ARMSE Omg-C2
MCUKF	$2n = 10$	10.41	18.84	40.38	27.29	4.532	4.613
3rd-degree MCSCKF	$2n = 10$	11.78	14.66	26.88	26.07	4.430	4.539
5th-degree E-MCCKF	$2n^2 + 1 = 51$	9.824	15.17	23.29	26.63	4.143	3.923
5th-degree S-MCCKF	$2n^2 + 1 = 51$	8.456	13.78	22.47	19.21	4.087	3.931
(Proposed) 5th-degree DD-MCCKF	$2n^2 + 1 = 51$	**5.956**	**11.85**	**15.50**	**17.42**	**3.492**	**3.869**

5. Conclusions

A high-degree robust CKF algorithm, based on a new cubature formula and MCC, was developed. First, according to the construction method of the fully symmetric cubature formula, different cubature-point coordinate generators were used to construct a new high-degree cubature formula and fine-tune its parameters to increase its accuracy. Subsequently, a new robust cost function was constructed by combining the MCC and WLS methods.

As a statistical measure of the similarity between random variables, MCC can extract information from all even moments under the appropriate kernel bandwidth. This helped us make better use of the higher-order moment information of the signal. Therefore, in filtering applications, MCC is more robust to non-Gaussian mixture noise than the conventional nonlinear Kalman filter, which can only use second-order information. The fixed-point iteration solution of the equation was embedded in the nonlinear Kalman filtering process to obtain a robust filtering algorithm. Finally, a new high-degree robust CKF algorithm was obtained by combining the new estimation process with the KF framework. The aim was to use as few cubature coordinate points as possible to achieve the highest filtering accuracy and stability in a non-Gaussian noise environment. The proposed method was applied to the target tracking of an unmanned surface vessel. The simulation experiment exhibited a smaller number of computations, higher filtering accuracy, and better numerical stability when compared to (or similar results to) other algorithms of the same order.

The new-proposed filter probably has a certain application value in practical application scenarios, such as the course tracking of unmanned surface vessels, tracking for maritime moving targets or surface vessel rescuing, and in the sea area with strong environmental interference (e.g., wind and waves). The autonomous navigation target of obstacle avoidance on the water surface may be achieved under the condition of large errors in sensor distance and angle measurement. However, owing to the uncertainty and complexity of the surface environment, the measurement process based on sensors and satellite positioning may be affected or temporarily unavailable. At the same time, some state estimation problems in more complex environments are also the focus of research, such as estimation with bias compensation, nonlinear filtering based on state constraints, robust constraints, state estimation in complex domain impulsive noise etc. As mentioned in References [28–45]. These issues need to be further studied in the following work.

Author Contributions: Writing—original draft preparation, T.W.; writing—review and editing, S.L. and L.Z.; validation and formal analysis, S.L., L.Z. and T.W. All authors have read and agreed to the published version of the manuscript.

Funding: Supported by the Fundamental Research Funds for the Central Universities (3072022JC0404).

Institutional Review Board Statement: Not applicable.

Informed Consent Statement: Not applicable.

Data Availability Statement: Data can be made available upon request to the authors.

Conflicts of Interest: The authors declare no conflict of interest.

References

1. Venkateswarlu, C.; Avantika, S. Optimal state estimation of multicomponent batch distillation. *Chem. Eng. Sci.* **2001**, *56*, 5771–5786. [CrossRef]
2. Ito, M.; Tsujimichi, S.; Kosuge, Y. Tracking a three-dimensional moving target with distributed passive sensors using extended Kalman filter. *Electron. Commun. Jpn.* **2010**, *84*, 74–85. [CrossRef]
3. Bucy, R.S.; Senne, K.D. Digital synthesis of non-linear filters. *Automatica* **1971**, *7*, 287–298. [CrossRef]
4. Julier, S.J.; Uhlmann, J.K. A new extension of the Kalman filter to nonlinear systems. In *AeroSense: The Ilth International Symposium on Aerospace/Defense Sensing, Simulation and Controls*; SPIE: New York, NY, USA, 1997; pp. 182–193.
5. Julier, S.J.; Uhlmann, J.K. Unscented filtering and nonlinear estimation. *Proc. IEEE* **2004**, *92*, 401–422. [CrossRef]
6. Arasaratnam, I.; Haykin, S. Cubature Kalman filters. *IEEE Trans. Autom. Control* **2009**, *54*, 1254–1269. [CrossRef]
7. Norgaard, M.; Poulsen, N.K.; Ravn, O. New developments in state estimation for nonlinear systems. *Automatica* **2000**, *36*, 1627–1638. [CrossRef]
8. Ito, K.; Xiong, K. Gaussian filters for nonlinear filtering problems. *IEEE Trans. Autom. Control* **2000**, *45*, 910–927. [CrossRef]
9. Arasaratnam, I.; Haykin, S.; Hurd, T.R. Cubature Kalman Filtering for Continuous-Discrete Systems: Theory and Simulations. *IEEE Trans. Signal Processing* **2010**, *58*, 4977–4993. [CrossRef]
10. Jia, B.; Xin, M.; Cheng, Y. High-degree cubature Kalman filter. *Automatica* **2013**, *49*, 510–518. [CrossRef]
11. Meng, D.; Miao, L.; Shao, H.; Shen, J. A Seventh-Degree Cubature Kalman Filter. *Asian J. Control* **2018**, *20*, 250–262. [CrossRef]
12. Zhang, X. Cubature Information Filters Using High-Degree and Embedded Cubature Rules. *Circuits Syst. Signal. Processing* **2014**, *33*, 1799–1818. [CrossRef]
13. McNamee, J.; Stenger, F. Construction of fully symmetric numerical integration formulas of fully symmetric numerical integration formulas. *Numer. Math.* **1967**, *10*, 327–344. [CrossRef]
14. Izanloo, R.; Fakoorian, S.A.; Yazdi, H.S.; Simon, D. Kalman filtering based on the maximum correntropy criterion in the presence of non-Gaussian noise. In Proceedings of the 2016 Annual Conference on Information Science and Systems (CISS), Princeton, NJ, USA, 16–18 March 2016.
15. Terejanu, G.; Singla, P.; Singh, T.; Scott, P.D. Adaptive Gaussian Sum Filter for Nonlinear Bayesian Estimation. *IEEE Trans. Autom. Control* **2011**, *56*, 2151–2156. [CrossRef]
16. Wang, G.; Li, N.; Zhang, Y. Maximum correntropy unscented Kalman and information filters for non-Gaussian measurement noise. *J. Frankl. Inst.* **2017**, *354*, 8659–8677. [CrossRef]
17. Meng, Q.; Li, X. Minimum Cauchy Kernel Loss Based Robust Cubature Kalman Filter and Its Low Complexity Cost Version With Application on INS/OD Integrated Navigation System. *IEEE Sens. J.* **2022**, *22*, 9534–9542. [CrossRef]
18. He, J.; Sun, C.; Zhang, B.; Wang, P. Variational Bayesian-Based Maximum Correntropy Cubature Kalman Filter With Both Adaptivity and Robustness. *IEEE Sens. J.* **2020**, *21*, 1982–1992. [CrossRef]
19. Zhao, L.; Wang, J.; Yu, T.; Chen, K.; Su, A. Incorporating delayed measurements in an improved high-degree cubature Kalman filter for the nonlinear state estimation of chemical processes. *ISA Trans.* **2019**, *86*, 122–133. [CrossRef] [PubMed]
20. Santos-León, C.; Orive, R.; Acosta, D.; Acosta, L. The Cubature Kalman Filter revisited. *Automatica* **2021**, *127*, 109541. [CrossRef]
21. Stroud, A.H. Approximate Calculation of Multiple Integrals. In *Englewood Cliffs*; Prentice Hall: Hoboken, NJ, USA, 1971.
22. Stroud, A.H.; Secrest, D. Approximate integration formulas for certain spherically symmetric regions. *Math. Comput.* **1963**, *17*, 105–135. [CrossRef]
23. Mysovskikh, I.P. *The Approximation of Multiple Integrals by Using Interpolatory Cubature Formulae*; DeVore, R.A., Scherer, K., Eds.; Quantitative Approximation Academic Press: New York, NY, USA, 1980; pp. 217–243.
24. Lu, J.; Darmofal, D.L. Higher-Dimensional Integration with Gaussian Weight for Applications in Probabilistic Design. *Siam J. Sci. Comput.* **2008**, *26*, 613–624. [CrossRef]
25. Wang, G.; Li, N.; Zhang, Y. Hybrid consensus sigma point approximation nonlinear filter using statistical linearization. *Trans. Inst. Measur. Control.* **2017**. [CrossRef]
26. Horn, R.A.; Johnson, C.R. *Matrix Analysis*, 2nd ed.; Cambridage University Press: Cambridge, UK, 2013.
27. He, J.; Sun, C.; Zhang, B.; Wang, P. Maximum Correntropy Square-Root Cubature Kalman Filter for Non-Gaussian Measurement Noise. *IEEE Access* **2020**, *8*, 70162–70170. [CrossRef]
28. Wang, Y.; Wang, H.; Li, Q.; Xiao, Y.; Ban, X. Passive Sonar Target Tracking Based on Deep Learning. *J. Mar. Sci. Eng.* **2022**, *10*, 181. [CrossRef]
29. Wang, X.; Liang, Y.; Pan, Q.; Zhao, C. Gaussian filter for nonlinear systems with one-step randomly delayed measurements. *Automatica* **2013**, *49*, 976–986. [CrossRef]

30. Ho, Y.C.; Lee, R.C.K.A. A Bayesian approach to problems in stochastic estimation and control. *IEEE Trans. Autom. Control* **1964**, *9*, 333–339. [CrossRef]
31. Liu, D.; Zhao, H. Bias-compensated subband adaptive filter algorithm based on maximum correntropy criterion. In Proceedings of the 2021 IEEE 16th Conference on Industrial Electronics and Applications (ICIEA), Chengdu, China, 1–4 August 2021; pp. 1202–1206. [CrossRef]
32. Xiang, W.; Zhao, H. Robust Filtering of Affine Projection Like Maximum Correntropy Algorithm with Bias Compensated. In Proceedings of the 2021 IEEE 16th Conference on Industrial Electronics and Applications (ICIEA), Chengdu, China, 1–4 August 2021; pp. 1207–1210. [CrossRef]
33. Feng, X.; Feng, Y.; Zhou, F.; Ma, L.; Yang, C.-X. Nonlinear Non-Gaussian Estimation Using Maximum Correntropy Square Root Cubature Information Filtering. *IEEE Access* **2020**, *8*, 181930–181942. [CrossRef]
34. Merabti, H.; Massicotte, D.; Zhu, W.-P. Nonlinearity-Robust Linear Acoustic Echo Canceller Using the Maximum Correntropy Criterion. In Proceedings of the 2019 IEEE International Symposium on Signal Processing and Information Technology (ISSPIT), Ajman, United Arab Emirates, 10–12 December 2019; pp. 1–5. [CrossRef]
35. Zou, Y.; Zou, S.; Tang, X.; Yu, L. Maximum Correntropy Criterion Kalman Filter Based Target Tracking with State Constraints. In Proceedings of the 2020 39th Chinese Control Conference (CCC), Shenyang, China, 27–29 July 2020; pp. 3505–3510. [CrossRef]
36. Liu, X.; Ren, Z.; Lyu, H.; Jiang, Z.; Ren, P.; Chen, B. Linear and Nonlinear Regression-Based Maximum Correntropy Extended Kalman Filtering. *IEEE Trans. Syst. Man Cybern. Syst.* **2021**, *51*, 3093–3102. [CrossRef]
37. Chen, S.; Zhang, Q.; Zhang, T.; Zhang, L.; Peng, L.; Wang, S. Robust State Estimation With Maximum Correntropy Rotating Geometric Unscented Kalman Filter. *IEEE Trans. Instrum. Meas.* **2022**, *71*, 1–14. [CrossRef]
38. Bhattacharjee, S.S.; Shaikh, M.A.; Kumar, K.; George, V.N. Robust Constrained Generalized Correntropy and Maximum Versoria Criterion Adaptive Filters. *IEEE Trans. Circuits Syst. II Express Briefs* **2021**, *68*, 3002–3006. [CrossRef]
39. Zhang, X.; Tian, Y.; Han, X.; Liu, Q.; Guo, L. Maximum Correntropy Criterion based Sparse Channel Estimation under Impulsive Noise in Complex Domain. In Proceedings of the 2019 IEEE International Conference on Signal, Information and Data Processing (ICSIDP), Chongqing, China, 11–13 December 2019; pp. 1–5. [CrossRef]
40. Cheng, L.; Wang, K.; Ren, M.; Yan, G. Adaptive Filter Approach for GPS Multipath Estimation Under Correntropy Criterion in Dynamic Multipath Environment. *IEEE Trans. Signal Processing* **2019**, *67*, 5798–5810. [CrossRef]
41. Fakoorian, S.; Izanloo, R.; Shamshirgaran, A.; Simon, D. Maximum Correntropy Criterion Kalman Filter with Adaptive Kernel Size. In Proceedings of the 2019 IEEE National Aerospace and Electronics Conference (NAECON), Dayton, OH, USA, 15–19 July 2019; pp. 581–584. [CrossRef]
42. Song, H.; Ding, D.; Dong, H.; Han, Q.-L. Distributed Maximum Correntropy Filtering for Stochastic Nonlinear Systems Under Deception Attacks. *IEEE Trans. Cybern.* **2022**, *52*, 3733–3744. [CrossRef] [PubMed]
43. Chu, L.; Li, W. Fusion-based multi-kernel learning filter with maximum correntropy criterion. In Proceedings of the 2020 IEEE 9th Data Driven Control and Learning Systems Conference (DDCLS), Liuzhou, China, 20–22 November 2020; pp. 556–561. [CrossRef]
44. Li, W.; Xiong, K.; Liu, Y. Diffusion Kalman filter by using maximum correntropy criterion. In Proceedings of the 2019 12th Asian Control Conference (ASCC), Kitakyushu, Japan, 9–12 June 2019; pp. 203–208.
45. Kulikova, M.V. Square-Root Approach for Chandrasekhar-Based Maximum Correntropy Kalman Filtering. *IEEE Signal Processing Lett.* **2019**, *26*, 1803–1807. [CrossRef]

Journal of
Marine Science and Engineering

MDPI

Article

Performance Simulation of Marine Cycloidal Propellers: A Both Theoretical and Heuristic Approach

Marco Altosole [1,†], Silvia Donnarumma [2,*,†], Valentina Spagnolo [3,†] and Stefano Vignolo [3,†]

1 Department of Industrial Engineering (DII), University of Naples Federico II, 80138 Napoli, Italy; marco.altosole@unina.it
2 Department of Electrical, Electronics and Telecommunication Engineering and Naval Architecture (DITEN), University of Genoa, 16145 Genoa, Italy
3 Department of Mechanical, Energy, Management and Transport Engineering (DIME), University of Genoa, 16145 Genoa, Italy; spagnolo.v86@gmail.com (V.S.); stefano.vignolo@unige.it (S.V.)
* Correspondence: silvia.donnarumma@unige.it
† These authors contributed equally to this work.

Abstract: The importance of mathematical and numerical simulation in marine engineering is growing together with the complexity of the designed systems. In general, simulation a makes it possible to improve the engineering design, reducing working time and costs of production as well. In this respect, the implementation of a simulation model for cycloidal propellers is presented. Cycloidal thrusters are being increasingly used in marine applications. Their best performance concerns low-speed applications, due to their ability to steer thrust in any direction. The proposed simulator is able to assess the performance of cycloidal propellers in terms of the generated thrust and torque, without resorting to consuming and demanding computational tools, such as CFD methods. This feature makes the presented model particularly suitable for the simulation in the time domain of the maneuverability of surface units, equipped with cycloidal propellers. In this regard, after embodying the implemented model in an already existing simulation platform for maneuverability, we show the most significant outputs concerning some simulated maneuvers, performed at cruise speed.

Keywords: marine propulsion; simulation-based design; cycloidal propellers

Citation: Altosole M.; Donnarumma, S.; Spagnolo, V.; Vignolo, S. Performance Simulation of MariPne Cycloidal Propellers: A Both Theoretical and Heuristic Approach. *J. Mar. Sci. Eng.* **2022**, *10*, 505. https://doi.org/10.3390/jmse10040505

Academic Editor: Tie Li

Received: 15 February 2022
Accepted: 29 March 2022
Published: 6 April 2022

1. Introduction

Until the past century, the only way that naval architects had to predict the behavior of the system (intended as the ship or a part of it, such as the propulsion plant, or the auxiliary systems) they were working on, was to build a prototype and test it. Nowadays, thanks to the development and the progress of computer science, it is possible to shape not prototypes but simulators, based on mathematical laws, that can predict in advance the behavior not only of a single subsystem, but also of the whole ship. Indeed, one of the main advantages of mathematical and numerical simulation is the possibility to compare different design choices, so improving the engineering design and reducing working time and costs as well.

For example, making use of mathematical models, the hull performance can be analyzed under any weather conditions [1–3], it is possible to assess whether the designed machinery can guarantee the needed power [4], the magnitude and the direction of propulsion and steering forces can be predicted [5,6], or any kind of failure conditions can be analyzed [7–9].

Among the several thruster types, cycloidal propellers (CPs) are widely used on water tractors, ferries and some naval vessels. CPs are made of a set of vertical blade protruding from the hull and performing two main rotations: one around the rotor axis and one around the axis passing through the blade pivoting centre. Depending on their eccentricity value e, namely the ratio between the distance of the steering centre from the propeller axis and the radius of the rotor, they can be classified into true cycloidal ($e = 1$), epicycloidal ($e < 1$) and trochoidal propellers ($e > 1$). Being able to generate almost the same thrust in every direction and combine both thrust and steering in a singel unit, CPs are very suitable for low-speed applications, such as dynamic positioning operations for instance.

To date, some investigations have been performed in order to model the behavior of such devices. In [10–13], some performances of cycloidal rudders have been shown, while propeller–hull interactions have been presented in [14], mainly in connection with maneuverability aspects. In this paper, the main features of a time domain simulation model for CPs are presented. The model is developed on a Matlab©Simulink platform and is able to calculate the time histories of the provided thrust and torque. The implemented model relies on a mixture of theoretical and empirical considerations. In particular, the propeller thrust and torque evaluation is based on the kinematics of each single blade, taking into account suitable correction factors in order to properly consider "dissipative" phenomena (such as interference between blades, the shielding induced by the half of the rotor which receives the oncoming flow, and the slight reduction of back thrust). The calibration of the simulator is carried out by comparing simulation outputs with real data found in open source. The final result is a simulation platform able to predict the performance of a CP in terms of generated thrust and torque, avoiding consuming and demanding computations, such as CFD methods. This is one of the significant aspects of the developed simulator that makes its use really effective when integrated into a platform for the simulation of ship motions.

In this regard, a first application of the developed simulator has been the validation of different thrust allocation logics of a DP system for a surface vessel equipped with a bow thruster and two cycloidal propellers at stern [15,16]. After that, in order to assess the reliability of the obtained simulation model, some maneuvers at cruise speed have been performed, embodying the CP simulator into a more complex dynamic model for maneuverability. In the following, the simulation outputs for one of these maneuvers are presented.

2. The Simulation Platform

As mentioned above, the present work is based on the simulation model which has been presented in [17,18] and concerns a 80 m long patrol vessel from the Italian Coast Guard. Such a simulation platform has been developed modular in order to be able to dealt with and possibly replace separately different sub-systems.

In modeling vessel dynamics, a crucial aspect is developing suitable mathematical models that reproduce the forces acting on the ship, as accurately as possible. In such a process, a complex system of coupled time-domain equations determines the evolution in time of each quantity which contributes to the ship dynamics. Usually, simulation platform are represented by means of flow–charts. In this work, the modular concept of the adopted platform is described by the mind map drawn in Figure 1, where the main part is represented by the motion equations of the ship (1) that mutually interact with the other sub-systems describing all the forces acting on the ship itself. Such forces mainly concern the interaction of the hull with the propulsion and steering systems as well as the environmental disturbs.

Whenever dealing with maneuvering problems, it is common to introduce two reference frames: the Earth-fixed reference frame $\{\Omega, \underline{n}_1, \underline{n}_2, \underline{n}_3\}$ and the body-fixed frame $\{O, \underline{b}_1, \underline{b}_2, \underline{b}_3\}$. Choosing the origin O as located on the mean water-free surface at midship, the main equations governing the ship motion are expressed as

$$\begin{cases} \Delta(\dot{u} - x_G r^2 - uv)) & = X \\ \Delta(\dot{v} - x_G \dot{r} + ur)) & = Y \\ I_z \dot{r} + m x_G(\dot{v} + ru) & = N \end{cases} \tag{1}$$

where $v_O = u\,\underline{b}_1 + v\,\underline{b}_2$ denotes the linear velocity of O expressed in the body-fixed basis and $\omega = r\underline{b}_3$ is the angular velocity, Δ is the vessel displacement, x_G is the longitudinal coordinate of gravity center w.r.t. $\{O, \underline{b}_1, \underline{b}_2, \underline{b}_3\}$, Δ is the vessel mass, I_z is the moment of inertia about $\underline{b}_3$-axis, $\underline{R} = X\,\underline{b}_1 + Y\,\underline{b}_2$ and $\underline{M} = N\,\underline{b}_3$ are the resultants of forces and the moments expressed in the $\underline{b}$-basis, respectively. In addition, $X = X_H + X_P + X_E$, $Y = Y_H + Y_P + Y_E$, $N = N_H + N_P + N_E$, where subscripts H, P, and E refer to hull, propellers, and environmental forces and moments respectively.

Figure 1. Vessel model layout.

3. Cycloidal Propeller Model

Cycloidal propellers allows precise and stepless thrust generation since propulsion and steering forces can be generated and varied simultaneously. As a result of the rotation around its vertical axis, the same amount of thrust can be provided almost over 360° by blades with hydrodynamically shaped profiles that assure a high degree of efficiency. In this section, a detailed description of the mathematical model developed for the computation of the thrust T and the torque Q_P delivered by a CP is presented.

3.1. Blade Motion

The rapid and precise thrust variation of CPs is based on the kinematics of the blades (usually from 4 to 6 and equally spaced from each other) that move along a circular path, centered at the rotor center, and at the same time perform a superimposed pivoting motion around a suitable vertical axis. When the steering center overlaps the center of the rotor casing, the blades are not angled with respect to the tangent to the blade circular trajectory and no thrust is originated in this circumstance. Instead, if the steering center is moved away from the center of the rotor casing, the blades are set at a variable angle with respect to the tangent of their circular path, and thrust is generated. From top to bottom, Figure 2 shows an example of the installation of two CPs on a ship, with some details of the machinery. During the revolution motion, the maximum angle reached by the blades increases with the eccentricity e, defined as:

$$e = \frac{OC}{D/2} \tag{2}$$

where OC is the distance between the center of the rotor O and the steering center C, and $D/2$ is the radius of the rotor. The motion of the pivot point (assumed as the center of mass) of the blade, relative to a stationary observer, results from the superimposition of the rotational movement of the rotor casing along a straight line representing the forward motion of the vessel. The pivot point follows the curve of a cycloid. The rolling radius of the cycloid is $D/2$ and the advance coefficient λ is:

$$\lambda = \frac{V_A}{\pi n D} \tag{3}$$

where V_A is the advance velocity and n is the rotor speed. During one revolution, the propeller travels a distance $\lambda D \pi$ in the direction of the vessel motion.

To generate thrust, the blades are angled with respect to the circular path described by their pivoting point. To achieve this, the steering center is moved from O to C. The resulting angle of attack leads to the generation of hydrodynamic lift and drag forces on each blade. The thrust provided by the propeller is the sum of such hydrodynamic forces, is always perpendicular to the line OC and its intensity increases with the distance OC. By shifting the steering center C, it is possible to produce thrust in any direction and of different intensities. Therefore, the thrust provided by an epicycloidal propeller can be represented as a function of two plane polar coordinates:

- the geometric or driving pitch (between 0 and $0.8R$ for constructive limits): that is the distance (expressed as a percentage of the rotor radius $R = D/2$) between the steering center C and the center of the rotor O;
- the steering pitch (between 0° and 360°): the angle between a fixed axis (with respect to the hull) and the line OC.

Figure 2. General overview of voith installation on a vessel on the top and some sketch of the machinery.

3.2. Thrust Generation

Changing the steering center, the blades acquire a certain attack angle, so generating corresponding lift and drag forces which give rise to the desired thrust. The hydrodynamic forces components acting transversely to the desired thrust direction cancel each other out. It is possible to produce thrust in any direction putting the steering center in the right position. The zero-thrust condition can be selected at any time, making the ship very safe to handle.

Each blade generates instantly a hydrodynamic force which is the sum of the lift (component of the hydrodynamic force, perpendicular to the oncoming flow) and the drag (parallel to the oncoming flow). The sum of all the hydrodynamic forces generated by all the blades gives rise to the corresponding total thrust. Also, each blade generates a corresponding torque which contributes to the total torque $\underline{M}$ acting on the propeller. For each choice of driving and steering pitch, there are corresponding curves of thrust and torque coefficients K_S and K_D as functions of the advance coefficient λ. The coefficients K_S and K_D are defined in analogy with the corresponding screw-propeller coefficients, respectively K_T and K_Q, by the formulae reported in Table 1.

Table 1. Cycloidal and screw propeller non-dimensional coefficients.

Coefficients	Cycloidal	Screw
advance coefficient	$\lambda = \frac{V_A}{\pi n D}$	$J = \frac{V_A}{nD}$
thrust coefficient	$K_S = \frac{T}{\frac{1}{2}\rho D L u^2}$	$K_T = \frac{T}{\rho n^2 D^4}$
torque coefficient	$K_D = \frac{4M}{\rho D^2 L u^2}$	$K_Q = \frac{Q_O}{\rho n^2 D^5}$
efficiency	$\eta_O = \frac{K_S}{K_D}\lambda$	$\eta_O = \frac{K_T}{K_Q}\frac{J}{2\pi}$

where V_A is the advance velocity, T is the propeller thrust, M and Q_O are the CP and screw propeller torque respectively, L is the blade length, D is the rotor diameter, ρ is the sea water density, L is the blade length, u is the tangential speed ($u = n\pi D$).

3.3. Kinematics of the Blade

In this section, the kinematical model of a blade is presented. In particular, a 2−dimensional plane model has been adopted, propellers have been modeled as counter-rotating and two distinguished coordinate systems have been introduced: the first one is the hull-fixed frame, while the second one $\{O, \underline{e}_1, \underline{e}_2, \underline{e}_3\}$ is rotated clockwise, about the vertical axis passing through O and parallel to $\underline{b}_3 = \underline{e}_3$, by an angle $\beta \in [0, 2\pi]$ (the steering pitch) which determines (the perpendicular of) the steering force direction. The angle β is related to the rudder pitch of the cycloidal propeller. The steering center C lies on the straight line passing through O and parallel to $\underline{e}_2$. The relationship between the bases $\{\underline{b}_i\}$ and $\{\underline{e}_i\}$, in accordance with Figure 3, is expressed as

$$\begin{cases} \underline{e}_1 &= \cos\beta\,\underline{b}_1 + \sin\beta\,\underline{b}_2 \\ \underline{e}_2 &= -\sin\beta\,\underline{b}_1 + \cos\beta\,\underline{b}_2 \\ \underline{e}_3 &= \underline{b}_3 \end{cases} \tag{4}$$

During the rotation, the projection P of the blade shaft on the plane $\langle O, \underline{b}_1, \underline{b}_2 \rangle$ describes a circumference having center O e radius R coinciding with the rotor radius. Such a circumference is parameterized by

$$P(\theta) : \begin{cases} x &= R\cos\theta \\ y &= R\sin\theta \\ z &= 0 \end{cases} \tag{5}$$

where θ denotes the angle (function of time) describing the revolution motion of the blade. The unit vector $\underline{t}$ tangent to the circular path of P has components in the vector basis $\{\underline{b}_i\}$ of the form

$$\underline{t}(\theta): \begin{cases} t_1 & = -\sin\theta \\ t_2 & = \cos\theta \\ t_3 & = 0 \end{cases} \tag{6}$$

Introducing the vector

$$(C - O) = s\,\underline{e}_2 = -s\,\sin\beta\,\underline{b}_1 + s\,\cos\beta\,\underline{b}_2, \quad s \in [0, 0.8R] \tag{7}$$

the vector joining the steering center C with the point P can be expressed as

$$(P - C) = (R\,\cos\theta + s\,\sin\beta)\,\underline{b}_1 + (R\,\sin\theta - s\,\cos\beta)\,\underline{b}_2 \tag{8}$$

The variable s is usually called driving pitch and controls the magnitude of the thrust. The unit vector orthogonal to $(P - C)$ and belonging to the plane $\langle O, \underline{b}_1, \underline{b}_2 \rangle$ identifies with the unit vector of the blade chord and it is given by

$$\frac{(P-C)^{\perp}}{|(P-C)^{\perp}|} = \frac{(-R\,\sin\theta + s\,\cos\beta)\,\underline{b}_1 + (R\,\cos\theta + s\,\sin\beta)\,\underline{b}_2}{\sqrt{(-R\,\sin\theta + s\,\cos\beta)^2 + (R\,\cos\theta + s\,\sin\beta)^2}} \tag{9}$$

The pivoting motion of the blade around its own vertical axis can be described by the angle α (function of time) between the unit vectors $\underline{t}$ and $\frac{(P-C)^{\perp}}{|(P-C)^{\perp}|}$. Due to the relation

$$\cos\alpha = \frac{(P-C)^{\perp}}{|(P-C)^{\perp}|}\cdot\underline{t} = \frac{R + s\,\sin(\beta - \theta)}{\sqrt{(-R\,\sin\theta + s\,\cos\beta)^2 + (R\,\cos\theta + s\,\sin\beta)^2}} \tag{10}$$

where the dot denotes the usual scalar product between vectors, choosing anticlockwise the positive direction of rotation around the blade shaft, the pivoting angle α can be defined as

$$\alpha = \pm\arccos\left(\frac{(P-C)^{\perp}}{|(P-C)^{\perp}|}\cdot\underline{t}\right) \quad \text{where} \quad \begin{array}{ll} + & \text{if}\cos(\theta - \beta) \geq 0 \\ - & \text{if}\cos(\theta - \beta) < 0 \end{array} \tag{11}$$

The above outlined kinematical model can be summarized by the following figure.

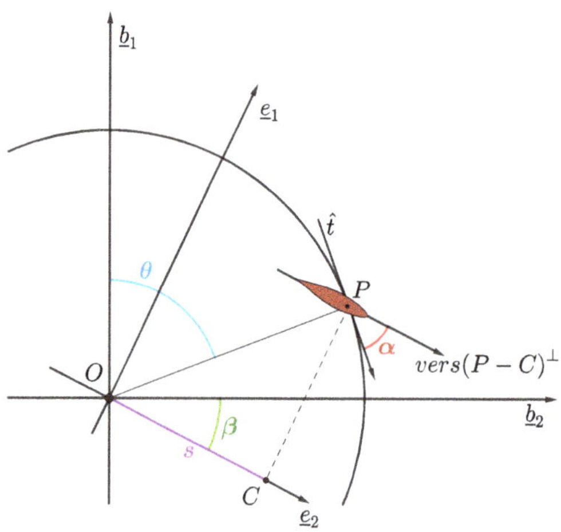

Figure 3. Kinematics of the blade.

Supposing now that the vessel is moving, let $\underline{v}_O = \hat{u}\underline{b}_1 + \hat{v}\underline{b}_2$ be the velocity of O (with respect to an Earth-fixed frame) expressed in the hull-fixed basis. Denoting by

$$\underline{v}'_p = -R\dot{\theta}\sin\theta\underline{b}_1 + R\dot{\theta}\cos\theta\underline{b}_2 \tag{12}$$

the velocity of the point P with respect to the body-fixed frame, the velocity of P with respect to the Earth-fixed frame is given by

$$\underline{v}_p = \underline{v}'_p + \underline{v}_O + \underline{\omega} \wedge (P - O) = \left[\hat{u} - R(\dot{\theta} + \dot{\psi})\sin\theta\right]\underline{b}_1 + \left[\hat{v} + R(\dot{\theta} + \dot{\psi})\cos\theta\right]\underline{b}_2 \tag{13}$$

where $\underline{\omega} = \dot{\psi}\underline{b}_3$ is the angular velocity of the vessel. The velocity of the oncoming flow experienced at P by a blade-fixed observer is then $-\underline{v}_P$; its unit vector $\hat{\underline{t}}$ is expressed as

$$\hat{\underline{t}} = -\frac{\underline{v}_P}{|\underline{v}_P|} = -\frac{\left[\hat{u} - R(\dot{\theta} + \dot{\psi})\sin\theta\right]\underline{b}_1 + \left[\hat{v} + R(\dot{\theta} + \dot{\psi})\cos\theta\right]\underline{b}_2}{\sqrt{\left[\hat{u} - R(\dot{\theta} + \dot{\psi})\sin\theta\right]^2 + \left[\hat{v} + R(\dot{\theta} + \dot{\psi})\cos\theta\right]^2}} \tag{14}$$

Making use of the unit vector $\hat{\underline{t}}$ it is possible to characterize the attack angle of the incident flow as

$$\hat{\alpha} = \pi - \arccos\left[\frac{(P - C)^{\perp}}{|(P - C)^{\perp}|} \cdot \hat{\underline{t}}\right] \tag{15}$$

3.4. Hydrodynamic Forces

Making use of some simplifying assumptions, a suitable model for evaluating the hydrodynamic forces generated by each blade is presented. It is supposed that the velocity of the incident flow is the same on the entire surface of the blade and coincides with $-\underline{v}_P$. Under such a condition, the lift and drag produced by each blade can be expressed as

$$\underline{L} = C_L\frac{1}{2}\rho_w A|\underline{v}_P|^2\hat{\underline{n}} \tag{16a}$$

$$\underline{D} = C_D\frac{1}{2}\rho_w A|\underline{v}_P|^2\hat{\underline{t}} \tag{16b}$$

where C_L and C_D are the lift and drag coefficients, respectively; ρ_w is the sea water density; A is the blade lateral area; $\underline{v}_P$ is the oncoming flow velocity; $\hat{\underline{t}}$ is the unit vector of the lift force (unit vector of the oncoming flow at P); and $\hat{\underline{n}}$ is the unit vector of the drag force (perpendicular to $\hat{\underline{t}}$).

The unit vector $\hat{\underline{n}}$ can be determined by the following procedure, in which two main scenarios are distinguished:

- the attack angle $\hat{\alpha}$ belongs to the interval $\left]0, \frac{\pi}{2}\right[$, namely the oncoming flow hits the blade from the front. In such a circumstance, the unit vector $\hat{\underline{n}}$ is determined according to the requirements

$$\hat{\underline{n}} = \begin{cases} \underline{b}_3 \wedge \hat{\underline{t}} & \text{when } \hat{\underline{t}} \wedge \frac{(P-C)^{\perp}}{|(P-C)^{\perp}|} \cdot \underline{b}_3 > 0 \\ -\underline{b}_3 \wedge \hat{\underline{t}} & \text{when } \hat{\underline{t}} \wedge \frac{(P-C)^{\perp}}{|(P-C)^{\perp}|} \cdot \underline{b}_3 < 0 \end{cases} \tag{17}$$

- $\hat{\alpha} \in \left]\frac{\pi}{2}, \pi\right[$, the oncoming flow hits the blade from the back. In this case, $\hat{\underline{n}}$ is singled out by the requests:

$$\hat{\underline{n}} = \begin{cases} -\underline{b}_3 \wedge \hat{\underline{t}} & \text{when } \hat{\underline{t}} \wedge \frac{(P-C)^{\perp}}{|(P-C)^{\perp}|} \cdot \underline{b}_3 > 0 \\ \underline{b}_3 \wedge \hat{\underline{t}} & \text{when } \hat{\underline{t}} \wedge \frac{(P-C)^{\perp}}{|(P-C)^{\perp}|} \cdot \underline{b}_3 < 0 \end{cases} \tag{18}$$

As remaining particular cases, if $\hat{\alpha} = 0$ or $\hat{\alpha} = \pi$ there is no lift while if $\hat{\alpha} = \frac{\pi}{2}$ then $\underline{\hat{n}} = \underline{\hat{t}}$. The above described procedure allows us to determine the lift and drag provided by each single blade. The resultant hydrodynamic force generated by the epicycloidal propeller can be computed as the sum of all contributions given by each blade.

3.5. Torque Acting on the Rotor

Figure 4 shows the implemented model for the propulsion system. Input data of the model are the desired engine speed RPM_{set}, the desired propeller pitch s, and the desired thrust angle β. The output is the array $\left[X_p, Y_p, N_p\right]^\top$ composed by the longitudinal and lateral propeller forces and the resulting moment.

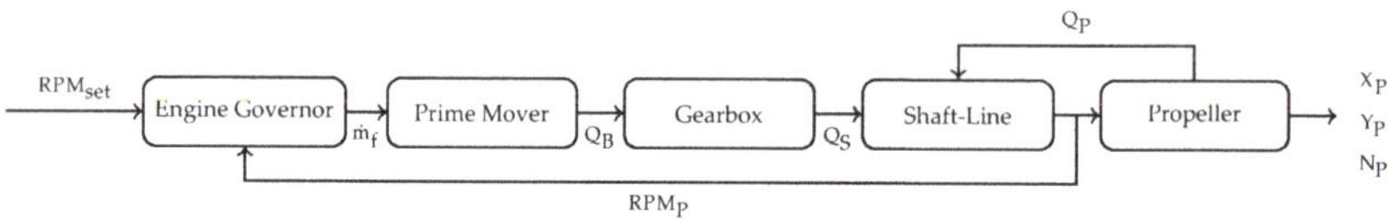

Figure 4. Layout of propulsion simulation platform.

The dynamics of the shaft line is described by the equation:

$$\frac{dn}{dt} = \frac{1}{I_{tot}}\left(Q_{eng} - Q_p - Q_{fric}\right) \tag{19}$$

where n is the shaft speed; I_{tot} is the total axial inertia taking into account: (i) engine, (ii) gears, (iii) shaft and (iv) propeller contributions; Q_{eng} is the engine torque; Q_{fric} represents frictions; and Q_p is the propeller torque.

In order to compute the torque acting on the rotor, the Newton-Euler moments equations for each single blade and for the rotor are taken into account separately. Developed in the hull-fixed reference frame and with respect to the point O (center of the rotor), the Newton-Euler moments equation for each blade can be expressed as

$$\underline{M}_O^H + \underline{M}_O^G + \underline{M}_O^R + \underline{M}_O^I = I_{G_B}^B(\underline{\dot{\omega}}) + \underline{\omega} \wedge I_{G_B}^B(\underline{\omega}) + m_B(G_B - O) \wedge \underline{a}_{G_B} \tag{20}$$

where $\underline{M}_O^H$, $\underline{M}_O^G$, $\underline{M}_O^R$, and $\underline{M}_O^I$ are the moments acting on the blade, respectively due to hydrodynamic, weight, reactive and inertial forces; $I_{G_B}^B$ is the inertia tensor of the blade w.r.t. its gravity center G_B (which is assumed to coincide with the pivot point P); $\underline{\omega} = (\dot{\theta} - \dot{\alpha})\underline{b}_3$ is the blade angular velocity w.r.t. the hull-fixed frame; $\underline{a}_{G_B}$ is the acceleration of G_B w.r.t. the hull-fixed frame; and m_B is the blade mass. The moment of the hydrodynamic force $\underline{M}_O^H$ is given by:

$$\underline{M}_O^H = (P - O) \wedge (\underline{L} + \underline{D}) \tag{21}$$

where the hydrodynamic force is described in terms of lift and drag. Expressing all vectors in the basis $\{\underline{b}_i\}$ as $(\underline{L} + \underline{D}) = f_1\underline{b}_1 + f_2\underline{b}_2$ and $(P - O) = R\underline{b}_1 \cos\theta + R\underline{b}_2 \sin\theta$, one has:

$$\underline{M}_O^H = (Rf_2 \cos\theta - Rf_1 \sin\theta)\underline{b}_3 \tag{22}$$

The weight force moment is given by:

$$\underline{M}_O^G = (P - O) \wedge m\underline{g} = m_B g(R \sin\theta \underline{b}_1 - R \cos\theta \underline{b}_2) \tag{23}$$

where $\underline{g} = g\underline{b}_3$ is the gravity acceleration. In order to evaluate the inertial force moment $\underline{M}_O^I$, it is necessary to assess the dragging and Coriolis forces acting on the blade and their associated moments $\underline{M}_O^S$ and $\underline{M}_O^C$. After that, the total inertial forces moment can be expressed as the sum:

$$\underline{M}_O^I = \underline{M}_O^C + \underline{M}_O^S \tag{24}$$

The moments $\underline{M}_O^S$ and $\underline{M}_O^C$ are calculated by means of the well-known formulae:

$$\underline{M}_O^C = \underline{M}_P^C + \underline{F}^C \wedge (O - P) \tag{25a}$$

$$\underline{M}_O^S = \underline{M}_P^S + \underline{F}^S \wedge (O - P) \tag{25b}$$

where $\underline{M}_P^C$ and $\underline{M}_P^S$ are the moments, w.r.t. the pivot point P, of the Coriolis and the dragging forces acting on the blade, while $\underline{F}^C$ and $\underline{F}^S$ denote the resultants of the Coriolis and dragging forces. The Coriolis force acting on any single blade is given by:

$$\underline{F}^C = -2\int_B k\underline{\omega}_\psi \wedge \underline{v}_Q \mathrm{d}\tau \tag{26}$$

where $\underline{\omega}_\psi = \dot{\psi}\underline{b}_3$ is the angular velocity of the vessel, $\underline{v}_Q = \underline{v}_P + \underline{\omega} \wedge (Q - P)$ (with $\underline{\omega} = (\dot{\theta} - \dot{\alpha})\underline{b}_3$) is the velocity of a generic point Q of the blade w.r.t. the hull-fixed frame and k is the mass density of the blade. After implementing calculations, one gets the expression:

$$\begin{aligned}\underline{F}^C &= -2m_B\underline{\omega}_\psi \wedge \underline{v}_P - 2m_B\underline{\omega}_\psi \wedge [\underline{\omega} \wedge (G_B - P)] = -2m_B\underline{\omega}_\psi \wedge \underline{v}_{G_B} \\ &= -2m_B(\dot{\psi}v_{G_B1}\underline{b}_2 - \dot{\psi}v_{G_B2}\underline{b}_1)\end{aligned} \tag{27}$$

where $\underline{v}_{G_B} = v_{G_B1}\underline{b}_1 + v_{G_B2}\underline{b}_2$ denotes the velocity of the gravity center of the blade w.r.t. the hull-fixed frame. By definition, the moment $\underline{M}_P^C$ of the Coriolis force with respect to the pivot point P is given by:

$$\begin{aligned}\underline{M}_P^C &= -2\int_B k(Q-P) \wedge \Big[\underline{\omega}_\psi \wedge [\underline{v}_P + \underline{\omega} \wedge (Q-P)]\Big]\mathrm{d}\tau \\ &= -2\int_B k(Q-P) \wedge \Big[\underline{\omega}_\psi \wedge \underline{v}_P\Big]\mathrm{d}\tau - 2\int_B k(Q-P) \wedge \Big[\underline{\omega}_\psi \wedge [\underline{\omega} \wedge (Q-P)]\Big]\mathrm{d}\tau\end{aligned} \tag{28}$$

Making use of the results shown in [19], the moment $\underline{M}_P^C$ can be expressed as:

$$\underline{M}_P^C = -2m_B(G_B - P) \wedge \Big[\underline{\omega}_\psi \wedge \underline{v}_P\Big] - \underline{\omega} \wedge I_P^B(\underline{\omega}_\psi) - \underline{\omega}_\psi \wedge I_P^B(\underline{\omega}) + I_P^B(\underline{\omega} \wedge \underline{\omega}_\psi) \tag{29}$$

where now I_P^B denotes the inertia tensor of the blade w.r.t. the pivot point P. Expression (29) holds in general. In our case, since the considered mathematical model is two-dimensional and in view of the assumption $G_B \equiv P$, it is easily seen that all terms appearing in Equation (29) vanish, so having $\underline{M}_P^C = 0$.

For what concerns the dragging force, by definition it reads as:

$$\begin{aligned}\underline{F}^s &= -\int_B k\Big[\underline{a}_O + \underline{\omega}_\psi \wedge \Big[\underline{\omega}_\psi \wedge (Q-O)\Big] + \underline{\dot{\omega}}_\psi \wedge (Q-O)\Big]\mathrm{d}\tau \\ &= -m_B\underline{a}_O - m_B\underline{\omega}_\psi \wedge \Big[\underline{\omega}_\psi \wedge (G_B - O)\Big] - m_B\underline{\dot{\omega}}_\psi \wedge (G_B - O) \\ &= -m_B(a_{O1}\underline{b}_1 + a_{O2}\underline{b}_2) + m_B\underline{\omega}_\psi \wedge (R\dot{\psi}\sin\theta\underline{b}_1 - R\dot{\psi}\cos\theta\underline{b}_2) + m_B(R\ddot{\psi}\sin\theta\underline{b}_1 - R\ddot{\psi}\cos\theta\underline{b}_2) \\ &= -m_B(a_{O1}\underline{b}_1 + a_{O2}\underline{b}_2) + m_B\dot{\psi}^2(R\cos\theta\underline{b}_1 + R\sin\theta\underline{b}_2) + m_BR\ddot{\psi}(\sin\theta\underline{b}_1 - \cos\theta\underline{b}_2)\end{aligned} \tag{30}$$

where $\underline{a}_O = a_{O1}\underline{b}_1 + a_{O2}\underline{b}_2$ is the acceleration of the rotor center O w.r.t. the Earth–fixed frame. The moment $\underline{M}_P^S$ of the dragging force w.r.t. the pivot point P is given by:

$$
\begin{aligned}
\underline{M}_P^S &= \int_{\mathcal{B}} k(Q-P)\wedge\Big[\underline{a}_O+\underline{\omega}_\psi\wedge\Big[\underline{\omega}_\psi\wedge(Q-O)\Big]+\underline{\dot{\omega}}_\psi\wedge(Q-O)\Big]\mathrm{d}\tau\\
&- \;\; m_B(G_B-P)\wedge\underline{a}_O-\int_{\mathcal{B}} k(Q-P)\wedge\Big[\underline{\omega}_\psi\wedge\Big[\underline{\omega}_\psi\wedge[(Q-P)+(P-O)]\Big]\Big]\mathrm{d}\tau+\\
&-\int_{\mathcal{B}} k(Q-P)\wedge\Big[\underline{\dot{\omega}}_\psi\wedge[(Q-P)+(P-O)]\Big]\mathrm{d}\tau\\
&= -m_B(G_B-P)\wedge\underline{a}_O-\underline{\omega}_\psi\wedge I_B\Big(\underline{\omega}_\psi\Big)-m_B(G_B-P)\wedge\underline{\omega}_\psi\wedge\Big[\underline{\omega}_\psi\wedge(P-O)\Big]+\\
&-I_B^P(\underline{\dot{\omega}}_\psi)-m_B(G_B-P)\wedge\Big[\underline{\dot{\omega}}_\psi\wedge(P-O)\Big]\\
&= -m_B(G_B-P)\wedge\Big[\underline{a}_O+\underline{\omega}_\psi\wedge\Big[\underline{\omega}_\psi\wedge(P-O)\Big]+\underline{\dot{\omega}}_\psi\wedge(P-O)\Big]-\underline{\omega}_\psi\wedge I_P^B\Big(\underline{\omega}_\psi\Big)-I_P^B\Big(\underline{\dot{\omega}}_\psi\Big)
\end{aligned}
\tag{31}
$$

Again, since the model is two-dimensional and $G_B \equiv P$, only the last term does not vanish, so yielding:

$$\underline{M}_P^S = -I_P^B(\underline{\dot{\omega}}_\psi) = -I_{33}\ddot{\psi}\underline{b}_3 \tag{32}$$

where I_{33} is the moment of inertia of the blade w.r.t. the vertical axis passing for P. In order to implement Equation (25), the terms $\underline{F}^S \wedge (O-P)$ and $\underline{F}^C \wedge (O-P)$ need to be calculated:

$$\underline{F}^S \wedge (O-P) = mR(a_{O1}\sin\theta - a_{O2}\cos\theta)\underline{b}_3 - mR^2\ddot{\psi}\underline{b}_3 \tag{33a}$$

$$\underline{F}^C \wedge (O-P) = -2mR(\dot{\psi}v_{G_B2}\sin\theta + \dot{\psi}v_{G_B1}\cos\theta)\underline{b}_3 \tag{33b}$$

Now, inserting all the obtained results into Equation (24), we end up with the final expression of the moment w.r.t. O of the inertial forces:

$$\underline{M}_O^I = -2mR(\dot{\psi}v_{G_B2}\sin\theta + \dot{\psi}v_{G_B1}\cos\theta)\underline{b}_3 - I_{33}\ddot{\psi}\underline{b}_3 + mR(a_{O1}\sin\theta - a_{O2}\cos\theta)\underline{b}_3 - mR^2\ddot{\psi}\underline{b}_3 \tag{34}$$

For our two-dimensional model with $G_B = P$, the first two terms on the right side of (20) are:

$$I_{G_B}^B(\underline{\dot{\omega}}) = I_{33}(\ddot{\theta} - \ddot{\alpha})\underline{b}_3 \tag{35a}$$

$$\underline{\omega} \wedge I_{G_B}^B(\underline{\omega}) = 0 \tag{35b}$$

$$m_B(G_B - O) \wedge a_{G_B} = m_B(R\underline{a}_{G2}\cos\theta - R\underline{a}_{G1}\sin\theta)\underline{b}_3 \tag{35c}$$

Inserting all the above calculated contributions into Equation (20), we obtain the explicit expression of the reactive moment:

$$\underline{M}_O^R = -\underline{M}_O^H - \underline{M}_O^G - \underline{M}_O^I + I_{G_B}^B(\underline{\dot{\omega}}) + \underline{\omega} \wedge I_{G_B}^B(\underline{\omega}) + m_B(G_B - O) \wedge \underline{a}_{G_B} \tag{36}$$

Inserting the reactive moments (36) acting on each single blade into the moments equation for the rotor, the engine torque can be calculated as:

$$M_O^E = \sum_{i=1}^{n}\Big(\underline{M}_O^R\Big)_i \cdot \underline{b}_3 - \underline{M}_O^I \cdot \underline{b}_3 + I_O^{rot}(\underline{\dot{\omega}}_r) \cdot \underline{b}_3 \tag{37}$$

where $\underline{M}_O^I$ is the inertial forces moment acting on the rotor, I_O^{rot} is the inertia tensor of the rotor w.r.t. its center O, $\underline{\omega}_r = \dot{\theta}\underline{b}_3$ is the angular velocity (w.r.t. the hull–fixed frame) of the rotor and n is the number of blades. The moment $\underline{M}_O^I$ can be calculated as already made for the blades, namely as the sum $\underline{M}_O^I = \underline{M}_O^C + \underline{M}_O^S$. In this case, since the center of the rotor O is fixed w.r.t. the hull, the same arguments as in [20] can be applied so obtaining the general explicit expressions:

$$\underline{M}_O^C = -\underline{\omega}_r \wedge I_O^{rot}(\underline{\omega}_\psi) - \underline{\omega}_\psi \wedge I_O^{rot}(\underline{\omega}_r) + I_O^{rot}(\underline{\omega}_r \wedge \underline{\omega}_\psi) \tag{38}$$

$$\underline{M}_O^S = -m_{rot}(G_{rot} - O) \wedge \underline{a}_O - \underline{\omega}_\psi \wedge I_O^{rot}(\underline{\omega}_\psi) - I_O^{rot}(\underline{\dot{\omega}}_\psi) \quad (39)$$

with I_O^{rot} the inertia tensor of the rotor w.r.t. its center O (which is assumed to coincide with the gravity center G_{rot}) and m_{rot} is the mass of the rotor. Again, since the model is two-dimensional and $G_{rot} = O$, we have $\underline{M}_O^C = 0$ and $\underline{M}_O^S = -I_O^{rot}(\underline{\dot{\omega}}_\psi)$.

Inserting now all the obtained results into Equation (37), we obtain the final expression for the engine torque:

$$M_O^E = \sum_{i=1}^{n} \left(\underline{M}_O^R\right)_i \cdot \underline{b}_3 + I_O^{rot}(\ddot{\psi}\underline{b}_3) \cdot \underline{b}_3 + I_O^{rot}(\ddot{\theta}\underline{b}_3) \cdot \underline{b}_3 \quad (40)$$

In conclusion, a detailed model for the evaluation of forces and torques acting on the CP has been expound. The relevance of the presented approach concerns the possibility to easily change the propeller characteristics (number, length and shape of blades as well as rotor diameter) and evaluate the corresponding performance variations.

4. Numerical Modeling and Validation: Free Running Test

The mathematical model illustrated in Section 3 has been used to develop a Matlab©Simulink simulator for cycloidal propellers. In this section, the main features and the validation of such simulator are presented.

In order to simplify the simulation platform, some hypotheses have been assumed: (*i*) the propeller has been considered in free-running conditions; (*ii*) the problem has been assumed stationary; (*iii*) the model has been implemented two dimensional; (*iv*) linear superposition of the contributions of each blade in terms of generated forces and moments has been adopted.

The propeller model input data are:

- the propeller geometry (length, chord and orbit diameter of the blade—see Table 2);
- the sea water characteristics (viscosity and density);
- the lift C_L and drag C_D coefficients of the blade;
- the rotor speed and maximum available pitch;
- the steering pitch angle β (0° in forward direction, 180° in astern condition) and the driving pitch s (expressed as a percentage of the rotor radius).

In the present case study, data are:

Table 2. Geometric parameters of the propeller.

Parameter	Value
Number of blades	5
Rotor diameter (m)	3.2
Blade length (m)	2.65
Blade chord (m)	0.7744
Maximum tip thickness (m)	0.242

The whole simulation model consists of a set of identical subsystems, each of them representing the behavior of a single blade. Making use of (16) and (40), the components of total thrust and torque are calculated in the basis $\{\underline{b}_i\}$.

Free-running characteristics have been evaluated through a simulation campaign where K_S and K_D coefficients have been evaluated, in accordance with Table 1, for several $\lambda \in (0, 0.6)$ and $s \in (30\%, 90\%)$ values. In particular, the evaluation of coefficients K_S and K_D has been performed in the pitch range from 30% to 90%, with steps of 20%. Results are reported in Figure 5 that shows the comparison between propeller manufacturer data (dashed lines) and the coefficients K_S and K_D obtained through simulation (continuous lines). The available data concern an existing cycloidal propeller, with the same geometry of the simulated one.

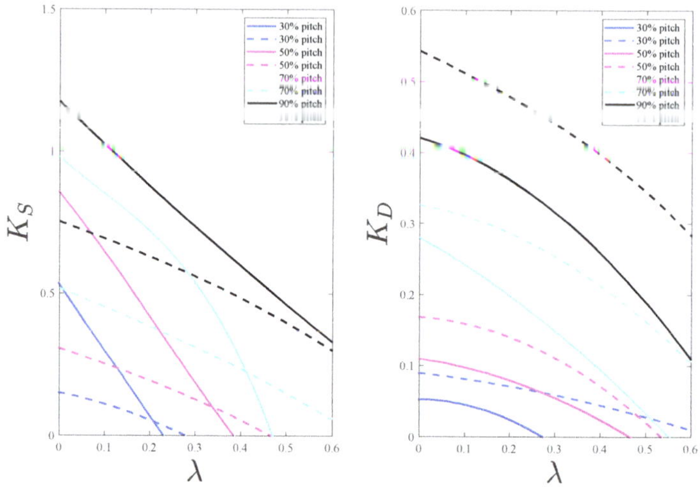

Figure 5. Thrust and torque propeller coefficients: model (solid lines) VS manufacturer data (dashed lines).

The discrepancies between real and simulated data appearing in Figure 5 have been supposed to be due to the stated simplifying assumptions about the interactions among the blades. In order to overcome this issue, some correction factors have been introduced, based on physical considerations: a shielding effect that can affect the oncoming flow for some blades and the interference of each blade with the others.

4.1. Shielding Correction

This correction concerns the shielding effect experienced by the blades that are in the half circumference not directly exposed to the oncoming water flow. Figure 6 gives a qualitative idea of how the flow is deviated by the blades. A corresponding correction factor, consisting in a matrix of corrective coefficients $w_s(s) < 1$, has been introduced in order to reduce the velocity of the oncoming flow in the part of the rotor not directly invested by the flow itself. The corrective factors depend only on the driving pitch values and not on the advance coefficient λ and are implemented as it follows

$$\hat{u} = \begin{cases} \hat{u}\, w_s(s) & \text{if} \cos\theta < 1 \\ \hat{u} & \text{otherwise} \end{cases} \tag{41}$$

where $\hat{u}$ is the velocity component of the rotor center along $\underline{b}_1$.

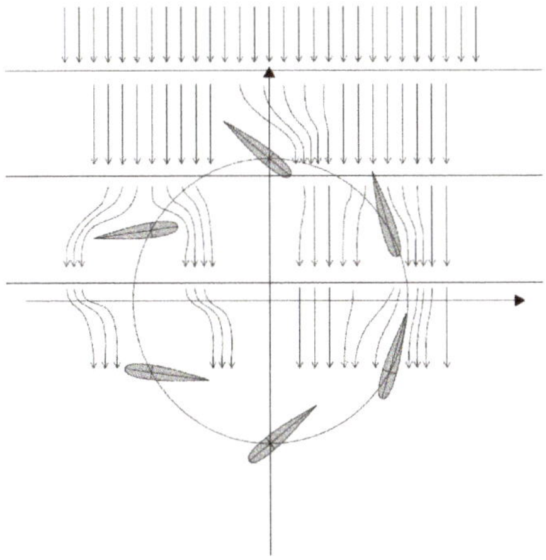

Figure 6. Shielding phenomena [21].

4.2. Interference Correction

As sketched in Figure 7, each blade influences the flow of the adjacent blade, so modifying the angle of attack of the incident flow itself. This interference among the blades has been modeled by reducing the attack angle of the incoming flow by a suitable quantity w_i. This correction depends on the advance coefficient λ and the pitch values

$$\hat{\alpha}_{corr} = \hat{\alpha}\, w_i(s, \lambda) \tag{42}$$

where $\hat{\alpha}_{corr}$ is the angle of attack defined in (15).

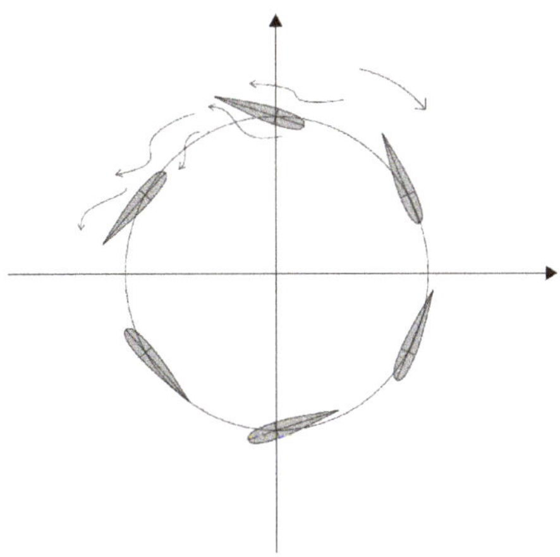

Figure 7. Interference phenomena [21].

Figure 8 shows the values of free-running propeller coefficients, obtained by applying the above mentioned corrections. Although the proposed corrections are purely empirical, the graph in Figure 8 exhibits a good agreement between simulated and real data.

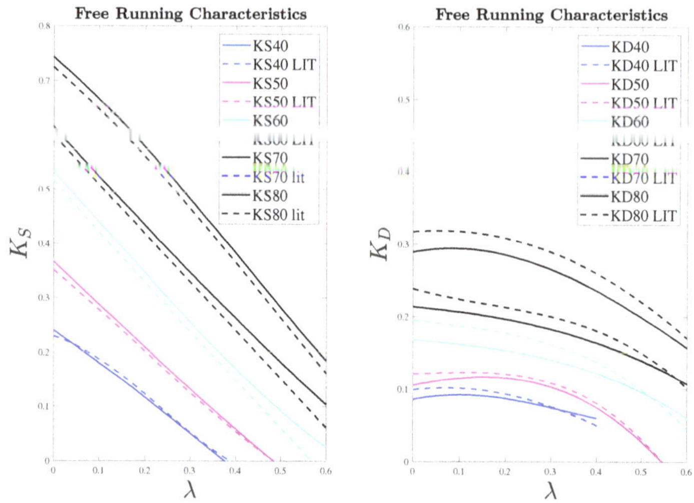

Figure 8. Thrust and torque propeller coefficients: model (solid lines) VS manufacturer (dashed lines) data.

5. Simulation Results

In this section results concerning a simulated maneuver are presented. In this contest, two counter rotating CPs have been fitted on a vessel, whose dynamics were known. In the proposed simulation, initially the vessel is moving forward, the propeller pitch is set at $s = 50\%$ and the shaft speed is constant. At a certain instant the steering pitch β is required to move from 0 to 20°. Figure 9 shows the required and delivered shaft speed. Generally, this kind of propellers work at constant shaft speed and the thrust is controlled through the pitch value s, as for controllable pitch screw propellers. Figure 10 shows the ship motion, made dimensionless with respect to the vessel length. From top to bottom, advance motion, side drift and heading are respectively shown. Figure 11 shows the vessel speed, the components of the linear and angular vessel velocities in the body-fixed basis, and the vessel drift angle. As expected, at instant 100s when a twenty-degree change in steering is required and kept constant for the rest of the maneuver, the ship begins to rotate and drift, it slows down, while drift and rotation velocities increase until they stabilize to constant values. The delivered forces and moment are reported in Figure 12, where they are expressed in the $\underline{b}$−basis. It is worth noticing that the model is able to evaluate the lateral forces (16) generated by each single propeller. Although two counter-rotating thrusters ensure the compensation of the lateral forces in the case of the forward navigation, this is no longer true during the maneuverings where the evaluation of such forces is an important aspect. Moreover, the implemented kinematic approach allows us to observe the asymmetry of X_p, that is the force component along $\underline{b}_1$, during the evolution. This is due to the fact that the two drifting thrusters are actually undergoing two different inlet flows. Such effect is also reflected in the torque. Indeed, time histories of required torque by each propeller at the shaft are reported in Figure 13. In this case, some differences can be evaluated during the vessel rotation where load on the portside shaft is different from the starboard one. The integration of the propeller models together with the propulsion system, Figure 14, allows us to assess the matching between available power, represented by the underlying area of the black curve, and the power required by the propellers at every time–step. Portside and starboard required powers (red and blue lines respectively) are clustered in a small area of the motor load diagram. Finally, the required fuel flow rate time history is reported in Figure 15, in terms of percentage of its maximum value, allowing us to compute fuel consumption.

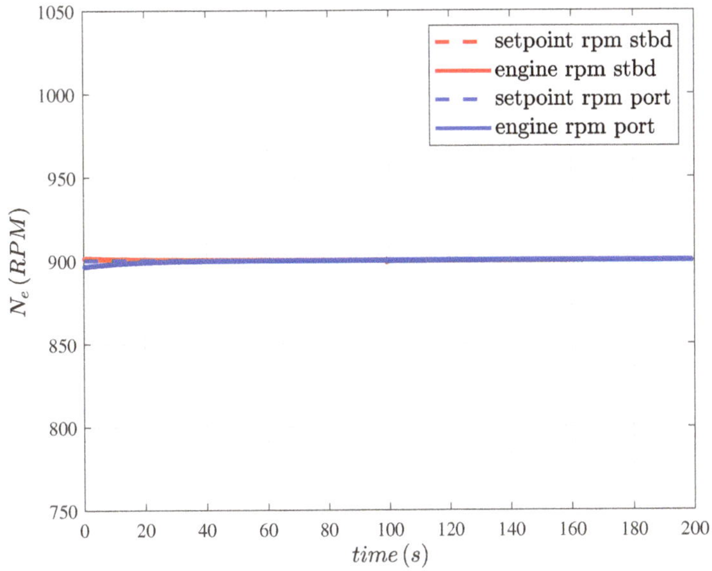

Figure 9. Engine shaft speed setpoint and feedback.

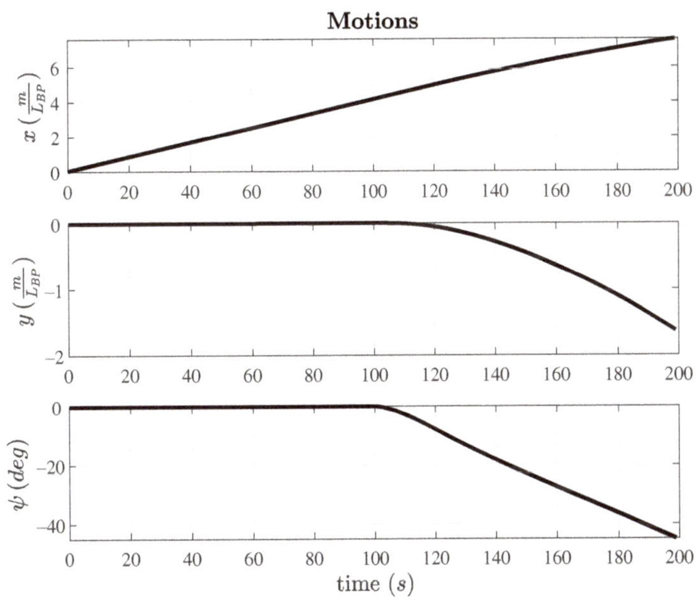

Figure 10. Vessel motions.

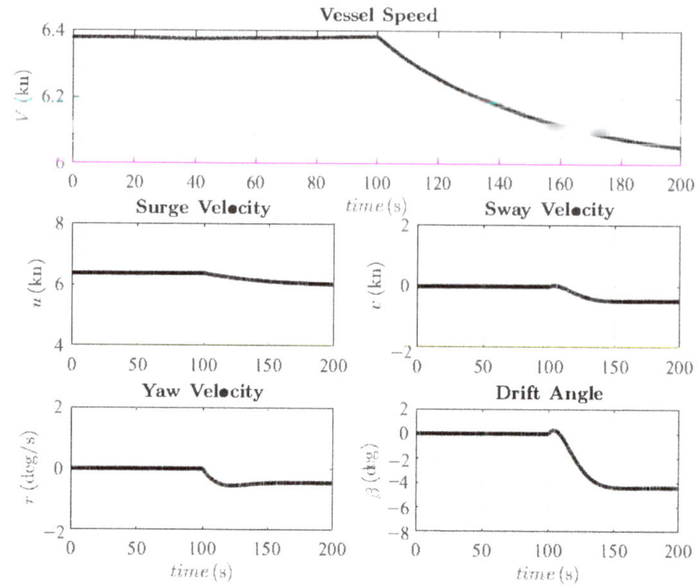

Figure 11. Vessel speed and velocity components in the body-fixed basis.

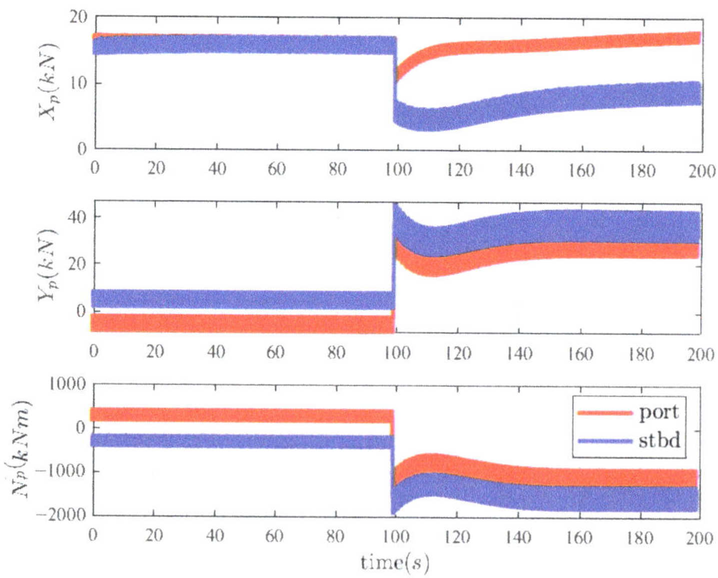

Figure 12. Propeller delivered forces and moment.

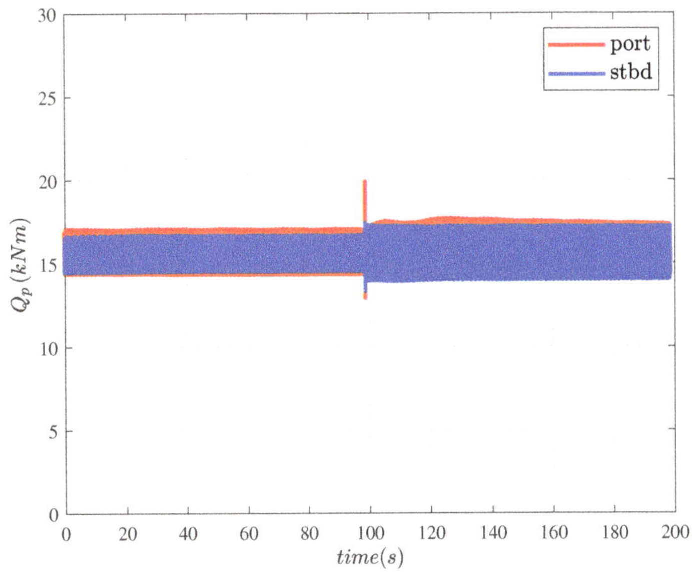

Figure 13. Propeller torque.

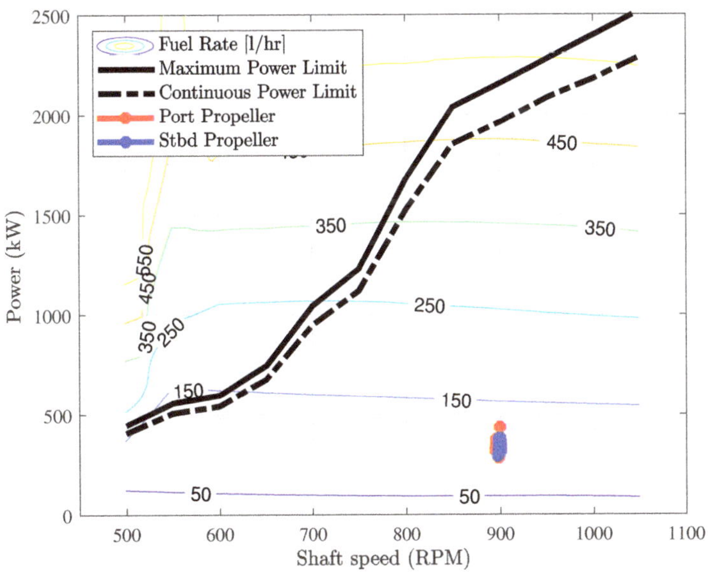

Figure 14. Propeller required power on the engine load diagram.

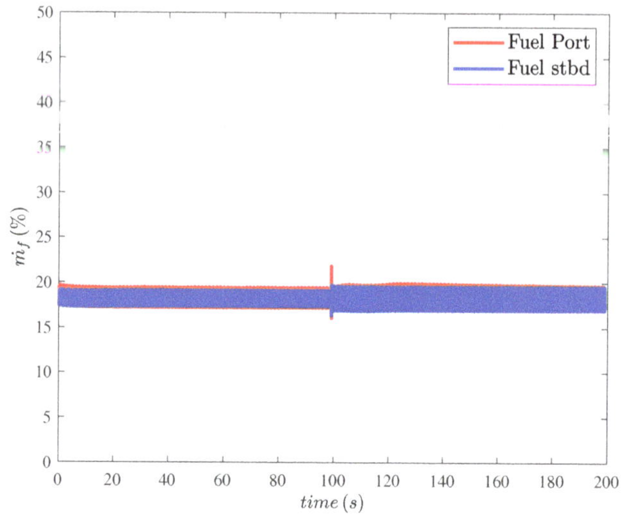

Figure 15. Fuel flow time history.

6. Conclusions

This work presents a simulation approach for both low and high speed manoeuvring of ships equipped with cycloidal propellers. The real strength comes from not having to calculate the propeller fluid dynamics, avoiding demanding computations that would make very difficult and complex an effective simulation of the whole ship propulsion plant behaviour (in the proposed approach, CFD method is just used for the evaluation of the lift and drag coefficients of the single blade). However, reliable results, concerning both the steady state and transient performance of the cycloidal propeller, are achieved. This is essentially due to the rigorous description of the motion of each blade and by introducing specific empirical correction factors that can be used for a preliminary performance estimation of several cycloidal propulsion units, characterized by different lengths and number of blades. In this sense, the propulsion simulator can be regarded as a parametric one. Indeed, open water diagrams can obtained for a wide range of cycloidal propellers only by changing the rotor diameter, number and length of blades. Through appropriate insights concerning the correction coefficients and therefore a dedicated hydrodynamic analysis, the simulator can reproduce the open water performance maps of other types of cycloidal thrusters, for which the literature and the industry provide very little information. Right because the lack of data, the proposed approach could be useful also to predict the behaviour of the ship during the design phase in terms of general performances, forces generation, response times and evaluation of energy/fuel consumptions. Moreover, a training platform for personnel could be implemented on this basis. Next developments should include the integration of the hydrodynamic interaction between the hull and the cycloidal propeller and vice versa, as well as the calculation of the hydrodynamic resistance of the cycloidal propellers intended as hull appendages, mainly by comparing empirical corrections with CFD results. Further improvements could include the study of different sizes of the main engine, represented by a thermodynamic model coupled to the cycloidal thruster model, in order to better analyse the engine-propeller dynamics (especially during transient conditions). Finally, the present research should be completed through a proper validation of the simulation approach, by means of experimental data of a cycloidal propulsion system installed on board a real ship.

Author Contributions: Conceptualization, M.A., S.V. and S.D.; methodology, M.A. and S.V.; software, V.S. and S.D.; validation, S.V. and S.D.; writing—original draft preparation, S.D.; writing—review and editing, M.A., S.V. and S.D. All authors have read and agreed to the published version of the manuscript.

Funding: This research received no external funding.

Institutional Review Board Statement: Not applicable.

Informed Consent Statement: Not applicable.

Data Availability Statement: Not applicable.

Conflicts of Interest: The authors declare no conflict of interest.

References

1. Acanfora, M.; Montewka, J.; Hinz, T.; Matusiak, J. Towards realistic estimation of ship excessive motions in heavy weather. A case study of a containership in the Pacific Ocean. *Ocean. Eng.* **2017**, *138*, 140–150. [CrossRef]
2. Acanfora, M.; Altosole, M.; Balsamo, F.; Micoli, L.; Campora, U. Simulation Modeling of a Ship Propulsion System in Waves for Control Purposes. *J. Mar. Sci. Eng.* **2022**, *10*, 36. [CrossRef]
3. Altosole, M.; Boote, D.; Brizzolara, S.; Viviani, M. Integration of Numerical Modeling and Simulation Techniques for the Analysis of Towing Operations of Cargo Ships. *Int. Rev. Mech. Eng.* **2013**, *7*, 1236–1245.
4. Altosole, M.; Figari, M.; Martelli, M.; Orrù, G. Propulsion control optimisation for emergency manoeuvres of naval vessels. In Proceedings of the INEC 2012-11th International Naval Engineering Conference and Exhibition, Edinburgh, UK, 15–17 May 2012.
5. Ghaemi, M.; Zeraatgar, H. Analysis of hull, propeller and engine interactions in regular waves by a combination of experiment and simulation. *J. Mar. Sci. Technol.* **2021**, *26*, 257–272. [CrossRef]
6. Saettone, S.; Tavakoli, S.; Taskar, B.; Jensen, M.V.; Pedersen, E.; Schramm, J.; Steen, S.; Andersen, P. The importance of the engine-propeller model accuracy on the performance prediction of a marine propulsion system in the presence of waves. *Appl. Ocean. Res.* **2020**, *103*, 102320. [CrossRef]
7. Altosole, M.; Campora, U.; Martelli, M.; Figari, M. Performance decay analysis of a marine gas turbine propulsion system. *J. Ship Res.* **2014**, *58*, 117–129. [CrossRef]
8. Zaccone, R.; Altosole, M.; Figari, M.; Campora, U. Diesel engine and propulsion diagnostics of a mini-cruise ship by using artificial neural networks. In Proceedings of the 16th International Congress of the International Maritime Association of the Mediterranean, Pula, Croatia, 21–24 September 2015; pp. 593–602. [CrossRef]
9. Campora, U.; Capelli, M.; Cravero, C.; Zaccone, R. Metamodels of a gas turbine powered marine propulsion system for simulation and diagnostic purposes. *J. Nav. Archit. Mar. Eng.* **2015**, *12*, 1–14. [CrossRef]
10. Halder, A.; Walther, C.; Benedict, M. Hydrodynamic modeling and experimental validation of a cycloidal propeller. *Ocean. Eng.* **2018**, *154*, 94–105. [CrossRef]
11. Hu, J.; Li, T.; Guo, C. Two-dimensional simulation of the hydrodynamic performance of a cycloidal propeller. *Ocean. Eng.* **2020**, *217*, 107819. [CrossRef]
12. Prabhu, J.; Nagarajan, V.; Sunny, M.; Sha, O. On the fluid structure interaction of a marine cycloidal propeller. *Appl. Ocean. Res.* **2017**, *64*, 105–127. [CrossRef]
13. Bakhtiari, M.; Ghassemi, H. CFD data based neural network functions for predicting hydrodynamic performance of a low-pitch marine cycloidal propeller. *Appl. Ocean. Res.* **2020**, *94*, 101981. [CrossRef]
14. Prabhu, J.; Dash, A.; Nagarajan, V.; Sha, O. On the hydrodynamic loading of marine cycloidal propeller during maneuvering. *Appl. Ocean. Res.* **2019**, *86*, 87–110. [CrossRef]
15. Altosole, M.; Donnarumma, S.; Spagnolo, V.; Vignolo, S. Marine cycloidal propulsion modelling for DP applications. In Proceedings of the VI International Conference on Computational Methods in Marine Engineering, Rome, Italy, 15–17 June 2017; pp. 206–219.
16. Altosole, M.; Donnarumma, S.; Spagnolo, V.; Vignolo, S. Simulation of a marine dynamic positioning system equipped with cycloidal propellers. In *Progress in Maritime Technology and Engineering*; CRC Press/Balkema: Boca Raton, FL, USA, 2018; pp. 257–264.
17. Donnarumma, S.; Figari, M.; Martelli, M.; Vignolo, S.; Viviani, M. Design and Validation of Dynamic Positioning for Marine Systems: A Case Study. *IEEE J. Ocean. Eng.* **2018**, *43*, 677–688. [CrossRef]
18. Donnarumma, S.; Figari, M.; Martelli, M.; Zaccone, R. Simulation of the Guidance and Control Systems for Underactuated Vessels. In Proceedings of the International Conference on Modelling and Simulation for Autonomous Systems, Palermo, Italy, 29–31 October 2019; Springer: Cham, Switzerland, 2020; Volume 11995, pp. 108–119. [CrossRef]
19. Martelli, M.; Viviani, M.; Altosole, M.; Figari, M.; Vignolo, S. Numerical modelling of propulsion, control and ship motions in 6 degrees of freedom. *Proc. Inst. Mech. Eng.* **2014**, *228*, 373–397. [CrossRef]
20. Martelli, M.; Figari, M.; Altosole, M.; Vignolo, S. Controllable pitch propeller actuating mechanism, modelling and simulation. *Proc. Inst. Mech. Eng. Part J. Eng. Marit. Environ.* **2014**, *228*, 29–43. [CrossRef]
21. Battistoni, L. Numerical Simulation Approach for the Preliminary Design of a Cycloidal Propeller. Master's Thesis, University of Genoa, Genoa, Italy, 2014.

Journal of
Marine Science and Engineering

MDPI

Article

Simulation Modeling of a Ship Propulsion System in Waves for Control Purposes

Maria Acanfora [1,*], Marco Altosole [1], Flavio Balsamo [1], Luca Micoli [1] and Ugo Campora [2]

[1] Department of Industrial Engineering (DII), School of Polytechnic and Basic Sciences, University of Naples "Federico II", 80125 Naples, Italy; marco.altosole@unina.it (M.A.); flavio.balsamo@unina.it (F.B.); luca.micoli@unina.it (L.M.)

[2] Department of Mechanical, Energy, Management and Transportation Engineering (DIME), Polytechnic School, University of Genoa, 16145 Genoa, Italy; ugo.campora@unige.it

* Correspondence: maria.acanfora@unina.it

Abstract: The article deals with a simulation approach to the representation of the ship motions in waves, interacting with the propulsion system behavior (diesel engine and propeller). The final goal is the development of a simulator, as complete as possible, that allows the analysis of the main engine thermodynamics in different sea conditions, also in the unfavorable event of dynamic instability of the hull, and the correct management of the other propulsion components. This latter aspect is particularly interesting in some of the last new energy solutions for decarbonization of ships, concerning, for example, auxiliary electric motors, powered by batteries, to support the traditional diesel-mechanical propulsion (especially in heavy weather conditions). From this point of view, a proper analysis of the engine dynamic performance, affected by particular sea states, is fundamental for a smart management and control of shaft generators/auxiliary electric motors, batteries, etc. To this end, the work presents and highlights the main features of a ship simulator, suitable for the study of the new propulsion solutions that are emerging in maritime transport. Some representative results will point out the complex non-linear behavior of the propulsion plant in waves. Moreover, a parametric roll scenario will be investigated, in order to highlight the capability of the conceived simulator in modeling the effects of the dynamic instability of the hull on the propulsion plant.

Keywords: ship propulsion; simulation; diesel engine; wave; ship motions; control

Citation: Acanfora, M.; Altosole, M.; Balsamo, F.; Micoli, L.; Campora, U. Simulation Modeling of a Ship Propulsion System in Waves for Control Purposes. *J. Mar. Sci. Eng.* **2022**, *10*, 36. https://doi.org/10.3390/jmse10010036

Academic Editor: Tie Li

Received: 13 December 2021
Accepted: 27 December 2021
Published: 31 December 2021

Publisher's Note: MDPI stays neutral with regard to jurisdictional claims in published maps and institutional affiliations.

1. Introduction

Marine time-domain simulation has traditionally found its main use in the field of ship maneuverability, especially for the development of training simulators, which have become increasingly effective with the advent of virtual reality [1]. However, the scientific literature shows that dynamic simulation techniques have also been significantly adopted in ship design, although the most representative industrial applications are just only from the last twenty years. This is essentially due to the enhancement of computers and the development of particular commercial software for a more user-friendly application. In this framework, Genoa University was a pioneer in the development of simulators for evaluating the ship propulsion performance and analyzing pertinent machinery control logics [2]. A first significant example of a ship propulsion simulation with a full-scale validation is reported in [3]. It concerns the mechanical diesel propulsion system of a cruise ferry, while [4] proposes again the same approach to the validation of the methodology in the case of a small naval vessel. After the presentation and validation studies of the simulation method, Genoa University developed the first simulators for industrial applications in the cases of the aircraft carrier Cavour [5,6] and the Italian Navy Multi-Mission Frigates (FREMM) [7–9]. In both projects, real-time simulation was used to design and test the propulsion control system. The collaboration with the Italian Navy continued in the development of a simulation study for the propulsion system refitting of the tall ship Amerigo Vespucci (in this case, the square sail's behavior was reproduced too [10,11]). These three applications represented a typical use of simulation in the design stage of a vessel, although it can also be used to evaluate the performance decay of already operating systems. In this regard, References [12,13] show simulation approaches to the representation of malfunctions/faults of marine engines and auxiliary systems for the thermal energy recovering from the exhaust gas of the engines. In the same research field, References [14–16] propose different simulation techniques (e.g., neural networks) for the marine engine diagnostics, while [17–19] show further examples of dynamic simulation approaches to the design of the ship energy systems

Focusing on marine propulsion systems, a good simulator is made up of several numerical sub-models that represent the main elements involved in ship propulsion. Surely, the main mover model can strongly characterize the complexity of the simulator as it can be faced with different approaches. Specific programming languages [20–25] or commercial software [26–29] can represent the engine behavior, with different levels of accuracy, according to the different purposes. On the contrary, the propeller is traditionally modeled by numerical tables, unless you want to describe further aspects, such as the dynamics of the rotation of the blades in the case of a controllable pitch propeller [30]. For other types of marine thruster, such as waterjet or cycloidal propellers, more complex models can be adopted [31–33]. Another important sub-model of the simulator certainly concerns the representation of the ship motions, which can be simulated with one [3,34], three (e.g., typical maneuverability simulators [35–37]), four [38] up to six degrees of freedom (DOF) [39]. Obviously, the complexity of the modeling grows in relation to the DOF number increasing. Moreover, different mathematical approaches may turn out to be more or less suitable for representing the motions of the ship at high or low speed (e.g., dynamic positioning applications [40]).

Due to the complexity of the whole ship dynamics, it is very difficult to develop a complete propulsion simulator taking into account both hull motions and main engine behavior. The whole ship dynamics in the time domain is usually studied by six DOF numerical models based on the equations of the rigid body motions. Several methods for the assessment of ship dynamic instabilities and maneuverability in waves have been extensively studied and compared [41,42]. However, all these numerical approaches disregard the propulsion power delivered by the main engine and rather assume ideal constant propeller revolutions or assume constant ship speed.

Several approaches of hull-propeller-engine interaction are available in the technical literature, although they mainly focus on the effects of the added resistance in waves, as underlined by [43]. In particular, in [43] the authors propose an interaction model of hull-propeller-engine by combining experimental data and numerical simulation in regular seas.

In [44], the simulation of the propeller performance accounts for the propeller emergence in waves, calculated beforehand from the estimation of ship motions in a head sea. The importance of developing a numerical model accounting for ship dynamics effects on the marine propulsion system performances has been recently pointed out in [45,46]. These research works account for simplified modeling of ship dynamics by means of a linear transfer function of heave and pitch and includes wake and propeller characteristic variations in regular waves.

Differently from the others, the present work shows a complete simulation approach to the description of the interaction between ship motions (six DoF) and machinery (i.e., prime mover and propeller), applicable in irregular sea, by combining the most sophisticated sub-models available in the state of the art, for a comprehensive simulation.

Indeed, the numerical model for ship dynamics is based on the so-called hybrid or blended non-linear approach, as described by [47,48].

Differently from [47], implemented in FORTRAN, the model adopted herein is developed and implemented in MATLAB/Simulink [49,50]. Similar to [47,51,52], it allows for the estimation of ship dynamic instabilities and maneuvering simulations in irregular waves with a fair level of accuracy. The main innovation of the present work is the interaction among the different propulsion components, in the presence of different sea states, accounting also for the possibility of simulating critical behaviors of the hull, such as dynamic instabilities. The weather conditions, in fact, greatly affect the hull motions, which, in turn, influence the behavior of the propeller and therefore the propulsion system. A constant wake factor in irregular sea is assumed, while ship speed changes because of the time-varying added resistance and propeller thrust.

The proposed simulation approach, described in its several sub-models, can become a useful tool for evaluating any critical issues on the engine, to be solved through appropriate control strategies. For the current application, a Ro-pax ship sailing in irregular sea is under investigation, assuming a thermodynamic model of a four-stroke medium diesel engine [53]. The results of several operational conditions are presented in order to point out the need for consistent interaction among the different sub-models for a proper depiction of the phenomena.

2. Numerical Model of the Simulator

The developed simulator aims at connecting existing sub-models of ship dynamics in irregular seas and a diesel engine, accounting for propeller actions in waves. The following subsections describe in brief each sub-model. The whole simulator is implemented in the MATLAB/Simulink platform.

2.1. Ship Dynamics in Wave

The numerical six DoF model for ship dynamics in waves (in the time domain) included all forces and moments acting on a sailing vessel in irregular sea. The hull surface was discretized by means of triangular panels up to the freeboard deck. Mass and inertia actions included all non-linear coupling terms of the rigid body dynamics. All non-linearities regarding hull geometry were accounted for in the calculation of Froude–Krylov and restoring actions.

Radiation forces and moments were implemented by means of the convolution integral technique, while diffraction forces and moments were obtained by linear superposition of regular wave components. The regular force components, which were the input data for both radiation and diffraction actions, were taken from potential strip theory calculations, carried out beforehand [54].

Additional details together with validations of the implemented numerical model, regarding ship motions and accelerations in waves, are available in [49]. It is worth noting, that the ship speed, in the reference applications, was set as constant, thus no attention was given to the effects of added resistance in waves (in addition to ship resistance in still water), which represents a demanding second-order problem [55].

In the current form of the simulation model, the ship resistance curve in still water was given as input data, while added wave resistance accounted only for first-order effects. In general, this approach could lead to approximate results, especially in the range of small wave components (i.e., smaller than ship length), where radiation and diffraction actions are predominant and the numerical modeling for these forces is not suited to the added wave resistance problem. On the other hand, the current approach provided quite accurate restoring and Froude–Krylov non-linear components, that for long waves (i.e., longer than ship length) became predominant.

Therefore, limiting the case studies to sea states characterized by a wavelength greater than ship length, this numerical model can be considered suited at the scope of the paper.

A brief validation of the model is presented in Table 1 for the hull KVLCC2, extensively studied for benchmark researches [56]. The experimental data (exp) available for this ship [56] concerned the speed reduction in regular waves given a fixed propeller revolution. The numerical simulations were carried out, including propeller actions but neglecting engine contribution. The obtained speed reduction in wave (sym) was compared against the experimental speed data.

As expected, the comparison of the results disclosed that the error on the speed reduction in waves decreased with wavelengths (λ) longer than ship length L. For this range of wavelength, the error remained below 10%.

Table 1. Comparison of experimental and numerical speed reduction in waves for KVLCC2.

λ/L	Exp (Knots)	Sim (Knots)	Error (Knots)	Err %
0.5	12.1	15.3	3.2	26%
0.7	10.9	15.1	4.2	39%
1	8.6	9.5	0.9	10%
1.2	10.2	9.2	−1	−10%
1.5	12.6	11.6	−1	−8%

2.2. Propeller Actions

The numerical model of the sailing ship in waves included screw propeller actions that were implemented by means of still water propeller coefficient K_T and K_Q, given as a function of the advance coefficient J. Propeller thrust has to balance ship resistance in waves, while propeller torque has to be balanced by the torque provided by the engine. The wake and the thrust deduction factors for modeling the propeller-hull interaction in irregular sea, refer to the still water condition.

The thrust T_{prop} and the torque Q_{prop} at the propeller were obtained as follows:

$$T_{prop} = \rho K_T^* N_{prop}^2 D_{prop}^4 \qquad Q_{prop} = \rho K_Q^* N_{prop}^2 D_{prop}^5 \tag{1}$$

In Equation (1), K_T^* and K_Q^* are modified propeller coefficients in order to account for a propeller operating very close to the free surface due to the wave effects. In particular, the simulation approach proposed by [57], for modeling generalized ventilation losses, was used. According to [57], thrust and torque ventilation loss factors (β_{tv} and β_{qv}, respectively) can be introduced:

$$K_T^* = \beta_{tv} K_T \qquad K_Q^* = \beta_{qv} K_Q$$

where β_{tv} depends on the propeller type, propeller loading, and on the relative blade submergence $2h/D_{prop}$, while $\beta_{qv} = (\beta_{tv})^m$, where m = 0.85 for open propeller type. When $2h/D_{prop} = -1$, it means that the propeller is fully emerged, thus $\beta_{tv} = \beta_{qv} = 0$.

In this paper, a hull equipped with twin-screw open propellers was chosen for the purpose of the applications; therefore the forces at the left-side and the right-side propellers were modeled independently. Indeed, due to the hull dynamics in waves (in particular due to the roll motions), the twin propellers can experience different submergences and loadings, resulting in different thrust and torque.

2.3. Diesel Engine

Whereas the diesel engine model is part of the whole ship dynamics, it should catch the engine behavior during transients with acceptable accuracy but without excessive computational work. A 0D model based on a filling and emptying approach can be a good compromise between these different requirements [20,22]. All the main engine components were assumed as blocks connected to each other representing algebraic and/or differential equations according to the principles of mass and energy conservation. The fluid that followed the so-called air path was considered as an ideal gas which composition varied when flowing through the engine components; in each, the fluid parameters (temperature, pressure, and composition) were supposed to be uniform, so neglecting the momentum conservation equation. The engine was a four-stroke turbocharged and the turbine and compressor were modeled using steady-state maps, where flow rate and isentropic efficiency are a function of pressure and speed, corrected as usual.

Engine speed and turbocharger speed were calculated from time to time by dynamic equilibrium equations and represent two of the main states of the engine model, as pressures in inlet and outlet manifolds. The heat release due to the fuel-injected and burned in the cylinders was modeled by a classical double zone Wiebe function. The inlet valves timing could be varied to control the airflow at different engine speeds and loads.

The model was fine-tuned by comparing the simulation results with the data extracted from the factory data sheet until a good agreement was reached between them. A detailed description of the engine model and data accuracy evaluation can be found in [23].

2.4. Engine Controller

The function of the engine governor is to keep the rotation speed of the diesel engine constant despite the variation of propeller load because of wind, current, waves, and fouling by allowing the engine to increase or decrease the torque generated. In the case of wave-induced propeller torque variations, whose characteristic time is between 5 and 15 s, quite lower than the other cases listed, the governor must operate quickly enough to avoid excessive engine speed variations that could cause an engine power failure and, consequently, represent a potential risk for ship safety.

The governor determines the torque supplied by the engine acting on the amount of fuel injected at each engine cycle by means of the so-called fuel rack; the engine response depends on many engine states, as crankshaft speed, turbocharger speed, pressure in the receivers, and other parameters.

In this paper, the engine governor was modeled as a conventional PI (proportional integral) controller, which action is proportional to the shaft speed error and its integral over time. The two values that represent the controller proportional (kp) and integral (ki) gains were determined by analyzing the engine response to a sudden variation of the fuel rack position, starting with the Ziegler Nichols method and followed by a further fine-tuning.

Even in its basic configuration, the engine governor must have more functions, as smoke limiters, torque limiter to avoid overload, and set point rate limiter to prevent the risk of compressor surging.

In order to check the effectiveness of the implemented regulator, the following numerical test was carried out. Given the engine working at a reduced load, an acceleration ramp was set for an increase of 1/3 of the revolutions. From the results presented in Figure 1, the control action of the regulator guaranteed the realization of such acceleration within a time range of approximately three minutes. This confirmed that the overall behavior of the regulator was acceptable for the purposes of the current paper. The engine revolutions were the result of the solution of the classic dynamic equation of the shaft line [2,5], dependent on the rotating inertia of the whole propulsive chain (i.e., engine, gearbox, shaft, and propeller).

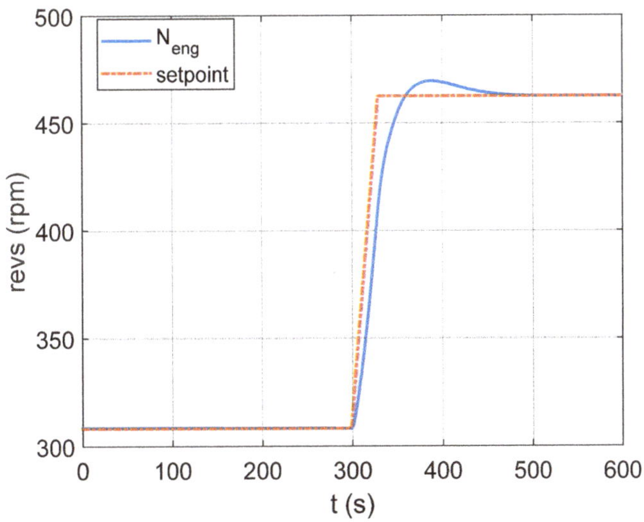

Figure 1. Revolution response to rpm ramp.

3. Case Study and Results

The hull chosen for the case study was a ro-pax ferry named SeatechD, which has been used for previous validations and applications of the method for ship dynamics in waves [49]. Since the hull model was developed mainly for research purposes on ship stability and maneuvering in waves [58], no precise information on propeller and engine at full scale are available. Therefore, at the scope of the case study, the maximum speed was set as 24 knots, (that is reasonable for the hull under investigation compared to other similar vessels) and two Wageningen B-series propellers (for which K_T and K_Q coefficients are available from [59]) were assumed in a twin-screw configuration.

These design choices allowed the use of two diesel engines of 12 MW each [60], for which the numerical model was developed in [53]. Table 2 provides a summary of the hull, propeller, and engine main data.

Table 2. Principal particulars of SeatechD.

Hull Seatech-D	
Length between perpendiculars, L (m)	158.00
Breadth, B (m)	25.00
Depth, D (m)	15.00
Draft forward, T_F (m)	6.10
Draft aft, T_A (m)	6.10
Displacement, Δ (tons)	13,766
Center of gravity above the keel, KG (m)	11.834
Long. coordinate of the center of gravity from aft perpendicular, LCG (m)	74.77
Transv. radius of gyration in air, k_{XX} (m)	10.06
Long. radius of gyration in air, k_{YY} (m)	39.36
Propeller Wageningen B-series	
Number of blades Z	4
A_e/A_0	0.750
D_{prop} (m)	4.8
$P/_{Drop}$	1.2
Engine	
Number of cylinders	12
Bore (m)	0.51
Stroke (m)	0.60
Engine revolution N_{eng} (rpm)	514
Engine power P_B (MW)	12

The sample irregular sea state for the simulation was obtained from the JONSWAP spectrum, given the significant wave height (Hs) and the zero-crossing period (Tz). The technique described in [49] was used for the generation of the irregular wave train.

The following application focuses on a realization of wave trains having a ratio of 1:2 of H_S, i.e., two similar wave profiles were generated, differing only in the H_S values that were $H_S = 6.5$ m and $H_S = 3.25$ m, respectively (see Figure 2). This analysis aimed at disclosing the non-linear behavior of the whole propulsion chain interacting with ship dynamics in waves. Moreover, an additional application will be carried out at a reduced speed in order to trigger a parametric roll that leads to non-symmetric loads on each propeller.

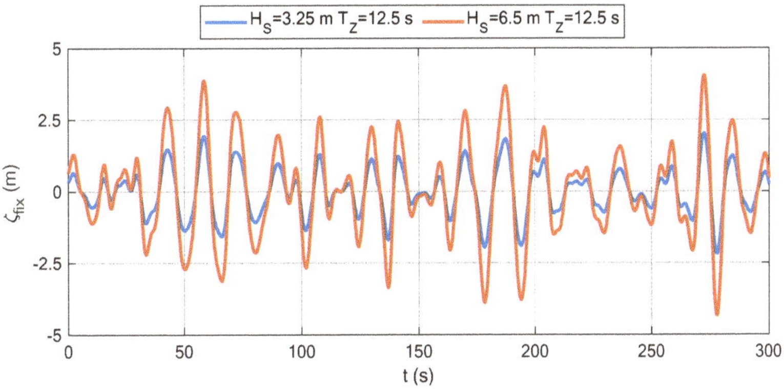

Figure 2. Time history of the two irregular wave trains (in the inertial reference frame).

3.1. Results

The results of this subsection regard the hull–propeller–engine performances for the ship sailing in the head sea in irregular waves characterized by T_Z = 12.5 s, H_S = 6.5 m, and T_Z = 12.5 s, H_S = 3.25 m. For the sake of synthesis, we limited the outcomes of the time history to 300 s.

Figure 3 shows ship dynamics and propeller behavior: green dash-dotted line refers to still water results; solid blue line refers to the sea state characterized by the smaller wave height; solid orange line refers to the wave train with the doubled wave height. It is possible observing that, as expected, ship motions (heave and pitch) exhibited larger amplitude in case of larger wave height, but they were not in phase. This was related to a visible reduction in ship speed (approximately 2 knots for Hs = 6.5 m) that significantly changed the encounter frequency of the hull. Moreover, ship dynamics modified the value of h/R, that is the head of water at the propeller h divided by the propeller radius R. Keeping in mind that h/R = −1 implies a fully emerged propeller; in both cases, the propeller stayed at least partially immersed. While for the case Hs = 3.25 m the propeller was scarcely affected by the wave effects (as observable also from revolutions N_{prop} and torque Q_{prop}), the larger wave amplitude Hs = 6.5 m induced a significant partial emergence of the propeller with h/R close to zero. This induced irregular profiles of propeller torque and revolution, with a reduction in torque and increase in the revolution when h/R tended to zero.

The propeller load directly influences the engine performances in terms of torque and revolutions, Q_{eng} and N_{eng}, respectively. Indeed, in Figure 4, it is possible to observe how the power profile provided by the engine P_B exhibited larger fluctuations for sea state Hs = 6.5 m, together with an increase in the specific fuel consumption SFCS. Figure 4 also shows the temperature T_{out} of the exhaust gases at the cylinder, in order to point out the capabilities of the model in simulating the thermodynamics of the whole turbocharged diesel engine.

At the same scope, Figure 5 illustrates compressor behavior given the compressor map. For the sea state with Hs = 3.25 m, the working conditions of the compressor remained close to the steady working condition (i.e., referring to still water), while they became scattered in the case of Hs = 6.5 m.

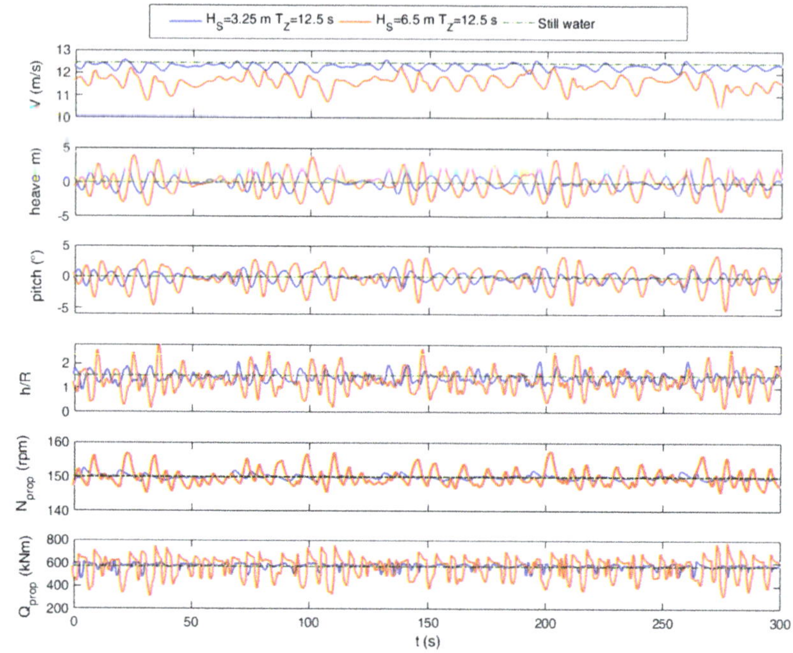

Figure 3. Time history of ship dynamics and propeller behavior in a head sea, for two proportional irregular sea states and still water.

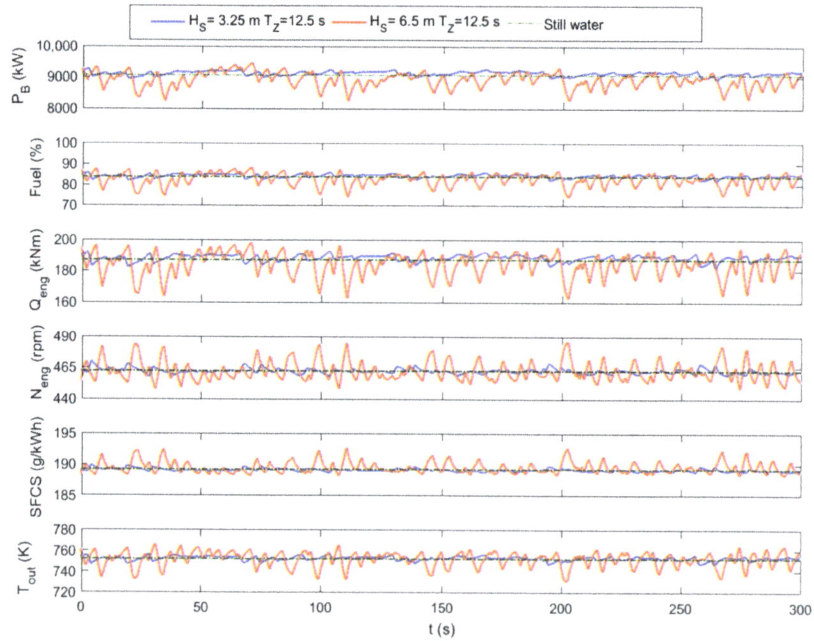

Figure 4. Time history of diesel engine performances in a head sea, for two proportional irregular sea states and still water.

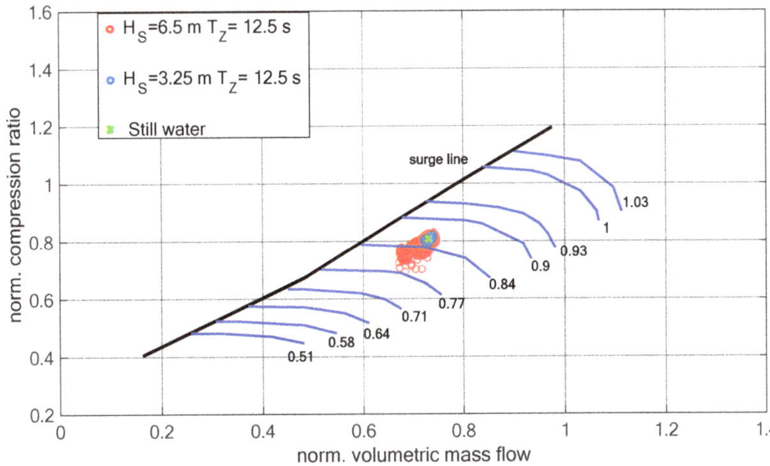

Figure 5. Comparison of compressor performances in a head sea, for two proportional irregular sea states and still water. Surge line in black, compression ratio at constant revolutions in blue.

3.2. *Results for Parametric Roll Scenario*

A ship sailing at maximum speed in a severe–moderate sea state is quite unrealistic, for safety reasons; indeed, it is quite usual that the vessel operates at reduced speed, i.e., at a lower engine load. Actually, this situation could be even more dangerous for ships prone to dynamic instabilities, such as the parametric roll, as in the case of the ro-pax ferry under investigation. The developed code also allowed simulating engine performances in these peculiar circumstances, related to the non-linearities of the immersed hull geometry in waves.

It was found that the ship developed a parametric roll in a head sea in the presence of a sea state characterized by Hs = 4.5 m and Tz = 9.0 s, for a speed of around 15 knots. In such a scenario, the mean wave encounter period was half of the natural roll period of the hull (around 17.5 s), as observable in Figure 6. This figure shows the ship dynamics in waves, focusing on the development of a parametric roll with maximum oscillations of around 20°. This situation induced a non-symmetric behavior of the two propellers and consequently of the two diesel engines. In Figure 7, the starboard propeller and engine outcomes are identified by solid blue lines, while the port propeller and engine outcomes are identified by solid orange lines. It is possible to notice that in presence of a small roll perturbation (i.e., absence of roll), the two propulsion drive chains behaved equally. Once parametric roll develops, the behaviors of starboard-side and port-side propellers and engines were almost 180° out of phase; actually, they were not completely opposite due to the effects of coupling of roll, heave, and pitch on propellers. This results in a large oscillation of engine powers due to the reduction in the propeller torque in correspondence with the propeller emergence, associated with a significant increase in revolutions.

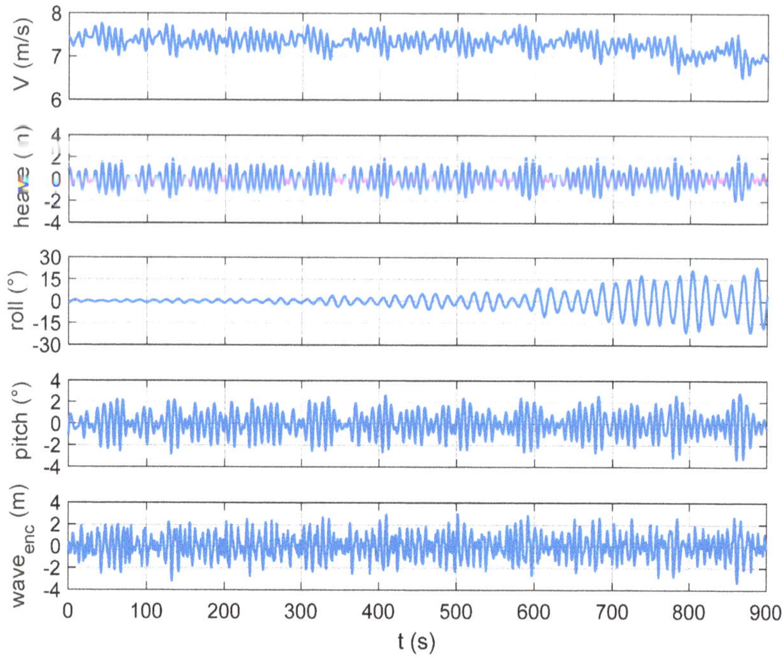

Figure 6. Time history of the ship dynamics in a head sea, Hs = 4.5 m Tz = 9.0 s: parametric roll development.

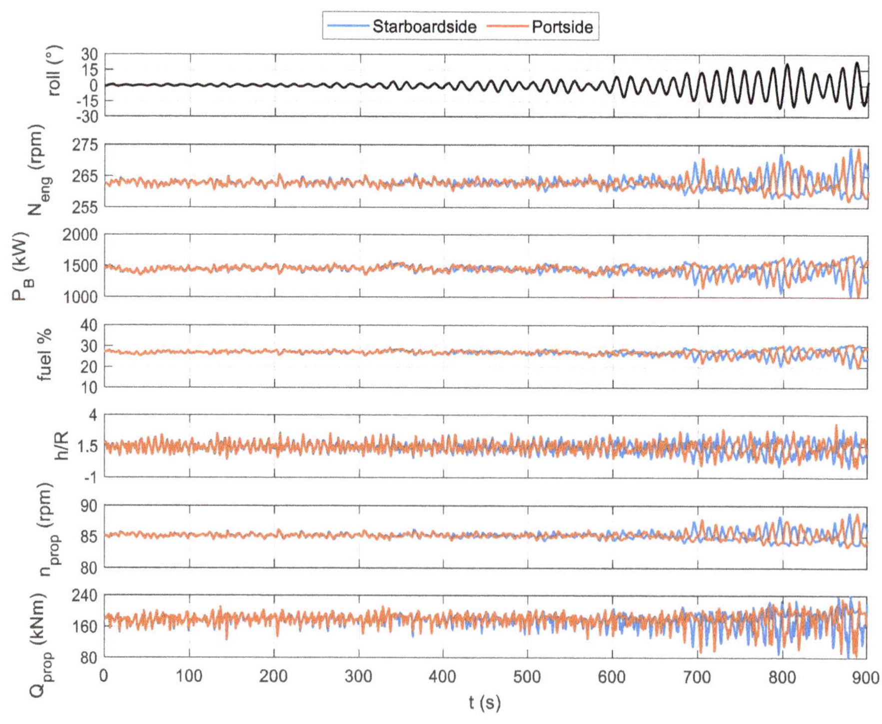

Figure 7. Time history of starboard-side and port-side propellers and engine in a head sea, Hs = 4.5 m Tz = 9.0 s, depending on ship roll motion (black line).

4. Conclusions

This paper concerned the development of a comprehensive numerical model for simulating engine performances accounting for ship dynamics in waves and its effects on propeller loading characteristics. This model represents a first step in the development of a numerical tool to analyze and control the ship propulsion system in unsteady conditions.

From the overall analysis of the numerical outcomes, it is possible to disclose that there was no proportional behavior between the sea state and hull–propeller–engine interaction. In other words, the variation in engine power and consumption did not show a linear behavior with the height of the sea state but depended on multiple interacting factors. Therefore, due to the complexity of phenomena, a simplified approach is not recommended for the assessment of engine performance in waves.

The numerical model presented herein is capable to account for non-linearities that arise from the hull dynamics in waves and interact with the propulsion system. In particular, in this research, the performances of the twin propellers and engines were simulated in event of parametric roll instability in head sea navigation (where usually no roll motion is expected). The main outcomes showed that this critical condition induced a somewhat nonsymmetrical behavior of the thermodynamic engine parameters with large oscillations of torque and revolutions

However, all obtained results can currently provide only a qualitative assessment of hull–propeller–engine behavior, since relevant validations of the complete model are still missing.

It is worth underlining that the main idea behind this research was the possibility to connect hull dynamics with any kind of propulsion system configurations, therefore allowing the modeling and the analysis of the performances of, among other, hybrid propulsion systems. In particular, by combining the prime mover (e.g., a diesel engine equipped with a PTO/PTI shaft generator) to batteries (or fuel cells or even renewable energy sources) for smoothing out the fluctuation of the propeller load, it is possible to optimize the ship propulsion efficiency in sizeable sea waves. The proposed simulator could test a specific control strategy, in order to achieve an averaged load of the diesel engine in severe waves, while the battery system, through the PTO/PTI modes of the shaft generator, will be discharged during short peaks and charged during short dips. Therefore, the engine load can be more constant, by improving fuel efficiency and reducing maintenance costs.

Author Contributions: Conceptualization, M.A. (Maria Acanfora) and M.A. (Marco Altosole); methodology, M.A. (Maria Acanfora), M.A. (Marco Altosole), F.B., L.M. and U.C.; software, M.A. (Maria Acanfora) and U.C.; formal analysis, M.A. (Maria Acanfora), M.A. (Marco Altosole), F.B. and L.M.; investigation, M.A. (Maria Acanfora), M.A. (Marco Altosole), F.B., L.M. and U.C.; resources, M.A. (Maria Acanfora) and U.C.; data curation, M.A. (Maria Acanfora); writing—original draft preparation, M.A. (Maria Acanfora), M.A. (Marco Altosole) and F.B.; writing—review and editing, M.A. (Maria Acanfora), M.A. (Marco Altosole), L.M. and F.B. All authors have read and agreed to the published version of the manuscript.

Funding: This research received no external funding.

Institutional Review Board Statement: Not applicable.

Informed Consent Statement: Not applicable.

Data Availability Statement: Not applicable.

Conflicts of Interest: The authors declare no conflict of interest.

Nomenclature

D_{prop}	Propeller diameter
h	Head of water on the propeller shaft
H_S	Significant wave height
J	Advance coefficient
K_Q	Propeller torque coefficient
K_T	Propeller thrust coefficient
L	Ship length
N_{eng}	Engine revolutions
N_{prop}	Propeller revolutions
Q_{eng}	Engine torque
Q_{prop}	Propeller torque
R	Propeller radius
SFCF	Specific fuel consumption
T_{out}	Temperature of the exhaust gases at the cylinder outlet
T_{prop}	Propeller thrust
T_Z	Zero crossing period
V	Ship speed
$wave_{enc}$	Wave profile encountered by the ship
ζ_{fix}	Wave profile in the inertial (fixed) frame
β_{qv}	Torque ventilation loss factors
β_{tv}	Thrust ventilation loss factors
λ	Wavelength
ρ	Density of the sea water

References

1. Okazaki, T.; Nishizaki, C. A study on situation awareness of ship maneuvering simulator training. *Int. J. Emerg. Trends Eng. Technol.* **2015**, *3*, 21–30.
2. Altosole, M.; Figari, M.; Martelli, M. Time-Domain Simulation for Marine Propulsion Applications. In Proceedings of the 2012—Summer Computer Simulation Conference, SCSC 2012, Part of SummerSim 2012 Multiconference, Genova, Italy, 8–11 July 2012; Volume 44, pp. 36–43.
3. Campora, U.; Figari, M. Numerical simulation of ship propulsion transients and full-scale validation. *Proc. Inst. Mech. Eng. Part M J. Eng. Marit. Environ.* **2003**, *217*, 41–52. [CrossRef]
4. Benvenuto, G.; Carrera, G.; Figari, M.; Campora, U. Interaction Between Ship Propulsion Plant Automation and Simulation. In Proceedings of the 12th International Congress of the International Maritime Association of the Mediterranean, IMAM 2005—Maritime Transportation and Exploitation of Ocean and Coastal Resources, Lisbon, Portugal, 26–30 September 2005; Volume 1, pp. 617–626.
5. Altosole, M.; Figari, M.; Bagnasco, A.; Maffioletti, L. Design and Test of the Propulsion Control of the Aircraft Carrier "Cavour" using Real-Time Hardware in the Loop Simulation. In Proceedings of the SISO European Simulation Interoperability Workshop 2007, EURO SIW 2007, Genova, Italy, 18–20 June 2007; pp. 67–74.
6. Altosole, M.; Benvenuto, G.; Figari, M.; Campora, U. Real-Time Simulation of a COGAG Naval Ship Propulsion System. In Proceedings of the 12th International Congress of the International Maritime Association of the Mediterranean, IMAM 2007—Maritime Industry, Ocean Engineering and Coastal Resources, Varna, Bulgaria, 2–6 September 2008; Volume 1, pp. 331–337.
7. Altosole, M.; Dubbioso, G.; Figari, M.; Viviani, M.; Michetti, S.; Trapani, A.M. Simulation of the Dynamic Behaviour of a Codlag Propulsion Plant. In *RINA, Royal Institution of Naval Architects—Warship 2010: Advanced Technologies in Naval Design and Construction, London, UK, 9–10 June 2010*; The Royal Institution of Naval Architects: London, UK, 2010; pp. 109–115.
8. Michetti, S.; Ratto, M.; Spadoni, A.; Figari, M.; Altosole, M.; Marcilli, G. Ship Control System Wide Integration and the use of Dynamic Simulation Techniques in the Fremm Project. In Proceedings of the International Conference on Electrical Systems for Aircraft, Railway and Ship Propulsion, ESARS 2010, Bologna, Italy, 19–21 October 2010.
9. Altosole, M.; Figari, M.; Martelli, M.; Orrù, G. Propulsion Control Optimisation for Emergency Manoeuvres of Naval Vessels. In Proceedings of the INEC 2012—11th International Naval Engineering Conference and Exhibition, Edinburgh, UK, 15 May 2012.
10. Altosole, M.; Figari, M.; Ferrero, C.; Giuffra, V.; Piva, L. Propulsion Retrofitting of the Tall Ship Amerigo Vespucci: Automation Design by Simulation. In Proceedings of the 2014 International Symposium on Power Electronics, Electrical Drives, Automation and Motion, SPEEDAM 2014, Ischia, Italy, 18–20 June 2014; pp. 313–318.
11. Altosole, M.; Piastra, F.; Canepa, E. Performance analysis of a motor-sailing propulsion system for control design purposes. *Ships Offshore Struct.* **2016**, *11*, 688–699. [CrossRef]

12. Altosole, M.; Campora, U.; Martelli, M.; Figari, M. Performance decay analysis of a marine gas turbine propulsion system. *J. Ship Res.* **2014**, *58*, 117–129. [CrossRef]
13. Altosole, M.; Campora, U.; Laviola, M.; Zaccone, R. Deterioration effects on the performance of a steam plant for the waste heat recovery from a marine diesel engine. *Ships Offshore Struct.* **2019**, *14*, 867–878. [CrossRef]
14. Zaccone, R.; Altosole, M.; Figari, M.; Campora, U. Diesel Engine and Propulsion Diagnostics of a Mini-Cruise Ship by using Artificial Neural Networks. In *Towards Green Marine Technology and Transport—Proceedings of the 16th International Congress of the International Maritime Association of the Mediterranean, IMAM 2015, Pula, Croatia, 21–24 September 2015*; CRC Press/Balkema: Boca Raton, FL, USA, 2015; pp. 593–602.
15. Campora, U.; Cravero, C.; Zaccone, R. Marine gas turbine monitoring and diagnostics by simulation and pattern recognition. *Int. J. Nav. Archit. Ocean Eng.* **2018**, *10*, 617–628. [CrossRef]
16. Campora, U.; Capelli, M.; Cravero, C.; Zaccone, R. Metamodels of a gas turbine powered marine propulsion system for simulation and diagnostic purposes. *J. Nav. Arch. Mar. Eng.* **2015**, *12*, 1–14. [CrossRef]
17. Altosole, M.; Campora, U.; Donnarumma, S.; Zaccone, R. Simulation Techniques for Design and Control of a Waste Heat Recovery System in Marine Natural Gas Propulsion Applications. *J. Mar. Sci. Eng.* **2019**, *7*, 397. [CrossRef]
18. Barone, G.; Buonomano, A.; Forzano, C.; Palombo, A. Implementing the dynamic simulation approach for the design and optimization of ships energy systems: Methodology and applicability to modern cruise ships. *Renew. Sustain. Energy Rev.* **2021**, *150*, 111488. [CrossRef]
19. Barone, G.; Buonomano, A.; Forzano, C.; Palombo, A.; Vicidomini, M. Sustainable energy design of cruise ships through dynamic simulations: Multi-objective optimization for waste heat recovery. *Energy Convers. Manag.* **2020**, *221*, 113166. [CrossRef]
20. Benvenuto, G.; Campora, U.; Carrera, G.; Casoli, P. A Two-Zone Diesel Engine model for the Simulation of Marine Propulsion Plant Transients. In Proceedings of the MARIND 98, Second International Conference on Marine Industry, Varna, Bulgaria, 28 September–2 October 1998.
21. Mrzljak, V.; Medica, V.; Bukovac, O. Quasi-dimensional diesel engine model with direct calculation of cylinder temperature and pressure | Kvazi-dimenzijski model dizelskog motora s direktnim izračunom tlaka i temperature u cilindru. *Tech. Gaz.* **2017**, *24*, 681–686.
22. Altosole, M.; Benvenuto, G.; Campora, U.; Laviola, M.; Zaccone, R. Simulation and performance comparison between diesel and natural gas engines for marine applications. *Proc. Inst. Mech. Eng. Part M J. Eng. Marit. Environ.* **2017**, *231*, 690–704. [CrossRef]
23. Altosole, M.; Campora, U.; Figari, M.; Laviola, M.; Martelli, M. A diesel engine modelling approach for ship propulsion real-time simulators. *J. Mar. Sci. Eng.* **2019**, *7*, 138. [CrossRef]
24. Mrzljak, V.; Medica, V.; Bukovac, O. Volume agglomeration process in quasi-dimensional direct injection diesel engine numerical model. *Energy* **2016**, *115*, 658–667. [CrossRef]
25. Mrzljak, V.; Žarković, B.; Poljak, I. Fuel mass flow variation in direct injection diesel engine—Influence on the change of the main engine operating parameters. *Pomorstvo* **2017**, *31*, 119–127.
26. Theotokatos, G.; Stoumpos, S.; Lazakis, I.; Livanos, G. Numerical Study of a Marine Dual-Fuel Four-Stroke Engine. In Proceedings of the 3rd International Conference on Maritime Technology and Engineering, MARTECH 2016, Lisbon, Portugal, 4–6 July 2016; Volume 2, pp. 777–786.
27. Tadros, M.; Ventura, M.; Guedes Soares, C. Data Driven In-Cylinder Pressure Diagram Based Optimization Procedure. *J. Mar. Sci. Eng.* **2020**, *8*, 294. [CrossRef]
28. Mocerino, L.; Soares, C.G.; Rizzuto, E.; Balsamo, F.; Quaranta, F. Validation of an Emission Model for a Marine Diesel Engine with Data from Sea Operations. *J. Mar. Sci. Appl.* **2021**, *20*, 534–545. [CrossRef]
29. Senčić, T.; Mrzljak, V.; Blecich, P.; Bonefačić, I. 2D CFD Simulation of Water Injection Strategies in a Large Marine Engine. *J. Mar. Sci. Eng.* **2019**, *7*, 296. [CrossRef]
30. Altosole, M.; Martelli, M.; Vignolo, S. A mathematical model of the propeller pitch change mechanism for the marine propulsion control design. Sustainable Maritime Transportation and Exploitation of Sea Resources. In Proceedings of the 14th International Congress of the International Maritime Association of the Mediterranean, IMAM 2011, Genova, Italy, 13–16 September 2011; Volume 2, pp. 649–656.
31. Altosole, M.; Benvenuto, G.; Figari, M. Performance Prediction of a Planing Craft by Dynamic Numerical Simulation. In Proceedings of the 7th Symposium on High Speed Marine Vehicles, 2005 Conference, HSMV 2005, Napoli, Italy, 21–23 September 2005; Volume 2005, pp. 105–111.
32. Altosole, M.; Benvenuto, G.; Figari, M.; Campora, U. Dimensionless Numerical Approaches for the Performance Prediction of Marine Waterjet Propulsion Units. *Int. J. Rotating Mach.* **2012**, *2012*, 321306. [CrossRef]
33. Altosole, M.; Donnarumma, S.; Spagnolo, V.; Vignolo, S. Marine Cycloidal Propulsion Modelling for DP Applications. In Proceedings of the 7th International Conference on Computational Methods in Marine Engineering, MARINE 2017, Rome, Italy, 29–31 May 2017; Volume 2017, pp. 206–219.
34. Tavakoli, S.; Najafi, S.; Amini, E.; Dashtimansh, A. Ship acceleration motion under the action of a propulsion system: A combined empirical method for simulation and optimization. *J. Mar. Eng. Technol.* **2021**, *20*, 200–215. [CrossRef]
35. Sutulo, S.; Soares, C.G. Mathematical Models for Simulation of Manoeuvring Performance of Ships. In *Marine Technology and Engineering*; Taylor & Francis Group: London, UK, 2011; pp. 661–698.

36. Altosole, M.; Boote, D.; Brizzolara, S.; Viviani, M. Integration of numerical modeling and simulation techniques for the analysis of towing operations of cargo ships. *Int. Rev. Mech. Eng.* **2013**, *7*, 1236–1245.
37. Ircani, A.; Martelli, M.; Viviani, M.; Altosole, M.; Podenzana-Bonvino, C.; Grassi, D. A simulation approach for planing boats propulsion and manoeuvrability. *Trans. R. Inst. Nav. Archit. Part B Int. J. Small Cr. Technol.* **2016**, *158*(Part B1), B-27–B-42.
38. Piaggio, B.; Viviani, M.; Martelli, M.; Figari, M. Z-Drive Escort Tug manoeuvrability model and simulation. *Ocean Eng.* **2019**, *191*, 106461. [CrossRef]
39. Martelli, M.; Viviani, M.; Altosole, M.; Figari, M.; Vignolo, S. Numerical modelling of propulsion, control and ship motions in 6 degrees of freedom. *Proc. Inst. Mech. Eng. Part M J. Eng. Marit. Environ.* **2014**, *228*, 373–397. [CrossRef]
40. Donnarumma, S.; Martelli, M.; Vignolo, S. Numerical Models for Ship Dynamic Positioning. In Proceedings of the VI International Conference on Computational Methods in Marine, Rome, Italy, 15–17 June 2015; pp. 1078–1088.
41. Spanos, D.; Papanikolaou, A. International Benchmark Study on Numerical Simulation Methods for the Prediction of Parametric Rolling of Ships in Waves. NTUA-SDL Report, Rev. 4. 2009. Available online: http://old.naval.ntua.gr/sdl/sibs/intact/BENCHMARK_REPORT_NTUA-rev.1.0.pdf (accessed on 12 December 2021).
42. Shigunov, V.; El Moctar, O.; Papanikolaou, A.; Potthoff, R.; Liu, S. International benchmark study on numerical simulation methods for prediction of manoeuvrability of ships in waves. *Ocean Eng.* **2018**, *165*, 365–385. [CrossRef]
43. Ghaemi, M.H.; Zeraatgar, H. Analysis of hull, propeller and engine interactions in regular waves by a combination of experiment and simulation. *J. Mar. Sci. Technol.* **2020**, *26*, 257–272. [CrossRef]
44. Benvenuto, G.; Campora, U.; Carrera, G.; Figari, M. Simulation of Ship Propulsion Plant Dynamics in Rough Sea. In Proceedings of the 8th International Conference on Marine Engineering Systems (ICMES 2000), New York, NY, USA, 22–23 May 2000.
45. Saettone, S.; Tavakoli, S.; Taskar, B.; Jensen, M.V.; Pedersen, E.; Schramm, J.; Steen, S.; Andersen, P. The importance of the engine-propeller model accuracy on the performance prediction of a marine propulsion system in the presence of waves. *Appl. Ocean Res.* **2020**, *103*, 102320. [CrossRef]
46. Tavakoli, S.; Saettone, S.; Steen, S.; Andersen, P.; Schramm, J.; Pedersen, E. Modeling and analysis of performance and emissions of marine lean-burn natural gas engine propulsion in waves. *Appl. Energy* **2020**, *279*, 115904. [CrossRef]
47. Matusiak, J. *Dynamics of a Rigid Ship*; Aalto University Publication Series: Espoo, Finland, 2013.
48. Matusiak, J. On the non-linearities of ship's restoring and the Froude-Krylov wave load part. *Int. J. Nav. Archit. Ocean. Eng.* **2010**, *3*, 151–159.
49. Acanfora, M.; Rizzuto, E. Time domain predictions of inertial loads on a drifting ship in irregular beam waves. *Ocean Eng.* **2019**, *174*, 135–147. [CrossRef]
50. Acanfora, M.; Balsamo, F. The Smart Detection of Ship Severe Roll Motions and Decision-Making for Evasive Actions. *J. Mar. Sci. Eng.* **2020**, *8*, 415. [CrossRef]
51. Taimuri, G.; Matusiak, J.; Mikkola, T.; Kujala, P.; Hirdaris, S. A 6-DoF maneuvering model for the rapid estimation of hydrodynamic actions in deep and shallow waters. *Ocean Eng.* **2020**, *218*, 108103. [CrossRef]
52. Acanfora, M.; Montewka, J.; Hinz, T.; Matusiak, J. Towards realistic estimation of ship excessive motions in heavy weather. A case study of a containership in the Pacific Ocean. *Ocean Eng.* **2017**, *138*, 140–150. [CrossRef]
53. Benvenuto, G.; Campora, U.; Laviola, M.; Terlizzi, G. Simulation Model of a Dual-Fuel Four Stroke Engine for Low Emission Ship Propulsion Applications. *Int. Rev. Mech. Eng.* **2017**, *11*, 817–824. [CrossRef]
54. Faltinsen, O.M. *Sea Loads on Ships and Offshore Structures*; Cambridge University Press: Cambridge, UK, 1990; Volume 1.
55. Yu, J.-W.; Lee, C.-M.; Choi, J.-E.; Lee, I. Effect of ship motions on added resistance in regular head waves of KVLCC2. *Ocean Eng.* **2017**, *146*, 375–387. [CrossRef]
56. Kim, D.J.; Yun, K.; Park, J.-Y.; Yeo, D.J.; Kim, Y.G. Experimental investigation on turning characteristics of KVLCC2 tanker in regular waves. *Ocean Eng.* **2019**, *175*, 197–206. [CrossRef]
57. Smogeli, O.N. Control of Marine Propellers from Normal to Extreme Conditions. Ph.D. Thesis, Faculty of Engineering Science and Technology, Trondheim, Norway, 2006.
58. Matusiak, J.; Stigler, C. Ship Roll Motion in Irregular Waves During a Turning Circle Maneuver. In Proceedings of the 11th International Conference on Stability of Ships and Ocean Vehicles, Athens, Greece, 23–28 September 2012; pp. 291–298.
59. Barnitsas, M.M.; Ray, D.; Kinley, P. *K_t, K_q and Efficiency Curves for the Wageningen B-Series Propellers*; University of Michigan: Ann Arbor, MI, USA, 1981.
60. MAN. MAN 51/60 DF IMO TIER II/IMO TIER III. In *Project Guide—Marine*; MAN: Augsburg, Germany, 2015.

Journal of
Marine Science and Engineering

MDPI

Article

Marine Dual-Fuel Engines Power Smart Management by Hybrid Turbocharging Systems

Marco Altosole [1,*], Flavio Balsamo [1], Ugo Campora [2] and Luigia Mocerino [1]

[1] Department of Industrial Engineering (DII), School of Polytechnic and Basic Sciences, University of Naples "Federico II", 80125 Naples, Italy; flavio.balsamo@unina.it (F.B.); luigia.mocerino@unina.it (L.M.)
[2] Department of Mechanical, Energy, Management and Transportation Engineering (DIME), Polytechnic School University of Genoa, 16145 Genoa, Italy; ugo.campora@unige.it
* Correspondence: marco.altosole@unina.it

Abstract: The performance of a marine dual-fuel engine, equipped with an innovative hybrid turbocharger producing electric power to satisfy part of the ship's electric load, is presented by a simulation comparison with the traditional turbocharging technology. The two distinct fuel types, combined with the hybrid turbocharger, involve a substantial change in the engine control modes, resulting in more flexible and efficient power management. Therefore, the investigation requires a numerical analysis depending on the engine load variation, in both fuelling modes, to highlight different behaviours. In detail, a dual-fuel engine simulation model is validated for a particular application in order to perform a complete comparison, reported in tabular and graphical form, between the two examined turbocharging solutions. The simulation analysis is presented in terms of the engine working data and overall energy conversion efficiency.

Keywords: marine engine; dual fuel; hybrid turbocharger; power management; efficiency; simulation

Citation: Altosole, M.; Balsamo, F.; Campora, U.; Mocerino, L. Marine Dual-Fuel Engines Power Smart Management by Hybrid Turbocharging Systems. *J. Mar. Sci. Eng.* **2021**, *9*, 663. https://doi.org/10.3390/jmse9060663

Academic Editor: Tie Li

Received: 15 May 2021
Accepted: 11 June 2021
Published: 15 June 2021

1. Introduction

Due to the need to mitigate the harmful impact deriving from marine engines powered by fossil fuels, the regulations implemented over the years by the International Maritime Organisation (IMO) have set a challenging horizon, which requires new solutions to minimise the environmental footprint of the maritime sector [1]. In particular, IMO rules are increasingly restrictive about the ship's polluting emissions of nitrogen and sulphur oxides (IMO Tier II and III, and ECA world areas [2]), while the energy efficiency design index (EEDI) has become mandatory for carbon dioxide [3,4]. This framework needs new measures to improve the energy conversion efficiency of propulsion engines and diesel–electric generators.

Currently, marine diesel engines have an efficiency of close to 50% but can be globally more efficient when combined with waste heat recovery (WHR) from their exhaust gas [5–10]. WHR steam plants allow achieving an overall efficiency improvement of $3 \div 5\%$, which corresponds to about the same percentage of carbon dioxide emission reduction. A further improvement in the ship's energy efficiency can be achieved by installing WHR systems for dual-fuel (DF) marine engines [11–13]. It should be noted that natural gas (NG), in comparison with the traditional heavy fuel oil (HFO), reduces significantly emissions from the engine, as well as being $1 \div 2\%$ more efficient at medium-high engine loads [14].

Mitsubishi Heavy Industries has recently developed an alternative method of exploitation regarding the engine's exhaust gas thermal power, based on the hybrid turbocharger (HTC) technology [15,16]. The compressor turbine's shaft line of the engine turbocharger integrates the high-speed electric motor/generator (EM/G). Thus, when the EM/G operates as an electric generator, it produces electricity to satisfy part of the ship's electric load, reducing the diesel generators' electric power, with a positive effect on fuel consumption and emissions. As far as the authors know, it is the only important application currently existing in the ship propulsion field. On the contrary, Formula One cars have been using HTC since 2014, and there are good application prospects for hybrid electric passenger cars [17,18] and fuel cell vehicles [19]. Scientific studies on marine applications are also rather limited: the study in [20] considers the HTC use for high-speed diesel engines aboard fast naval combatants, while in [21], the authors compare the simulation of a four-stroke DF marine engine equipped with its original turbocharger (TC) and, alternatively, with the HTC. The comparison between the two turbochargers demonstrated the validity of the HTC solution to increase the overall efficiency of the energy conversion system in the NG fuel mode. Obviously, this innovative turbocharging configuration requires controlling some important operational data, paying due attention especially to the different combustion control methods for the two fuel modes. Therefore, this second article extends the analysis of the two turbocharging solutions (TC and HTC) at different loads and speeds of the engine, including the HFO mode, for optimal management of the engine power in all its possible operating conditions. The main idea would be to provide, by numerical simulation, a series of operational data in the widest possible working range of the engine, for the evaluation of new smart ship propulsion solutions to complement the existing applications or studies [22–24]. In fact, the proposed innovation combines the advantages of thermal energy recovery with a new concept of hybrid power generation, differentiating itself from the traditional meaning of hybrid propulsive applications [22,25–28]. To this end, simulation results, concerning a MAN 51-60 DF 12V marine engine (Germany), are discussed after a brief description of a DF engine's working principles and numerical modelling.

2. Materials and Methods

2.1. DF Engines

A marine DF engine is substantially derived from a compression-ignition (CI) engine by adding the capability to operate as a positive-ignition (PI) engine burning a gaseous fuel. The two working modes are completely different. In the diesel mode, the fuel is injected into the cylinder in a liquid phase and it ignites by compression; the air–fuel ratio (AFR) varies with the load. In NG mode, the fuel is injected and mixed with the air before the inlet valve and the ignition occurs as in a spark-ignition engine. In DF engines, the spark plug is replaced by a small liquid fuel injection, whose combustion is promoted by compression. The natural gas, depending on methane contents, has good resistance to autoignition, particularly when the mixture is lean; therefore, the engine compression ratio can be as high as in CI engines. The pilot fuel flame behaves as multiple sources of high-temperature points that move in the combustion chamber and fire up gaseous mixture burning. The value of air–fuel ratio (AFR), to avoid knocking and to achieve a stable combustion, varies between 1.7 and 2.1, depending on brake mean effective pressure (b.m.e.p.) and methane number; hence, in NG mode, a more refined and precise AFR regulation is required, mainly during transients. The airflow towards the cylinders can be controlled by acting on the pressure in the air inlet manifold; a fuel controller determines the right amount of gaseous fuel to be injected according to the value of the air receiver pressure, which becomes the main parameter to manage the NG-powered engine load. The turbocharger system must be designed to supply the proper airflow to the engine at the different loads, and this requires accurate control of the energy transferred from the exhaust gas to the charge air. Due to the increased efficiency of current turbochargers for marine engines, at higher loads, only a part of the exhaust flow is needed to guarantee a stable operation of the compressor and to avoid a dangerous increase of the boost pressure. A waste gate valve (WGV) on the exhaust line can control the boost pressure, bypassing the turbine and therefore reducing the energy to the compressor. The WGV system can be overcome by a variable geometry turbine that allows the pressure in the exhaust to be controlled manifold. In fact, the turbine cross-sectional area variation corresponds to a change in the turbine aspect ratio (AR) and, consequently, in the turbine performance map. There are different technical solutions to obtain such AR variation, including variable nozzle turbine, variable volute turbine, variable sliding ring turbine. In this study, a variable nozzle turbine area (VTNA) is considered. Although the peak efficiency of an AR variable turbine is lower than a corresponding fixed aspect ratio one, the overall turbocharger system efficiency is generally higher [29]. A further system to control pressure in the air inlet manifold is the adoption of a bleed valve (BV) at the intercooler outlet to discharge outside part of the airflow at the compressor outlet. Blowing off part of the air allows a faster and more precise control of the pressure in the air receiver, suitable for application requiring a constant set value of AFR during the transient. From the energetic point of view, blow-off operation, as in the case of WGV or VTNA, dissipates part of the compressor work without the possibility of further recovery.

2.2. *Hybrid Turbochargers*

Since the mid-1980s, the non-negligible energy content of the engine exhaust gases has been used to produce mechanical or electrical energy by means of auxiliary turbines [30]. More recently, taking advantage of the improved efficiency of marine turbochargers, a parallel configuration has also been adopted by deriving a portion of the exhaust gas flow (generally about 10%) to the auxiliary turbine. In this context, Mitsubishi Heavy Industries has developed an HTC, i.e., a turbocharging system, integrating an electric motor/generator (EM/G) [15]. The shaft of the turbogenerator is equipped with EM/G in a very compact installation so that a unique turbine drives the compressor and the electric motor. Installing an HTC gives the possibility to control the turbocharging system dynamics (i.e., turbocharger speed) and the compressor work to avoid BV use. Thus, the HTC generator mode can favourably convert the energy, otherwise dissipated by the blow-off process, into electric power. Alternatively, the HTC electrical machine can operate as a motor to support the compressor during transients at low loads, when the exhaust gas energy from the engine cylinders decreases considerably. Due to the very fast dynamics of electrical machines, this solution can improve greatly the overall system dynamics, particularly useful in gensets' application, where sudden and large load variations can be common. Adding a VTNA gives the turbocharging system two degrees of freedom to optimise the working point for each load condition. The electric machine should be designed compactly enough to be mounted on the turbocharger shaft (usually on the compressor silencer side [15] or in the TC shaft [20]) and to withstand the very high rotational speeds (12 ÷ 20 krpm). In addition, the electrical design has to take into account the problems related to the high-frequency currents and magnetic flux variations. Finally, the HTC speed depends on the engine load condition; therefore, a bidirectional AC/AC converter is required to adapt current frequency and voltage to the ship's electric power grid [21].

3. Simulation Case Study

The present analysis shows the behaviour of a marine DF engine in the TC/HTC alternative configurations, in both fuel modes. The examined engine is a four-stroke MAN 51-60 DF 12V [31], whose main characteristics are listed in Table 1 [21].

In the original version (i.e., TC equipment), the engine uses a BV to control the pressure in the air inlet manifold, determining the airflow rate to the cylinders. On the contrary, in the HTC configuration, illustrated in Figure 1, the BV is no longer essential for controlling the manifold air pressure, as previously stated. The compressor and turbine characteristics for the numerical modelling are assumed identical in the two different configurations.

Table 1. Engine dimensions and design data [21].

MAN 51-60 DF 12V	HFO/NG
engine length (mm)	10,254
height (mm)	5517
width (mm)	4713
dry weight (t)	189
cylinders number	12V
bore (mm)	510
stroke (mm)	600
fuel type	HFO/NG
brake power (kW)	12,000
speed (rpm)	514
b.m.e.p. (bar)	19.1
s.f.c. (g/kWh)	189/157
charge air pressure (barg)	4.29/3.77

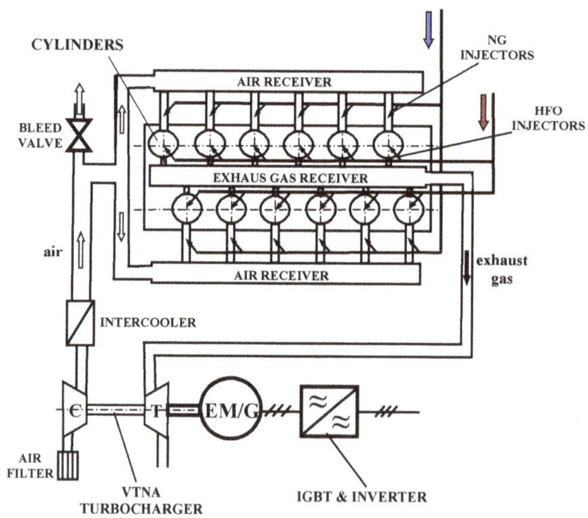

Figure 1. Engine equipment scheme in HTC configuration.

3.1. Numerical Modelling

The HTC behaviour in the two possible operating modes is assessed through a complete thermodynamic simulation model of the engine, based on previous work [32]. Later, the authors updated the code by adding the HTC modelling [21] to show a performance comparison between the two different turbocharging systems, although limited to the NG fuel mode. As a next step, the present study aims to extend the comparison analysis to include the case of the HFO-powered engine in order to have a valid overview of the influence of the different working conditions and related combustion control methods on the behaviour of the engine in the two distinct turbocharging configurations. There is a large scientific literature on modelling of marine diesel engines, while simulation studies on DF engines are still limited. The models can be developed through specific programming languages [32,33] or commercial software [34,35], according to the different purposes [36–39]. The present DF engine simulator is arranged in a modular form, by using MATLAB-Simulink software. It is shown in Figure 2, where each engine module is represented [21].

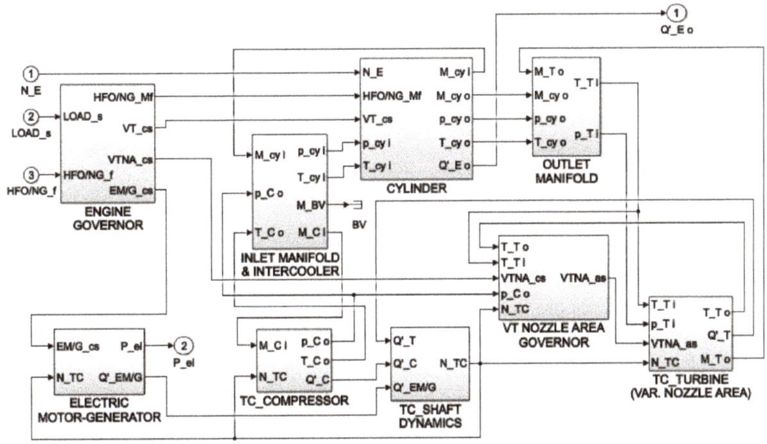

Figure 2. HTC engine simulator in Simulink environment [21].

The simulation follows a filling and emptying approach, in which each block provides features and performance of the pertinent engine components by means of algebraic, differential equations, and steady-state performance maps. The ideal gas equation is used to model the working fluid, while the specific internal energy and enthalpy are evaluated according to the temperature and fluid composition. The engine simulator requires the following main data: cylinders number and pertinent geometric characteristics; intake and exhaust valve data with opening/closing timing logics; compressor and turbine performance maps; VTNA variation; BV control details and ambient conditions.

As shown in Figure 2, the engine working conditions are represented by the rotational speed (N_E), the load signal (LOAD_s), and fuel type (HFO/NG_f). The actual engine cycle is calculated by the 'cylinder' module, in which the thermodynamic processes are modelled through a single zone approach, depending on the crank angle value. The cylinder inlet and exhaust mass flow rates through the poppet valves are given by the gas-dynamic equations adopted for the compressible fluid through a flow restriction (both subsonic and choked flow conditions are considered).

The combustion heat release fraction, depending on the crank angle and proportional to the fuel mass burned, is calculated through the double Wiebe equation, while the in-cylinder pressure variation is determined by integrating, at each crank angle step, the energy differential equation of a closed system.

However, the authors have already provided a broad explanation of the numerical model in [21,32]; therefore, this article prefers to delve into the main aspects of combustion control and power management, depending on the specific operating modes of the engine. The engine operates in the Miller cycle mode, being equipped with variable timing of the opening and closing phases of the intake valve. In this regard, different settings depend on fuel type (HFO or NG) and engine load, managed by the 'engine governor' module (Figure 2) giving the valves timing control signal (VT_cs) as output.

To optimise the engine performance under all operating conditions, the turbine of the turbocharger is characterised by a variable nozzle area (VTNA). This could represent a problem in the case of DF engines, because of the significant difference between the fuels' lower heating value (42,700 kJ/kg for HFO, 49,000 kJ/kg for NG), the substantial difference in the combustion processes, and in the cylinder air–fuel ratios. For these reasons, the optimal turbocharger performance is different depending on the fuel used.

According to manufacturer indications [40], the VTNA is adopted only when using NG as fuel, while maximum turbine nozzle area is selected for the engine powered by HFO, under all operating conditions. Indeed, the exhaust gas flow rate is significantly higher over the entire operating range of the engine in HFO mode; therefore, the turbine in fully open condition can exploit all the exhaust thermal power. On the contrary, the gas mass flow rate is reduced by about 20% in NG mode [31], and therefore, it requires an appropriate adjustment of the turbine nozzle area, as the engine load decreases. The simulation of the VTNA turbine is based on the typical steady-state turbine performance maps [33]. As shown in Figure 3, reduced mass flow rate ($M\sqrt{T_{oi}}/p_{oi}$) and efficiency (η_T) are functions of the expansion ratio. The VTNA is described by different curves, depending on the variable nozzle turbine actuator signal (VTNA_as), which is an output of the 'VT nozzle area governor' block in Figure 2. The influence of this parameter is evident in the generic turbine performance curves reported in Figure 3.

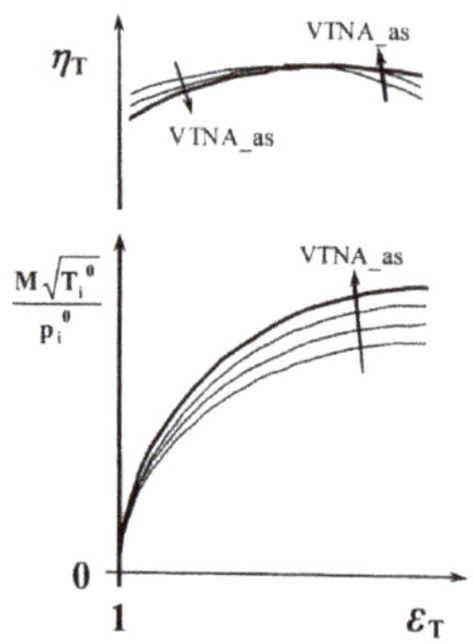

Figure 3. Typical VTNA turbine performance map.

The TC turbine variable geometry nozzle ring is controlled, according to the logic indicated by the manufacturer [40], taking into account inlet and outlet turbine temperature (T_Ti and T_To), VTNA control signal (VTNA_cs) (depending in turn on the used fuel), compressor outlet pressure (p_Co), and turbocharger shaft speed (N_TC).

3.2. Engine Load Control

In the original version of the considered DF engine [31], equipped with a traditional TC, the engine load, proportional to the delivered torque, is controlled in different ways, depending on the fuel type used. When the engine is fuelled by HFO, the reference thermodynamic cycle is the diesel one; therefore, the classical approach of graduating the fuel mass injected into the cylinders is adopted. The engine load control considers also the BV opening, according to a logic depending on the engine speed and the fuel mass injected into the cylinders, as reported in [31]. In NG mode, the engine torque is controlled only by the BV, since the NG injection system keeps an almost constant value for the air–fuel ratio (about twice the stoichiometric value), in accordance with the Otto cycle. Similar to the HFO mode, an increase of the BV opening determines an engine torque decrease and vice versa. The amount of the bled air at the different operating points represents the capability of the control system to increase more or less quickly the boost pressure, depending on the manifold volume; when the BV is completely closed, there is the maximum increase in the inlet manifold pressure. In the engine simulator, the BV is modelled in the same way as a cylinder poppet valve [32]. On the contrary, in HTC engine configuration, the EM/G governor can vary its delivered electric power to control the turbocharger shaft speed [21], while the BV remains always closed. In fact, the HTC speed is obtained hrough he shaft dynamics equation as follows:

$$d\omega/dt = (1/J)\cdot(Q'_T - Q'_C - Q'_{EM/G}) \quad (1)$$

where ω is the HTC shaft angular speed, J the rotor inertia, Q'_T, Q'_C, and $Q'_{EM/G}$, respectively, the torque of HTC turbine, compressor, and electric motor/generator. Equation (1) shows that the HTC speed can depend on the EM/G torque ($Q'_{EM\text{-}G}$), which can be used to control the turbocharger speed in both fuelling modes. In addition, the electric power generated by the EM/G is available to satisfy part of the ship's electric load. The engine dynamic performance can also be improved by allowing the electrical machine to operate as a motor during the load transients. In the Simulink model, the 'engine governor' block is able to manage the engine load in all the possible engine control configurations (use of BV or HTC EM/G) and working conditions (i.e., HFO or NG fuel, constant or variable speed load). In particular, in the HTC engine configuration, the 'engine governor' block is scheduled to maintain the same working conditions of cylinders and turbocharging turbine considered in the case of the original TC, corresponding at the same engine load and speed. This is possible through its output EM/G control signal ('EM/G_cs' in Figure 2), which controls the EM/G delivered electric power ('P_el', output of the 'electric motor-generator' block). The 'P_el' parameter has an influence on the $Q'_{EM/G}$ value, therefore on the HTC angular velocity (ω), according to Equation (1).

4. Results and Discussion

In this section, the simulator results obtained for the TC engine configuration are compared with the engine data sheet [31], referred to as steady-state working conditions between 25% and 100% of the load, both at constant (514 rpm) and at variable speed.

Table 2 shows the results of this comparison as percentage errors, for both possible fuels (HFO or NG). The table (where C_o and T_o represent the compressor and turbine outlet sections) shows good accuracy of the engine simulator results, particularly at medium-high engine loads, for both types of used fuels. No significant differences between constant and variable speed running conditions are detected.

Table 2. Errors (%) comparison between reference data and simulation.

Delivered Engine Power	Results at Constant Engine Speed (514 rpm): Without Brackets Results at Variable Engine Speed: In Brackets [ ]									
% (rpm)	s.f.c		C_o Air Mass Flow		Charge Air pr.		C_o Temp.		T_o Temp.	
	HFO	NG	HFO	NG	HFO	NG	HFO	NG	HFO	NG
100 (514 rpm)	0.21 [0.17]	0.49 [0.36]	0.06 [−0.11]	0.32 [0.21]	0.12 [0.08]	−0.23 [0.07]	1.37 [−0.56]	0.46 [0.18]	0.00 [−0.32]	1.05 [0.82]
85 (514 rpm)	−0.79 [−0.65]	−0.57 [−0.46]	−2.18 [1.82]	0.78 [0.85]	−0.48 [−0.36]	1.02 [0.74]	−1.03 [1.08]	0.01 [0.02]	1.23 [1.15]	1.36 [0.93]
75 (501 rpm)	0.29 [−0.31]	−0.45 [−0.48]	−0.97 [−0.45]	−1.52 [−1.08]	−1.97 [−2.04]	2.21 [1.51]	−1.23 [1.74]	0.00 [0.02]	0.87 [−1.06]	1.39 [1.78]
50 (442 rpm)	−0.19 [0.43]	0.24 (0.71]	0.31 [−0.19]	−0.65 [−1.02]	−1.21 [−0.84]	1.52 [1.03]	−1.42 [1.05]	1.38 [1.53]	−2.11 [−1.73]	2.42 [1.67]
25 (402 rpm)	−1.01 [−0.67]	−0.34 [−0.43]	–	–	–	–	–	–	–	–

As regards the engine efficiency (η_E), the simulation results reported in Figure 4a (HFO-powered TC engine) and Figure 4b (NG-powered TC engine) depend on engine power, fuel mass flow rate, and fuel lower heating value. The data are normalised with respect to the TC engine configuration, working at the maximum continuous rating (MCR) of the HFO-powered engine. The figure shows for both fuel types a very good agreement between calculated and reference data, in every operating condition (constant/variable speed). The NG mode is more efficient than HFO at high engine loads; on the contrary, the engine efficiency is greater in HFO mode at very low loads and at variable speed.

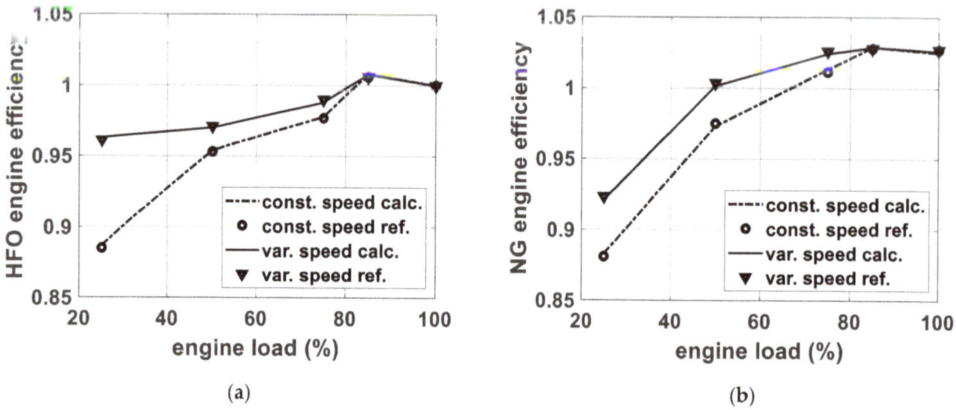

Figure 4. Comparison between calculated and reference efficiencies (normalised with respect to the HFO-powered TC engine) in diesel fuel mode (**a**) or gas mode (**b**).

In order to investigate the influence of some important parameters on engine power management, the VTNA device performance is assessed in Figure 5. In addition to the good agreement between simulation and reference data [40], in both conditions of VTNA fully opened (HFO mode) and for different nozzle areas (i.e., variable VTNA, NG-powered engine), the best efficiency is provided at 70% of the engine load.

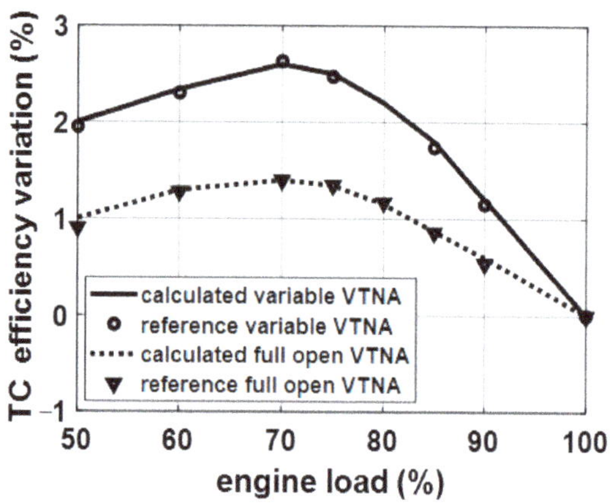

Figure 5. VTNA turbocharger efficiency comparison between simulation and reference data.

The good accuracy of the overall simulation results makes the simulator suitable for the examined case study.

4.1. Turbochargers' Influence on Engine Performance

The simulation model is mainly used to show the engine performance differences due to the use of the HTC as an alternative to the original TC. The comparison is carried out in both HFO and NG modes, for different engine loads at constant and variable speed working points. In detail, Tables 3 and 4 and refer to the engine loads equal to 100%, 85%, 75%, 50%, and 25% of MCR.

Table 3. TC and HTC comparison for different fuels and engine working points at constant speed.

Line	Engine Parameters	Engine Loads and Fuel Type									
1	Engine power (%)	100		85		75		50		25	
2	Engine power (kW)	12,000		10,260		9000		6000		3000	
3	Engine speed (rpm)	514		514		514		514		514	
4	Fuel type	HFO	NG	HFO	NG	HFO	NG	HFO	NG	HFO	NG
5	EM/G electric power (kW)	526	498	426	551	320	671	203	562	87	351
6	EM/G electric power (%)	4.38	4.15	4.15	5.37	3.55	7.45	3.38	9.37	2.91	11.70
7	Turbocharger speed (Δ%)	−5.29	−4.86	−5.81	−7.51	−4.88	−10.97	−11.31	−28.41	−10.52	−23.68
8	Air compression ratio (Δ%)	−9.73	−7.70	−5.72	−10.95	−3.64	−15.15	−5.34	−24.79	−1.37	−17.69
9	Compressor temperature (Δ%)	−3.26	−3.85	−2.84	−4.44	−2.09	−5.46	−3.13	−6.52	−1.43	−6.89
10	Compressor air flow rate (Δ%)	−8.19	−7.70	−10.53	−10.31	−10.03	−14.47	−15.23	−22.70	−14.68	−16.95
11	Compressor efficiency (Δ%)	2.09	−0.12	1.65	0.65	1.86	1.60	1.21	0.97	2.68	1.06
12	Cylinder air flow rate (Δ%)	0.06	0.01	−0.18	−0.01	−0.31	0.01	0.03	0.01	0.08	0.01
13	Cylinder air temperature (Δ%)	−0.29	−0.51	−0.25	−0.68	−0.23	−0.61	−0.06	−0.29	−0.07	−0.79
14	Cylinder exhaust gas flow (Δ%)	0.06	0.01	−0.18	0.02	−0.32	0.01	0.04	0.02	0.08	0.02
15	Turbine expansion ratio (Δ%)	−0.03	0.22	−0.51	1.17	−0.24	0.58	−0.17	2.62	−0.15	5.42
16	Turbine inlet pressure (Δ%)	−0.03	0.21	−0.18	0.39	−0.28	0.61	−0.22	0.25	0.07	0.32
17	Turbine inlet temperature (Δ%)	−0.11	−0.12	0.10	−0.17	0.18	−0.34	−0.07	−0.21	−0.06	−0.58
18	Turbine outlet temperature (Δ%)	−0.10	−0.11	0.13	−0.14	−0.21	−0.27	−0.01	−0.10	−0.05	−0.24
19	Turbine efficiency (Δ%)	0.00	0.00	0.00	0.01	0.01	0.01	−0.01	−0.01	−0.01	0.01
20	Engine i.m.e.p. (Δ%)	0.02	−0.02	−0.01	−0.02	−0.01	−0.03	0.03	−0.01	0.00	−0.02

Table 4. TC and HTC comparison for different fuels and engine working points at variable speeds.

Line	Engine Parameters	Engine Loads and Fuel Type									
1	Engine power (%)	100		85		75		50		25	
2	Engine power (kW)	12,000		10,260		9000		6000		3000	
3	Engine speed (rpm)	514		514		501		462		402	
4	Fuel type	HFO	NG	HFO	NG	HFO	NG	HFO	NG	HFO	NG
5	EM/G electric power (kW)	526	498	426	551	293	571	98	576	27	226
6	EM/G electric power (%)	4.38	4.15	4.15	5.37	3.25	6.34	1.63	9.60	0.90	7.53
7	Turbocharger speed (Δ%)	−5.29	−4.86	−5.81	−7.51	−3.59	−11.85	−22.21	−29.12	−17.15	−21.36
8	Air compression ratio (Δ%)	−9.73	−8.19	−5.72	−10.95	−3.84	−14.11	−3.17	−21.69	−1.15	−18.68
9	Compressor temperature (Δ%)	−3.26	−3.85	−2.84	−4.44	−1.92	−5.46	−1.23	−6.52	−0.61	−6.89
10	Compressor airflow rate (Δ%)	−8.19	−7.70	−10.53	−10.31	−7.41	−13.27	−7.34	−20.66	−6.89	−17.95
11	Compressor efficiency (Δ%)	2.09	0.12	1.65	0.65	2.21	1.71	2.44	1.29	1.09	2.02
12	Cylinder airflow rate (Δ%)	0.06	0.01	−0.18	0.02	0.24	0.01	−019	−0.01	−0.23	0.01
13	Cylinder air temperature (Δ%)	−0.29	−0.55	−0.25	−0.63	−0.19	−0.59	−0.14	−0.43	−0.02	−0.66
14	Cylinder exhaust gas flow (Δ%)	0.06	0.01	−0.18	−0.01	0.23	0.01	−0.20	−0.01	−0.23	0.02
15	Turbine expansion ratio (Δ%)	−0.03	0.21	−0.51	1.18	−0.49	0.58	−0.22	2.43	0.00	3.29
16	Turbine inlet pressure (Δ%)	−0.03	0.22	−0.18	0.16	−0.04	0.53	−0.17	0.34	−0.08	0.27
17	Turbine inlet temperature (Δ%)	−0.11	−0.14	0.10	−0.16	−0.23	−0.39	0.08	−0.26	0.12	−0.45
18	Turbine outlet temperature (Δ%)	−0.10	−0.11	0.13	−0.14	−0.19	0.26	0.11	0.15	0−09	−0.33
19	Turbine efficiency (Δ%)	0.00	0.00	0.00	−0.01	0.00	0.00	0.01	0.01	−0.01	0.01
20	Engine i.m.e.p. (Δ%)	0.02	−0.01	−0.01	−0.03	−0.04	−0.01	−0.02	−0.02	−0.03	−0.02

The electric power generated by the HTC is also reported in Table 3 as a percentage of the mechanical engine power (line 6) and shows the significant power rate recovered by the HTC, for both fuels, especially at high loads (i.e., 100% and 85%).

At the engine medium loads (i.e., 75% and 50%), the recovered power increases in gas mode, while it reduces progressively in the HFO-powered engine condition. At a very low engine load (25%), the EM/G power decreases but remains quite substantial in the case of NG fuel, while it drops considerably for HFO. The different values of the EM/G power between the two fuels are mainly due to the following:

- A greater NG lower heating value, compared to that of HFO (49,000 kJ/kg for NG; 42,700 kJ/kg for HFO);
- A substantial difference of the combustion processes;
- A different equivalence ratio in the cylinders (this last quantity is defined by dividing the AFR value by the stoichiometric ratio).

As regards the latter, Figure 6 reports the equivalence ratio values versus the engine load, at constant and variable speed conditions, for both fuels.

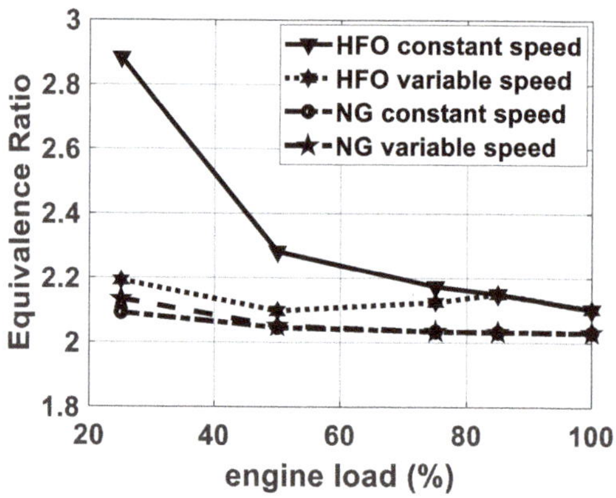

Figure 6. Cylinder equivalence ratio for different fuel types and engine speed conditions.

The equivalent ratio, in the case of constant engine speed and HFO mode, strongly decreases when the engine load increases, while it remains almost constant in the other operating conditions.

The quantities in Table 3, from line 7 to line 20, are shown in a form expressed through the following generic equation:

$$\Delta x\% = (x_{HTC} - x_{BV}) \cdot 100 / x_{BV} \quad (2)$$

where x is the generic variable, and the subscripts HTC and BV indicate, respectively, the engine adopting the HTC and the conventional TC using BV. The HTC use, in comparison with the original TC, involves not negligible and sometimes high variations of the parameters, considered from lines 7 to 11 as the TC speed and the compressor behaviour. On the other hand, the cylinder and the turbine parameters (lines 12–20), as well as the engine indicated mean effective pressure (i.m.e.p.), show negligible variations in these parameters consequent to the HTC adoption. This is mainly due to the abovementioned 'engine governor' setting of the simulator. Table 4 reports the same simulation analysis, although referring to the engine working at variable speeds. A similar behaviour is noticed between the two engine speed conditions.

Figure 7 reports the compressor working points for the HFO-powered engine, according to a normalised representation.

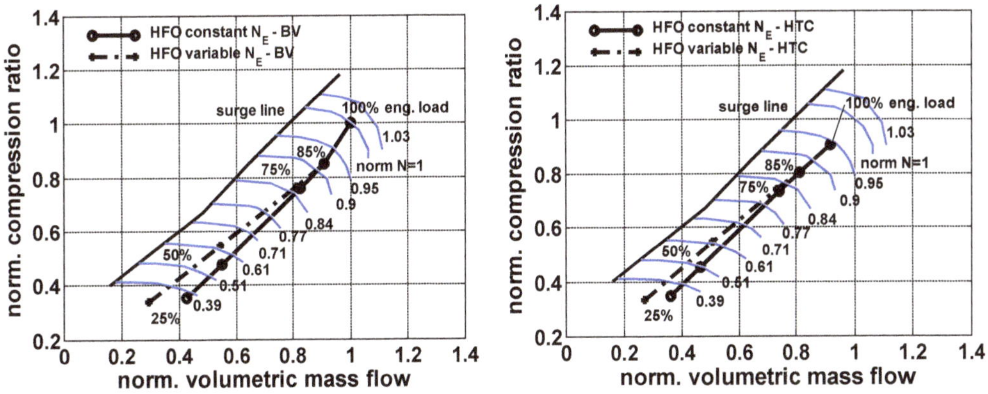

(**a**) (**b**)

Figure 7. Working points in the compressor map for the HFO-powered engine, with original TC **(a)** and HTC **(b)**, at constant and variable engine speeds N_E.

In particular, Figure 7a, pertaining to the TC engine configuration, shows the working conditions in the compressor map, for different engine loads (100%, 85%, 75%, 50%, and 25% of the MCR power) at both constant and variable engine speed settings. The comparisons with Figure 7b, concerning the HTC compressor map, show a reduction in the compression ratio, volumetric mass flow, and TC speed in all the engine load conditions. The main reason for these differences is that the whole air mass flow rate is introduced into the HTC engine cylinders, while in the case of the original TC, a part of the airflow is discharged into the environment by the bleed valve (see Figure 1).

The compressor outlet pressure is higher in the TC application, while by using HTC (at the same engine power and speed), the conditions of the flow in the intake manifold (i.e., pressure, temperature) are about the same in the two cases. In fact, with the BV partially opened to control the TC engine, the air pressure in the intake manifold becomes lower than the compressor outlet pressure. This does not occur for the HTC, because in this case, the BV is always closed and the pressure in the intake manifold is practically equal to that of the turbocharger compressor outlet (except for a small pressure loss due to friction). In the engine simulation model, the BV effect on the inlet manifold flow conditions is assessed by the mass and energy dynamic equations [33].

Figure 8a shows again the compressor working points corresponding to the same engine loads and speeds of Figure 7a but for the NG-powered TC engine. For every engine load condition, the comparison shows a reduction in the compression ratio, volumetric mass flow, and TC speed. This difference between the two fuels is mainly due to the same reasons regarding the power recovered by HTC. The comparison between Figure 8a,b, with the latter pertaining to the engine equipped with HTC, shows that similar to the HFO mode and for the same reasons, the HTC for the NG-powered engine also involves a reduction in the compressor compression ratio, volumetric mass flow, and TC speed.

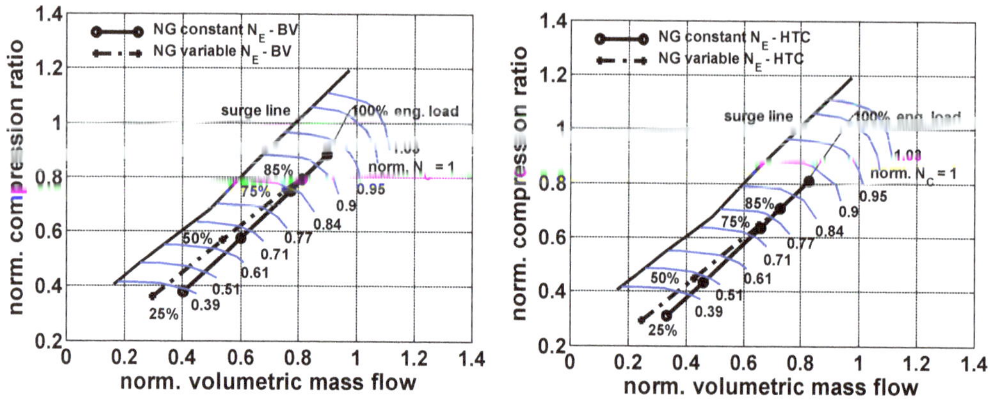

(a) (b)

Figure 8. Working points in the compressor map for the NG-powered engine, equipped with original TC (**a**) and HTC (**b**), at constant and variable engine speeds N_E.

4.2. Turbochargers' Influence on the Engine Overall Efficiency

Figure 9 reports the normalised engine efficiency versus engine load and working conditions (constant/variable speed) in HFO (Figure 9a) and NG (Figure 9b) engine modes.

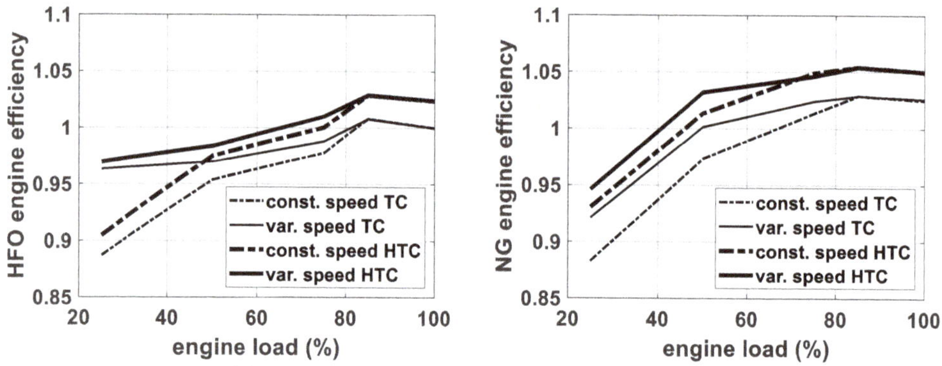

(a) (b)

Figure 9. Normalised efficiency of the engine powered by HFO (**a**) and NG (**b**) for TC/HTC engine conditions (constant/variable speed).

The HTC engine efficiency is defined as:

$$\eta E\ HTC\% = ((PE + Pel\ EM/G) \cdot 100)/(Mf\ FLHV) \quad (3)$$

where $P_{el\ EM/G}$ is the electric power delivered by the electric motor/generator.

Both Figure 9a,b show that the use of the HTC increases the overall engine efficiency by about 2.5% at high loads (85–100% of the MCR power), in both constant and variable engine speed conditions.

For the HFO-powered engine, Figure 9a shows also that the difference between the efficiency of the HTC engine and TC engine remains about constant; a slight reduction is mainly observed at minimum load (25% of the MCR power) when the engine is running at a constant speed. The reduction in the efficiency difference between TC and HTC is more evident at the variable speed condition, especially at 25% of the engine load; this is due to the strong reduction in the HTC EM/G electric power produced at the low loads, as reported in line 5 of Table 4.

Figure 9b, relating to the NG engine mode, shows the efficiency advantage already observed for all engine loads (25 ÷ 100% of the MCR power), with an increase of up to about 3% at 50% engine load, at variable speed. At constant speed (dash and dot lines in Figure 9b), the efficiency advantage of the HTC engine increases progressively as the engine load decreases, to almost 5% at 25% of the engine load.

This different behaviour of the engine powered by HFO or NG is due to several factors as the VTNA turbocharger for the NG-powered engine (see Figure 5), the different combustion modes, and reference thermodynamic cycles of the two fuels (i.e., Diesel cycle for HFO, Otto for NG). Moreover, the two fuels require different cylinder equivalence ratio values, depending on the engine load (see Figure 6).

5. Conclusions

The present simulation analysis aims to investigate the behaviour of a marine DF engine for distinct turbocharging configurations and fuel modes. The particular engine load control, due to the use of an innovative hybrid turbocharger, combined with a proper combustion control method (depending on whether gas or diesel oil is used), can lead to significant advantages in the system's overall efficiency and hence should be considered for better power management in ships. To this end, an engine simulator has allowed testing control devices and performance for the examined application. The comparative analysis of the simulation results represents the main purpose of the work, as the shown numerical tables and graphs collect useful information for efficient use of the innovative hybrid system. In this regard, although the HTC use, in comparison with the traditional turbocharger, leads to significant variations of the compressor operating conditions, no relevant changes in the engine hot components (cylinder and TC turbine), and consequently, in the efficiency of the thermodynamic cycle, are present. Again, the mass flow rate of the exhaust gases and their temperature downstream of the turbocharging turbine do not depend on the type of turbocharger used (original TC or HTC). This last aspect could be important because, for example, the turbocharger configuration would not affect the possible installation of WHR steam plants to improve the ship's energy efficiency further.

Finally, with a view to the development of adequate smart control strategies, it is useful to notice a more significant increase in the overall efficiency of the NG-powered HTC engine, especially at low and medium loads.

Author Contributions: Conceptualization, M.A., U.C., F.B. and L.M.; methodology, M.A., U.C., F.B. and L.M.; software, U.C.; validation, U.C.; investigation, M.A., U.C., F.B. and L.M.; writing—original draft preparation, M.A. and U.C.; writing—review and editing, M.A., F.B. and L.M.; supervision, M.A. All authors have read and agreed to the published version of the manuscript.

Funding: This research received no external funding.

Institutional Review Board Statement: Not applicable.

Informed Consent Statement: Not applicable.

Data Availability Statement: Not applicable.

Acknowledgments: The authors are grateful to MAN Diesel & Turbo and Andrea Rossetto by Ranieri Tonissi S.p.A. Genova (Italy) for their precious cooperation.

Conflicts of Interest: The authors declare no conflict of interest.

Abbreviations

Nomenclature

AFR	Air–fuel ratio
AR	Aspect ratio
b.m.e.p.	Brake mean effective pressure
BV	Bleed valve
CI	Compression ignition
DF	Dual fuel
ECA	Emission control areas
EEDI	Energy efficiency design index
EM/G	Electric motor/generator
FLHV	Fuel lower heating value
HFO	Heavy fuel oil
HTC	Hybrid turbocharger
i.m.e.p	Indicated mean effective pressure
IMO	International maritime organisation
J	Rotor inertia
m or M	Mass
MCR	Engine Maximum Continuous Rating
NG	Natural gas
P	Power
PI	Positive ignition
Q'	Torque
s.f.c.	Specific fuel consumption
T	Temperature
TC	Turbocharger
V	Volume
VTNA	Variable turbine nozzle area
WGV	Waste gate valve
WHR	Waste heat recovery
x	Generic variable
xb	Fuel mass burned fraction
η	Efficiency
ω	Angular speed
Subscripts	
a	Ambient
b	Burned
BV	Bleed valve
C	Compressor
E	Engine
el	Electric
EM/G	Electric motor/generator
f	Fuel
HTC	Hybrid turbocharger
i	Inlet
o	Outlet
T	Turbine
TC	Turbocharger
Symbols	
EM/G_cs	Electric motor/generator control signal
HFO/NG_f	Engine fuel type
HFO/NG_Mf	Cylinder fuel mass flow rate
LOAD_s	Engine load signal
M_BV	Bleed valve mass flow rate
M_C i	Compressor inlet mass flow rate
M_cy	Cylinder inlet mass flow rate
M_cy o	Cylinder outlet mass flow rate

M_T o	Turbocharger turbine outlet mass flow rate
N_E	Engine speed
N_TC	Turbocharger speed
p_C o	Compressor outlet pressure
p_cy i	Cylinder inlet pressure
p_T i	Turbocharger turbine inlet pressure
P_el	Electric motor/generator electric power
Q′_C	Turbocharger compressor torque
Q′_E o	Engine torque
Q′_EM/G	Electric motor/generator torque
Q′_T	Turbocharger turbine torque
T_C o	Compressor outlet temperature
T_cy i	Cylinder inlet temperature
T_cy o	Cylinder outlet temperature
T_T i	Turbocharger turbine inlet temperature
T_T o	Turbocharger turbine outlet temperature
VT_c	Cylinder valves timing control signal
VTNA_as	Turbocharger variable turbine nozzle area actuator signal
VTNA_cs	Turbocharger variable turbine nozzle area control signal

References

1. Fan, A.; Wang, J.; He, Y.; Perčić, M.; Vladimir, N.; Yang, L. Decarbonising inland ship power system: Alternative solution and assessment method. *Energy* **2021**, *226*, 120266. [CrossRef]
2. Marine Environment Protection Committee (MEPC). 66th Session, 31 March to 4 April 2014. Available online: https://www.imo.org/en/MediaCentre/MeetingSummaries/Pages/MEPC66.aspx (accessed on 15 June 2021).
3. MEPC (Marine Environment Protection Committee). *Guidelines to the Method of Calculation of the Attained Energy Efficiency Design Index (EEDI) for New Ships*; Resolution, MECP 212 (63/23), ANNEX 8; MEPC: Abingdon, UK, 2012.
4. IMO. *Energy Efficiency Ship Operation, Module 2—Ship Energy Efficiency Regulation and Related Guidelines*; IMO: London, UK, 2016.
5. Ioannidis, J. Waste heat recovery from Diesel engines. In Proceedings of the 3rd International Congress of International Maritime Association of the East Mediterranean (IMAEM), Athens, Greece, 28 May–1 June 1984.
6. Ito, K.; Akagi, S. An optimal planning method for a marine heat and power generation plant by considering its operational problem. *Energy Res.* **2007**, *10*, 75–85. [CrossRef]
7. Theotokatos, G.; Livanos, G.A. Exhaust gas waste heat recovery in marine propulsion plants. In Proceedings of the 14th Conference on International Maritime Association of Mediterranean (IMAM), Genova, Italy, 13–16 September 2011.
8. Byung, C.C.; Young, M.K. Thermodynamic analysis of a dual loop heat recovery system with trilateral cycle applied to exhaust gases of internal combustion engine for propulsion of the 6800 TEU container ship. *Energy* **2013**, *58*, 404–416.
9. Altosole, M.; Benvenuto, G.; Campora, U.; Laviola, M.; Trucco, A. Waste heat recovery from marine gas turbines and diesel engines. *Energies* **2017**, *10*, 718. [CrossRef]
10. Barone, G.; Buonomano, A.; Forzano, C.; Palombo, A.; Vicidomini, M. Sustainable energy design of cruise ships through dynamic simulations: Multi-objective optimization for waste heat recovery. *Energy Convers. Manag.* **2020**, *221*, 113166. [CrossRef]
11. Livanos, G.A.; Theotokatos, G.; Patagonis, D.N. Techno-economical investigation of alternative propulsion plant for ferries and roro ships. *Energy Convers. Manag.* **2014**, *79*, 640–651. [CrossRef]
12. Altosole, M.; Campora, U.; Donnarumma, S.; Zaccone, R. Simulation techniques for design and control of a waste heat recovery system in marine natural gas propulsion applications. *J. Mar. Sci. Eng.* **2019**, *7*, 397. [CrossRef]
13. Altosole, M.; Benvenuto, G.; Zaccone, R.; Campora, U. Comparison of saturated and superheated steam plants for waste-heat recovery of dual-fuel marine engines. *Energies* **2020**, *13*, 985. [CrossRef]
14. Altosole, M.; Benvenuto, G.; Campora, U.; Laviola, M.; Zaccone, R. Simulation and performance comparison between diesel and natural gas engines for marine applications. *J. Eng. Marit. Environ.* **2017**, *231*, 690–704. [CrossRef]
15. Yoshihisa, O.; Keiichi, S.; Yukio, Y. Application of a large hybrid turbocharger for marine electric-power generation. *Mitsubishi Heavy Ind. Tech. Rev.* **2012**, *49*, 29.
16. Keiichi, S.; Yoshihisa, O.; Yukio, Y.; Musashi, S. Energy Savings Through Electric-Assist Turbocharger for Marine Diesel Engines. *Mitsubishi Heavy Ind. Tech. Rev.* **2015**, *52*, 36–41.
17. Alshammari, M.; Alshammari, F.; Pesyridis, A. Electric Boosting and Energy Recovery Systems for Engine Downsizing. *Energies* **2019**, *12*, 4636. [CrossRef]
18. Dong, H.; Zhao, Z.; Fu, J.; Liu, J.; Li, J.; Liang, K.; Zhou, Q. Experiment and simulation investigation on energy management of a gasoline vehicle and hybrid turbocharger optimization based on equivalent consumption minimization strategy. *Energy Convers. Manag.* **2020**, *226*, 113518. [CrossRef]

19. Metz, D.; Werner, J.; Münz, S.; Becker, M. Charging system for fuel cell applications. *MTZ Worldw.* **2013**, *74*, 40–43. [CrossRef]
20. Rusman, J.D. Charge air configurations for propulsion diesel engines aboard fast naval combatants. In Proceedings of the 14th International Naval Engineering Conference & Exhibition (INEC), Glasgow, UK, 2–4 October 2018.
21. Altosole, M.; Benvenuto, G.; Campora, U.; Silvestro, F.; Terlizzi, G. Efficiency improvement of a natural gas marine engine using a hybrid turbocharger. *Energies* **2018**, *11*, 1924. [CrossRef]
22. Geertsma, R.D.; Negenborn, R.R.; Visser, K.; Hopman, J.J. Design and control of hybrid power and propulsion systems for smart ships: A review of developments. *Appl. Energy* **2017**, *194*, 30–54. [CrossRef]
23. Coppola, T.; Micoli, L.; Turco, M. State of the art of high temperature fuel cells in maritime applications. In Proceedings of the International Symposium on Power Electronics, Electrical Drives, Automation and Motion, Sorrento, Italy, 24–26 June 2020; pp. 430–435.
24. Sapra, H.; Stam, J.; Reurings, J.; van Biert, L.; van Sluijs, W.; de Vos, P.; Visser, K.; Vellayani, A.P.; Hopman, H. Integration of solid oxide fuel cell and internal combustion engine for maritime applications. *Appl. Energy* **2021**, *281*, 115854. [CrossRef]
25. Geertsma, R.D.; Negenborn, R.R.; Visser, K.; Hopman, J.J. Parallel Control for Hybrid Propulsion of Multifunction Ships. *IFAC-Pap. Line* **2017**, *50*, 2296–2303. [CrossRef]
26. Capasso, C.; Veneri, O.; Notti, E.; Sala, A.; Figari, M.; Martelli, M. Preliminary design of the hybrid propulsion architecture for the research vessel "G. Dallaporta". In Proceedings of the International Conference on Electrical Systems for Aircraft, Railway, Ship Propulsion and Road Vehicles & International Transportation Electrification Conference, Toulouse, France, 2–4 November 2016.
27. Altosole, M.; Campora, U.; Vigna, V. Energy efficiency analysis of a flexible marine hybrid propulsion system. In Proceedings of the International Symposium on Power Electronics, Electrical Drives, Automation and Motion, Sorrento, Italy, 24–26 June 2020; pp. 436–441.
28. Mauro, F.; Ghigliossi, E.; Bucci, V.; Marinò, A. Design of hybrid-electric megayachts: The impact of operative profile and smart berthing infrastructures. *J. Mar. Sci. Eng.* **2021**, *9*, 186. [CrossRef]
29. Feneley, A.J.; Pesiridis, A.; Andwari, A.M. Variable Geometry Turbocharger Technologies for Exhaust Energy Recovery and boosting-A Review. *Renew. Sustain. Energy Rev.* **2017**, *71*, 959–975. [CrossRef]
30. Woodyard, D. *Pounder's Marine Diesel Engines and Gas*, 9th ed.; Elsevier Ltd.: Oxford, UK, 2009.
31. *MAN 51/60 DF IMO TIER II/IMO TIER III Marine Project Guide*; MAN: Munich, Germany, 2015.
32. Benvenuto, G.; Campora, U.; Laviola, M.; Terlizzi, G. Simulation model of a dual-fuel four stroke engine for low emission ship propulsion applications. *Int. Rev. Mech. Eng.* **2017**, *11*, 817–882. [CrossRef]
33. Benvenuto, G.; Campora, U.; Carrera, G.; Casoli, P. A two-zone Diesel engine model for the simulation of marine propulsion plant transients. In Proceedings of the MARIND 98, Second International Conference on Marine Industry, Varna, Bulgaria, 28 September–2 October 1998.
34. Theotokatos, G.; Stoumpos, S.; Lazakis, I.; Livanos, G. Numerical study of a marine dual-fuel four-stroke engine. *Marit. Technol. Eng. 3* **2016**, *2*, 777–786.
35. Tadros, M.; Ventura, M.; Guedes Soares, C. Data driven in-cylinder pressure diagram based optimization procedure. *J. Mar. Sci. Eng.* **2020**, *8*, 294. [CrossRef]
36. Mrzljak, V.; Medica, V.; Bukovac, O. Quasi-dimensional diesel engine model with direct calculation of cylinder temperature and pressure. *Technical Gazette* **2017**, *24*, 681–686.
37. Altosole, M.; Campora, U.; Figari, M.; Laviola, M.; Martelli, M. A diesel engine modelling approach for ship propulsion real-time simulators. *J. Mar. Sci. Eng.* **2019**, *7*, 138. [CrossRef]
38. Tavakoli, S.; Saettone, S.; Steen, S.; Andersen, P.; Schramm, J.; Pedersen, E. Modeling and analysis of performance and emissions of marine lean-burn natural gas engine propulsion in waves. *Appl. Energy* **2020**, *279*, 115904. [CrossRef]
39. Tadros, M.; Ventura, M.; Guedes Soares, C. Simulation of the performance of marine genset based on double-wiebe function. Sustainable Development and Innovations in Marine Technologies. In Proceedings of the 18th International Congress of the International Maritime Association of the Mediterranean, IMAM, Varna, Bulgaria, 9–11 September 2019; pp. 292–299.
40. MAN. *VTA—Variable Turbine Area. MAN Diesel & Turbo Report*; MAN: Munich, Germany, 2010.

Journal of
Marine Science and Engineering

MDPI

Article

Parameter Identification of a Model Scale Ship Drive Train

Arthur Vrijdag [1] and Michele Martelli [2,*]

[1] Faculty of Mechanical, Maritime and Materials Engineering, Delft University of Technology, 2628 CD Delft, The Netherlands; a.vrijdag@tudelft.nl
[2] Department of Electrical, Electronic, Telecommunications Engineering and Naval Architecture, Polytechnic School, University of Genoa, 16126 Genoa, Italy
* Correspondence: michele.martelli@unige.it

Citation: Vrijdag, A.; Martelli, M. Parameter Identification of a Model Scale Ship Drive Train. *J. Mar. Sci. Eng.* **2021**, *9*, 268. https://doi.org/10.3390/jmse9030268

Academic Editor: Kostas A. Belibassakis

Received: 19 January 2021
Accepted: 22 February 2021
Published: 2 March 2021

Abstract: Simulation models of the ship propulsion system play an increasingly important role, for instance in controller design and condition monitoring. However, creation of such simulation models requires significant time and effort. In this paper, the application of deterministic identification techniques on a DC-electric ship drive train is explored as an alternative for data-driven identification techniques that require extensive measured data sets collected over long periods of ship operation. First, a nonlinear and a linear simulation model that represent the dynamic behavior of the propulsion plant are developed, and the main parameters to be identified are defined. Then, a set of experiments on a model scale boat in the bollard pull condition are conducted using an ad hoc experimental setup and data acquisition system. Subsequently, various types of identification techniques are applied, aiming to determine the unknown model parameters. Eventually, a comparison is made between experimental and simulated results, using the different sets of the estimated parameters. The value of the demonstrated approaches lies in the fast determination of unknown system parameters. These parameters can be used in simulation models, which in turn can be used for various purposes such as system controller development and tuning. Furthermore, periodic determination of system parameters can support condition monitoring to detect faults or degradation of the system. The latter point directly deals with the condition-based maintenance issue; in fact, monitoring the propulsion plant parameters over time could allow for better management (and timing) of maintenance. Although the developed ideas are far from ready to be used on the full-scale, the authors believe that the methodologies are promising enough to be developed further towards a full-scale application.

Keywords: marine propulsion system; parameter identification; data-driven ship propulsion model; condition-based maintenance; digital twin

1. Introduction

Simulation models of the ship propulsion system play an increasingly important role, for instance in controller design [1,2] and condition monitoring [3]. The drawback of using simulation models, however, is that the required parameters are often unknown or very uncertain. Therefore, building a simulation model and determination or estimation of its parameters can be a time-consuming task, which often requires significant experience (see for recent examples [4–6]). After building and verifying the model, its validity can sometimes be quantified, at least for a specific domain of applications [7]. Periodic re-validation is not commonly reported, while it is known that many of the physical parameters that play a role in the performance of the ship propulsion plant are time-variant. Examples of time-variant factors are fouling of the hull and propeller, turbocharger contamination, and so on.

A comprehensive description of identification techniques is given by Ljung [8]. Since the 1990s artificial neural network techniques have been widely used to identify electric motor parameters [9–11] as well as linear and nonlinear least-squares algorithms [12,13].

Despite the abundant literature on identification techniques, publication of their application to determine marine propulsion plant parameters is not widespread. A noteworthy exception is the research effort that has been put into the identification of parameters of a dynamic thruster performance model for remotely operated underwater vehicles, which attempts to capture the dynamic response of propeller thrust and torque to the applied electric motor torque [14–25]. The following observations are made regarding these papers:

- They do not all adhere to general accepted ship propulsion theory and notation. In various papers the system parameters are lumped together such that direct comparison between different cases becomes difficult, and in the opinion of the authors of this paper, the physical viewpoint is easily lost;
- Various papers account for the axial and/or rotational acceleration of the water flowing through the propeller disc. The assumptions and modeling approach, however, differ. Although the effect of flow dynamics on propulsion performance is very interesting, this effect is not included in this paper and does not seem to lead to poor agreement between simulation and measurement;
- None of the papers includes a differential equation for motor current, which is included in this paper to ensure that all relevant electric parameters are captured in the model.
- Examples are given of the use of various input signals for identification purposes, such as the triangular wave, square wave, and single sinusoidal wave. In this paper the use of multiple sinusoidal waves and band-limited white noise as input signals will be treated as well, aiming to ensure good agreement between model and measurement over a wide frequency domain.

Data-driven modeling approaches such as those reported by Coraddu et.al. [26] might offer benefit in the sense that by making use of large amounts of historical data in combination with advanced algorithms, a "superfit" model can be generated. Drawbacks of using such a black box approach are the amount of required data, the time over which the data are to be collected, and the lack of insight on the physical behavior of the underlying system.

Although the data-driven approaches based on huge datasets will, without doubt, play an important role in the future, in this paper multiple identification techniques are proposed to obtain the propulsion system parameters, based on short (but information-rich) controlled performance tests, and are tested on model scale. The potential benefit of application of these approaches on full scale is that they can be used to, in a relatively short time span (possibly in real time), quantify system performance during sea acceptance trials, after periodic maintenance or following a system upgrade. Comparison of this fingerprint with sister ships or with previous fingerprints could potentially be used to understand the state of decay of components giving a significant contribution to a condition-based approach to ship maintenance operations [27].

To demonstrate the idea, a model scale ship available at Delft University of Technology (DUT) and Genoa University (UNIGE) is used. First, the non-linear system model of its propulsion plant including electric DC-motor, gearbox, and propeller is derived and subsequently linearized. Both models contain the same unknown parameters. Note that this paper focuses on bollard pull conditions, although the ideas can be extended to free sailing conditions as well.

Subsequently, multiple identification methods are explained and applied, making use of data collected during various types of experiments. The resulting parameter sets are implemented in the non-linear and linear simulation models, and their behavior is validated in both time and frequency domains.

At the end of the paper, a possible path is given for the development of full-scale ship propulsion "fingerprinting" techniques through system performance tests. Such a path includes simulation-based research and both model-scale and full-scale experimental research.

2. Ship Drive Train and Its Mathematical Model

The ship propulsion simulation model is based on a model scale ship called "Tito Neri", which is shown in Figure 1. A detailed picture of its azimuthing thrusters is shown in Figure 2, and its main particulars are given in Table 1. A schematic representation of one of its two drive trains is given in Figure 3. It consists of a DC motor that drives an azimuthing thruster with a ducted fixed pitch propeller. Although not shown in the figure, the upper shaft is supported by a shaft bearing.

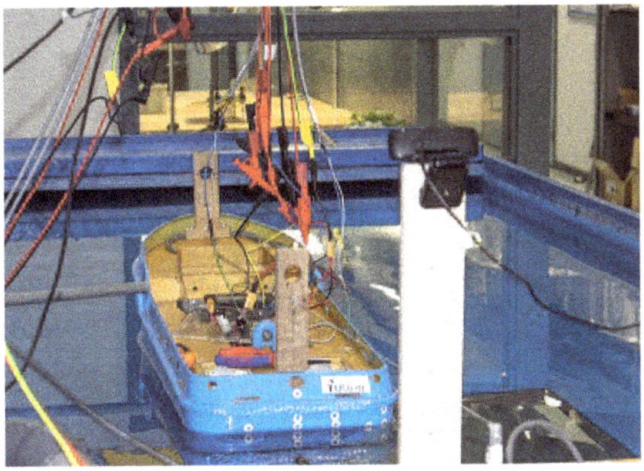

Figure 1. Tito Neri overview.

Figure 2. Tito Neri azimuthing thrusters from astern.

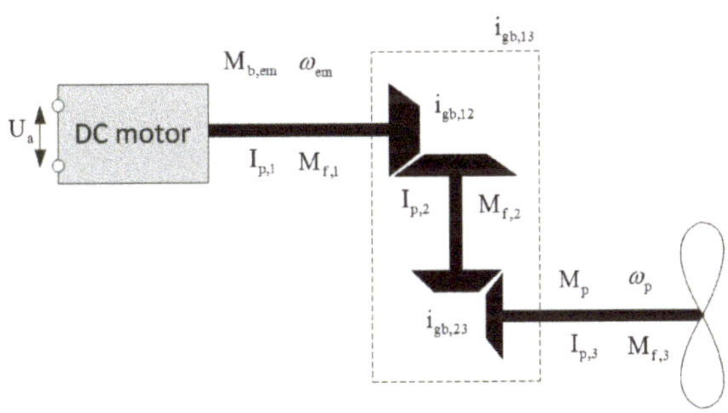

Figure 3. Schematic representation of Tito Neri drive train.

Table 1. Main particulars of Tito Neri.

L_{oa}	0.97	[m]
B_{oa}	0.32	[m]
draft forward/aft	0.10/0.13	[m]
displacement Δ with/without battery	15.4/13.5	[kg]
upper bevel gear teeth ratio	13:39	[-]
total gear reduction ratio $i_{gb,13}$	3	[-]
Propeller diameter D	0.065	[m]

The system behavior is governed by two differential equations that interact with each other. One is related to the (faster) electrical circuit, and the other related to the (slower) mechanical part of the drive train. The differential equation commonly used to model an electric DC motor circuit is given by

$$L_a \frac{di_a}{dt} = U_a - K_e \omega_{em} - R_a i_a \tag{1}$$

in which L_a is the inductance, i_a the current, U_a the supply voltage, K_e the motor coefficient, ω_{em} the motor speed (in rad/s), and R_a the resistance.

The reduction ratio between motor shaft and intermediate vertical shaft $i_{gb,12}$ between intermediate vertical shaft and propeller shaft $i_{gb,23}$ and the resulting total reduction ratio $i_{gb,13}$ are defined by

$$i_{gb,12} = \frac{\omega_{em}}{\omega_{int}}, \quad i_{gb,23} = \frac{\omega_{int}}{\omega_p}, \quad i_{gb,13} = i_{gb,12}\, i_{gb,23}$$

The differential equation for electric motor speed, assuming constant friction torque on all three shafts, is given by

$$I_p \frac{d\omega_{em}}{dt} = M_{b,em} - M_f - \frac{M_p}{i_{gb,13}} \tag{2}$$

in which brake motor torque $M_{b,em}$ is given by

$$M_{b,em} = K_e i_a \tag{3}$$

and in which the total polar moment of inertia is given by

$$I_p = I_{p,1} + \frac{I_{p,2}}{i_{gb,12}^2} + \frac{I_{p,3}}{i_{gb,13}^2}$$

and in which the total friction is given by:

$$M_f = M_{f,1} + \frac{M_{f,2}}{i_{gb,12}} + \frac{M_{f,3}}{i_{gb,13}}$$

The propeller torque M_p and thrust T are modeled following Carlton [28], making use of the torque and thrust coefficients K_Q and K_T at advance ratio $J = 0$:

$$M_p = \frac{Q}{\eta_R} = \frac{K_{Q,J=0}\, \rho\, \omega_{em}^2 D^5}{\eta_R 4\pi^2\, i_{gb,13}^2} \tag{4}$$

in which Q is the open water propeller torque, η_R is the relative rotative efficiency, ρ is the water density, and D is the propeller diameter. Although not further used in this paper, propeller thrust T and bollard pull force F_{BP} are modeled by

$$T = \frac{K_{T,J=0}\,\rho\,\omega_{em}^2 D^4}{4\pi^2 i_{gb,13}^2} \tag{5}$$

and

$$F_{BP} = k_p T(1-t) \tag{6}$$

in which k_p is the number of operating propellers, and $(1-t)$ corrects for thrust deduction.

To summarize, the following system of differential equations describes the nonlinear system dynamics:

$$\begin{cases} L_a \frac{di_a}{dt} = U_a - K_e\omega_{em} - R_a i_a \\ I_p \frac{d\omega_{em}}{dt} = K_e i_a - M_f - \frac{M_p}{i_{gb,13}} \end{cases} \tag{7}$$

in which M_p is given by Equation (4). The unknown parameters of this model are inductance L_a, resistance R_a, motor coefficient K_e, polar moment of inertia I_p, friction torque M_f, propeller torque coefficient $K_{Q,J=0}$, and relative rotative efficiency η_R. However, $K_{Q,J=0}$ and η_R are observationally equivalent, meaning that (with the sensors in this experimental setup) they cannot be distinguished from each other. Therefore, propeller torque coefficient and relative rotative efficiency are combined into a single combined unknown parameter $\frac{K_{Q,J=0}}{\eta_R}$, leaving a total of six unknown parameters.

Note that the unknown parameters $K_{T,J=0}$ and $(1-t)$ are not further considered in this paper due to difficulties in measuring the small thrust force during the experiment.

Linearized Propulsion System Model and Step-Responses

In this section the ship propulsion system model (7) is linearized, and its analytical step responses are given. Later these will be shown to be useful tools for the identification of the unknown parameters.

The linearization process of the ship propulsion plant in free sailing mode is described in detail in [29,30], although in both papers no electric circuit including DC-motor was included. Note that in the main text of this paper only the main results are given, and details of the notation and the full derivations are available in Appendixes A–C. The normalized and linearized versions of (7) are given by

$$\tau_{em}\frac{di_a^*}{dt} = \frac{U_{a,0}}{R_a i_{a,0}}\delta U_a^* - \frac{K_e\omega_{em,0}}{R_a i_{a,0}}\delta\omega^* - \delta i_a^* \tag{8}$$

$$\tau_\omega \frac{d\omega^*}{dt} = \delta i_a^* - 2\eta_{trm,0}\delta\omega^* \tag{9}$$

in which the delta-asterisk notation indicates normalized difference as follows:

$$\delta i_a^* = \frac{\delta i_a}{i_{a,0}} = \frac{i_a - i_{a,0}}{i_{a,0}}$$

such that for example a value of $\delta U_a^* = 0.05$ means a +5% perturbation from the nominal value $U_{a,0}$. The two integration constants are defined as

$$\tau_{em} = \frac{L_a}{R_a}, \quad \tau_\omega = \frac{I_p\omega_{em,0}}{M_{b,em,0}} = \frac{I_p\omega_{em,0}}{K_e i_{a,0}} \tag{10}$$

The transmission efficiency $\eta_{trm,0}$ is related to the friction torque M_f by

$$\eta_{trm,0} = \frac{M_{b,em,0} - M_f}{M_{b,em,0}}$$

When Equations (8) and (9) are put in state space notation, this results in the following linear system:

$$\begin{bmatrix} \frac{di_a^*}{dt} \\ \frac{d\omega^*}{dt} \end{bmatrix} = \begin{bmatrix} -\frac{1}{\tau_{em}} & -\frac{1}{\tau_{em}}\frac{K_e\omega_{em,0}}{R_a i_{a,0}} \\ \frac{1}{\tau_\omega} & -\frac{2\eta_{trm,0}}{\tau_\omega} \end{bmatrix} \begin{bmatrix} \delta i_a^* \\ \delta\omega^* \end{bmatrix} + \begin{bmatrix} \frac{1}{\tau_{em}}\frac{U_{a,0}}{R_a i_{a,0}} \\ 0 \end{bmatrix} \delta U_a^* \tag{11}$$

The benefit of this notation is that it can easily be programmed and analyzed in software like MATLAB. Alternatively, the Laplace transfer function can be used. As derived in Appendix B, the two transfer functions from the supply voltage δU_a^* to the two state variables electric current δi_a^* and rotation speed $\delta\omega^*$ are

$$\frac{\delta i_a^*(s)}{\delta U_a^*(s)} = \frac{(\tau_{\omega,e}s+1)\frac{U_{a,0}}{R_a i_{a,0}}}{\tau_{em}\tau_{\omega,e}s^2 + (\tau_{em}+\tau_{\omega,e})s + 1 + \frac{1}{2\eta_{trm,0}}\frac{K_e\omega_{em,0}}{R_a i_{a,0}}} \tag{12}$$

$$\frac{\delta\omega^*(s)}{\delta U_a^*(s)} = \frac{\frac{1}{2\eta_{trm,0}}\frac{U_{a,0}}{R_a i_{a,0}}}{\tau_{em}\tau_{\omega,e}s^2 + (\tau_{em}+\tau_{\omega,e})s + 1 + \frac{1}{2\eta_{trm,0}}\frac{K_e\omega_{em,0}}{R_a i_{a,0}}} \tag{13}$$

in which $\tau_{\omega,e} = \frac{\tau_\omega}{2\eta_{trm,0}}$. The transfer function for current is recognizable as a summation of a bandpass system and a (lowpass) second-order system, while the transfer function for motor speed is (only) a second-order lowpass system.

Analytic expressions for the two poles s_1 and s_2, the single zero z_1, and the two DC-gains of the transfer functions are derived in Appendix B.

As derived in Appendix C the approximate response of motor speed to a unit step in supply voltage is given by

$$\delta\omega^*(t) \approx K\left(1 - e^{s_2 t}\right) \tag{14}$$

in which $K = \frac{U_{a,0}}{K_e\omega_0 + 2\eta_{trm,0}R_a i_{a,0}}$. The response of current to a unit step in supply voltage is

$$\delta i_a^*(t) \approx K_{LP}\left(1 - e^{s_2 t}\right) + K_{BP}\left(\frac{1}{s_2 - s_1}e^{s_2 t} - \frac{1}{s_2 - s_1}e^{s_1 t}\right) \tag{15}$$

with $K_{LP} = \frac{U_{a,0}2\eta_{trm}}{K_e\omega_0 + 2\eta_{trm,0}R_a i_{a,0}}$ and $K_{BP} = \frac{U_{a,0}}{L_a i_{a,0}}$.

3. Applied Identification Techniques

Once both the non-linear and the linearized plant models have been formulated, measurement data can be used to determine the unknown model parameters by making use of parameter identification techniques.

Many different identification techniques can be used, such as for instance the various possibilities that are included in the "system identification" toolbox of MATLAB. A possibility is to search for an optimal parameter vector $\boldsymbol{\theta}$ by minimizing the (weighed) sum of squared errors given by the cost function V_t:

$$V_t(\boldsymbol{\theta}) = \frac{1}{N}\sum_{t=1}^{N} \mathbf{e}(t,\boldsymbol{\theta})^T \mathbf{W}(\boldsymbol{\theta})\mathbf{e}(t,\boldsymbol{\theta}) \tag{16}$$

where the error $\mathbf{e}$ is the difference between the vectors of measurement and simulation, containing all output signals that are to be taken into account:

$$\mathbf{e}(t) = \mathbf{y}(t)_{measured} - \mathbf{y}(t)_{model} \tag{17}$$

Another approach, which prevents usage of all time samples in the minimization algorithm and which ensures a balanced representation of system behavior throughout the frequency domain, is to define the cost function V_ω directly in the frequency domain:

$$V_\omega(\boldsymbol{\theta}) = \frac{1}{N}\sum_{\omega=1}^{N} \mathbf{e}(\omega,\boldsymbol{\theta})^T \mathbf{W}(\boldsymbol{\theta})\mathbf{e}(\omega,\boldsymbol{\theta}) \tag{18}$$

in which the error is defined as the Euclidean norm of the error in the complex frequency response:

$$\mathbf{e} = ||\frac{Y(\omega)}{X(\omega)} - G(\boldsymbol{\theta},\omega)|| \tag{19}$$

in which $\frac{Y(\omega)}{X(\omega)}$ indicates the measured frequency response data (FRD), model and $G(\boldsymbol{\theta},\omega)$ indicates the modeled frequency response.

Within the two main groups "time domain approach" and "frequency domain approach", there are various possible refinements and alternatives. For an in-depth review, reference is made to Ljung [8].

The "goodness of fit" of a model with a given parameter set can be expressed in various ways. The quality metrics "FitPercent" and mean squared error "MSE" are used here:

$$FitPercent = 100\left(1 - \frac{||\mathbf{y}_{measured} - \mathbf{y}_{model}||}{||\mathbf{y}_{measured} - \overline{\mathbf{y}}_{measured}||}\right) \tag{20}$$

$$MSE = \frac{1}{N}\sum_{t=1}^{N} \mathbf{e}^T(t)\mathbf{e}(t) \tag{21}$$

Equivalent versions of quality metrics can be defined for the goodness of fit in the frequency domain.

From the following non-exhaustive list of possible identification techniques, in this paper three different parameter identification procedures (1, 4, and 5) are applied to the "Tito Neri" drive train in the bollard pull condition:

1. a time domain identification approach based on multiple steady operating points and a step response;
2. a time domain approach, aiming at minimization of cost function (16) by fitting the linear model (11);
3. a time domain approach as 2, but now fitting the non-linear model defined by (7) and (4);
4. a frequency domain identification approach based on experimental FRD generated by processing the experimental time domain data obtained with multiple single frequency input voltage signals with a correlation algorithm;
5. a discrete transfer function estimation based on the Welch method combined with a modified periodogram method [31].

Note that the frequency domain approaches 4 and 5 only differ in the way that they generate the experimental FRD. The subsequent parameter identification procedure is the same and aims to minimize cost function (18). 2 and 3 are not taken into account in the present work since they are investigated in open literature.

3.1. Time Domain Identification: 1

In this first method, for the sake of computational simplicity, the procedure to obtain parameters is split into two parts, assuming that the parameters do not change during the experimental time.

First, the stationary condition $\frac{d\omega_e}{dt} = 0$ and $\frac{di_a}{dt} = 0$ is considered to reduce the number of unknowns, and a least-squares algorithm is applied. Starting from Equation (7) it is possible to obtain

$$\begin{cases} 0 = K_e i_a - M_f - \frac{M_p}{i_{gb,13}} \\ 0 = U_a - K_e \omega_{em} - R_a i_a \end{cases} \tag{22}$$

These equations are rearranged in matrix notation as follows, separating known from unknown variables:

$$\begin{pmatrix} 1 & -i_a & 0 & c \\ 0 & \omega_{em} & i_a & 0 \end{pmatrix} \begin{pmatrix} M_f \\ K_e \\ R_a \\ \frac{K_{Q,j=0}}{\eta_R} \end{pmatrix} = \begin{pmatrix} 0 \\ U_a \end{pmatrix} \tag{23}$$

where c is obtained from Equation (4):

$$c = \frac{\rho \omega_{em}^2 D^5}{4\pi^2 i_{gb,13}^3} \tag{24}$$

In this way the system consists of two equations and four unknown variables (M_f, K_e, R_a, and $\frac{K_{Q,j=0}}{\eta_R}$), such that ∞^2 solutions exist. However, if measurements at n different steady state operating points are available, n quadruplets have to satisfy the system of Equation (23), resulting in the following over-determined system:

$$\begin{pmatrix} 1 & -i_{a,1} & 0 & c_1 \\ \dots & \dots & \dots & \dots \\ 1 & -i_{a,n} & 0 & c_n \\ 0 & \omega_{em,1} & i_{a,1} & 0 \\ \dots & \dots & \dots & \dots \\ 0 & \omega_{em,n} & i_{a,n} & 0 \end{pmatrix} \begin{pmatrix} M_f \\ K_e \\ R_a \\ \frac{K_{Q,j=0}}{\eta_R} \end{pmatrix} = \begin{pmatrix} 0 \\ U_a \end{pmatrix} \tag{25}$$

The last can be written in general form, as follows:

$$A\mathbf{x} = \mathbf{b} \tag{26}$$

The system (25) cannot be solved in principle since it is overdetermined. Although an exact solution does not exist, an approximate solution to (25) can be determined by means of, for instance, a (weighed) least-squares approach; in our case we used unweighted least squares. The final goal, according to notation reported in (26), is to evaluate the vector $\mathbf{x}$ that minimizes the squared $l^2 norm$ of the residual, naming $A, \mathbf{x}, \mathbf{b}$, the coefficient matrix, the unknown vector, and the constant terms vector, respectively. The quantity $S(\mathbf{x})$ to be minimized is written as follows, in matrix notation:

$$S(\mathbf{x}) = ||\mathbf{b} - A\mathbf{x}||^2 \tag{27}$$

Differentiating the above equation, and setting to zero the solution, it is possible to obtain the Normal Equation:

$$A^T A\mathbf{x} = A^T(b) \tag{28}$$

If $A^T A$ is non-singular, the solution is given by solving the linear algebraic system (28).

Once M_f, K_e, R_a, and $\frac{K_{Q,j=0}}{\eta_R}$ are known, the second part of the procedure is deterministic. By using Equation (10), it is possible to evaluate the inertia I_p and the motor inductance L_a :

$$\begin{aligned} L_a &= \tau_{em} R_a \\ I_p &= \tau_\omega \frac{K_e i_a}{\omega_{em,0}} \end{aligned} \tag{29}$$

To obtain the two parameters, knowledge of the time constants τ_{em} and τ_{ω} is necessary, and the step response of both current and motor speed reported in Appendix C is used. From Equation (14) and fixing whichever time, t^* (authors suggest to use the t^* when the response is at 63.2%), since parameters are time independent, it is possible to obtain s_2, as follows:

$$s_2 \approx \frac{1}{t^*} ln\left(1 - \frac{\delta\omega^*(t^*)}{K}\right) \tag{30}$$

Substituting the value of s_2 into Equation (A34) and remembering the difference between $\tau_{\omega,e}$ and τ_{ω} gives

$$\tau_{\omega} \approx \frac{-C}{s_2}\eta_{trm,0} \tag{31}$$

The evaluation of τ_{ω}, as it can be intuitive from the last equations, it is an approximate solution.

The electric time-constant τ_{em} is more challenging to estimate. As reported in Appendix C the step response of current could be obtained as a summation of two terms. The first term is represented by a low-pass filter in its simplified form and the second by a bandpass filter as reported in Equation (15). The total response is known from the experiment, and all terms describing the low-pass filter are known at this stage; so, numerically, it is possible to obtain the shape of the bandpass filter response over time. A specific time called t^* should be fixed, and at that time the value of $\delta i^*_{a,BP}(t^*)$ can be obtained. After some adjustment the following relation is obtained:

$$\delta i^*_{a,BP}(t^*) - \frac{U_{a,0}}{\tau_{em} R_a i_{a,0}}\left(\frac{1}{-\frac{C}{\tau_{\omega,e}} + \frac{1}{\tau_{em}}}\right)\left(e^{-\frac{C}{\tau_{\omega,e}}t^*} - e^{-\frac{1}{\tau_{em}}t^*}\right) = 0 \tag{32}$$

From the previous equation, it is not possible to obtain a solution in closed form for τ_{em}, and numerical methods must be used (i.e., bisection methods or Newton–Raphson method). Eventually, using Equation (29) L_a can be obtained.

3.2. Frequency Domain Approach Using Sinusoidal Input Voltage Signals 4

The idea of the this method is to generate a sinusoidal voltage of a specific frequency and amplitude, to superimpose it on a constant voltage value $U_{a,0}$, and to apply the resulting signal as a voltage input to the system, while recording the response of current i_a and electric motor speed ω_{em}. Based on the input and response at each frequency, the gain and phase of the transfer functions of the system are estimated with a correlation-based single-frequency approach [8,32], in line with Figure 4. Since this method does not deliver the unknown parameters of the model directly, it is called a non-parametric identification method. The non-parametric frequency response data (FRD) model can however be used as a basis to determine the parameters of the model.

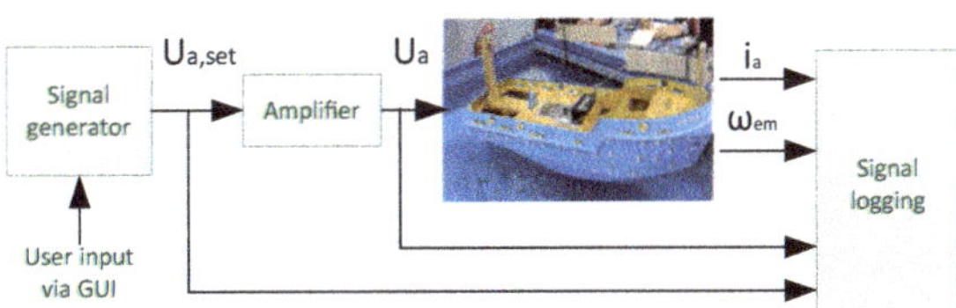

Figure 4. Block diagram of experimental setup.

In more detail the basis of the method is to generate two harmonic signals:

$$x(t) = X\sin(\omega t)$$

and the out-of-phase signal

$$x'(t) = X\sin(\omega t + \pi/2) = X\cos(\omega t)$$

of which the first signal is used to excite the system. The response of the system is

$$y(t) = Y\sin(\omega t + \varphi) + n(t)$$

in which n(t) is a noise signal which is assumed uncorrelated with input and output signals.

Both input signals $x(t)$ and $x'(t)$ in combination with the output $y(t)$ are used to determine the cross-correlations and auto-correlation according to

$$R_{xy} = \frac{1}{T}\int_0^T X\sin(\omega t)\, Y\sin(\omega t + \varphi)dt + R_{xn} = \frac{XY}{2}\cos\varphi + R_{xn} \tag{33}$$

$$R_{x'y} = \frac{1}{T}\int_0^T X\cos(\omega t) Y\sin(\omega t + \varphi)dt + R_{x'n} = \frac{XY}{2}\sin\varphi + R_{x'n} \tag{34}$$

$$R_{xx} = \frac{1}{T}\int_0^T X\sin(\omega t) X\sin(\omega t)dt = \frac{X^2}{2} \tag{35}$$

where X is the amplitude of the input signal (in this case the amplitude of voltage δU_a^*), and Y is the amplitude of the output signal under consideration (in this case the amplitude of motor current δi_a^* or motor speed $\delta\omega_{em}^*$). R_{xn} is the cross-correlation between input and noise, which reduces to zero with increasing measurement time. Division of Equation (33) by Equation (35) delivers the in phase (real) component of the frequency response:

$$\frac{R_{xy}}{R_{xx}} = \frac{Y}{X}\cos\varphi \tag{36}$$

while division of Equation (34) by Equation (35) gives the out-of-phase (imaginary) part of the response:

$$\frac{R_{x'y}}{R_{xx}} = \frac{Y}{X}\sin\varphi \tag{37}$$

Based on the real and imaginary components the gain and phase of the transfer function are calculated by

$$\frac{Y}{X} = \sqrt{\left(\frac{R_{xy}}{R_{xx}}\right)^2 + \left(\frac{R_{x'y}}{R_{xx}}\right)^2} \tag{38}$$

$$\varphi = \arctan\left(\frac{R_{x'y}}{R_{xy}}\right) \tag{39}$$

By using this approach, the gain and phase can be determined experimentally for multiple appropriately spaced frequencies, resulting in a non-parametric frequency-response data (FRD) model. The results of the procedure are presented in Table 2. Subsequently, the procedure to derive the unknown system parameters from the obtained FRD model is based on minimization of the cost function (18).

The main advantage of the correlation method to determine an FRD model is the inherent high noise immunity.

Table 2. Experimental FRD based on single sinusoidal testing, followed by processing with the correlation approach.

ω [rad/s]	0.3	1	5	10	20	100	500	1000	5000
$\lvert\frac{\delta i^*}{\delta U_a^*}\rvert$ [-]	0.71	0.79	1.62	2.31	2.94	3.08	2.99	2.90	2.11
$\angle\frac{\delta i^*}{\delta U_a^*}$ [deg]	8.6	19.4	37.7	31.3	19.6	3.5	−4.5	−8.8	−26.8
$\lvert\frac{\delta \omega^*}{\delta U_a^*}\rvert$ [-]	1.24	1.18	1.05	0.81	0.54	0.12	0.02	0.01	0.00
$\angle\frac{\delta \omega^*}{\delta U_a^*}$ [deg]	−2.2	−7.0	−26.2	−45.4	−64.3	−95.5	−144.9	146.3	20.7

3.3. Noise Input Testing: 5

An FRD model of a system can also be determined from the measured system response to a random input signal. This approach is often practical for processes that cannot be taken off-line for dedicated testing, but due to their nature do contain measurable random input disturbances. In this paper, a sequence of random supply voltage will be superimposed on the nominal supply voltage. The method is based on the relation between the transfer function $H(j\omega)$, power spectral density of the input $S_{xx}(j\omega)$, and cross-spectral density $S_{xy}(j\omega)$ [8,31]:

$$H(j\omega) = \frac{S_{xy}(j\omega)}{S_{xx}(j\omega)} \tag{40}$$

The estimation of both the input power spectral density S_{xx} and the cross-spectral density S_{xy} requires sufficient length of data, and can be improved by application of suitable "windowing" and "smoothing", which can be done by averaging the spectrum derived from multiple segments of the total time-trace. Secondly, it is possible to increase the number of portions of a given time-trace by allowing a specific percentage of overlap between the parts.

For this method to work well, it is essential to ensure that the input signal is persistently exciting, which indicates that the signal power is sufficiently large for all frequencies of interest.

When using this method, the coherency γ usually is presented side by side with the estimated transfer function. It expresses the correlation between the input and output signal of the system with a value between 0 and 1, where 0 means no correlation and 1 means full correlation, thereby giving an idea of the quality of the estimated transfer function at different frequencies. Note that operations such as windowing, smoothing, and quantization of signals due to A-D conversion in the measurement system and noise in the measurement influence the coherency negatively.

4. Experimental Campaign

4.1. Setup and Experimental Matrix

The schematic experimental setup used is shown in Figure 4. The signal generator is operated via a customized graphical user interface and delivers the required voltage signal $U_{a,set}$ to the amplifier, which in turn feeds the electric motor of the model scale ship with the supply voltage U_a. Two sensors are installed: a current sensor just before the electric motor and a 15 pulse encoder mounted on the motor shaft. The two sensor signals i_a and ω_{em}, together with the voltages $U_{a,set}$ and U_a, are recorded with a data acquisition system. Although not discussed in detail in this paper, the transfer function of the amplifier itself could be determined experimentally, showing that the amplifier only causes a small drop in voltage (<1%), and a small phase lag (<2°) over the frequency range of interest.

Several experiments with varying sequences of voltage $U_{a,set}$ have been done. The sampling rate of the data acquisition system was established based on the goals and duration of the specific experiment.

Trials were performed with the following input voltage signals: one staircase, nine sinusoidal waves with the different amplitudes and frequencies, a band-limited white noise input signal, and at the end a mix of the previous signals. Each identification technique uses data from a specific (set of) experiments. The final "mixed" test is used for validation purposes, as reported in Table 3.

Table 3. Experimental test matrix.

Test	Identification	Validation
Staircase	yes (1)	yes (4)
9 × Sinusoidal	yes (4)	yes (1,4,5)
White noise	yes (5)	
Mix of signals		yes (1,4)

4.2. *Inspection of Current and Motor Speed Signals*

Initial measurements of the current revealed some unexpected behavior. The current signal showed a considerable amount of noise, and the reason was investigated. In particular, specific higher-order frequencies appeared when inspecting the FFT of the current signal. It was hypothesized that these higher-order frequencies, which are not captured by the linear or non-linear system model, could be caused by unmodeled system behavior. Examples could be, for instance, the gear-meshing frequency, shaft misalignment, unbalance in the shafting system, propeller blade passing frequencies, or cogging of the electric motor due to a discrete number of permanent magnets and the gaps in between them.

To obtain insight into the cause of the higher-order frequencies, an order tracking of current in the motor speed range from 220 to 1995 rpm was carried out, as shown in Figures 5 and 6. The figures reveal that although many harmonic frequencies were present in the current signal, the 6th and 12th harmonics of motor speed were particularly dominant. A similarly strong 6th and 12th harmonic were found when carrying out the test with disconnected gearwheels. Manual rotation of the motor shaft revealed a strong cogging effect at 6 times the motor shaft rate. Based on this it is concluded that the root cause of the higher-order frequencies lies in the interaction between rotor and stator of the electric motor.

Filtering has been considered to reduce the visually disturbing effect of cogging-related harmonics from the plotted current signal. However, by filtering additional phase lag would be introduced, which would result in less steep current increase following a step in voltage, and it could reduce the amplitude of the current signal following a sinusoidal voltage input. In the end, it was decided to show the unfiltered current measurements.

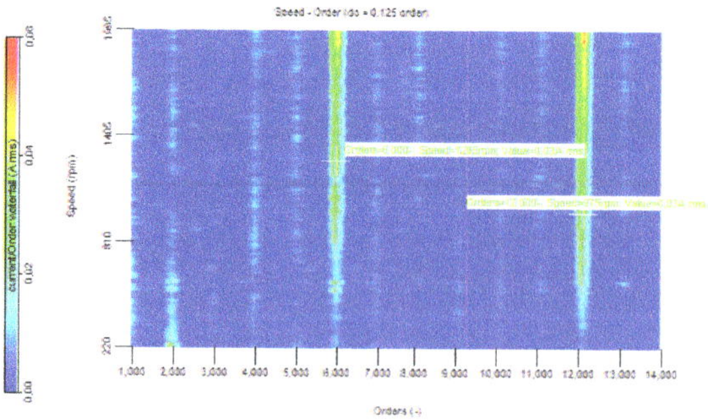

Figure 5. Order waterfall plot.

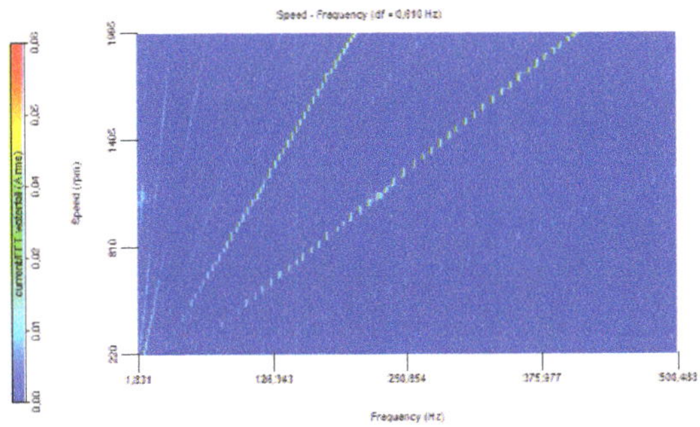

Figure 6. FFT waterfall plot.

In addition, the motor speed signal showed unexpected behavior, which appeared to be caused by the sensor. A sketch of the encoder disk used in the experiments is shown in Figure 7. It is a round disk with 15 holes, which causes 15 pulses per revolution, generated by a photosensitive sensor. The motor speed is derived from the time interval between two upcoming flanks of the pulses. The resulting motor speed signal as shown in Figure 8 shows a repeating sequence of 15 motor speed values, indicating that the angle $\Delta\Psi_{i,j}$ between the holes varied slightly around 360/15 = 24°. No further corrections have been made to the signal, which explains the relatively "noisy" motor speed signal presented in the following sections.

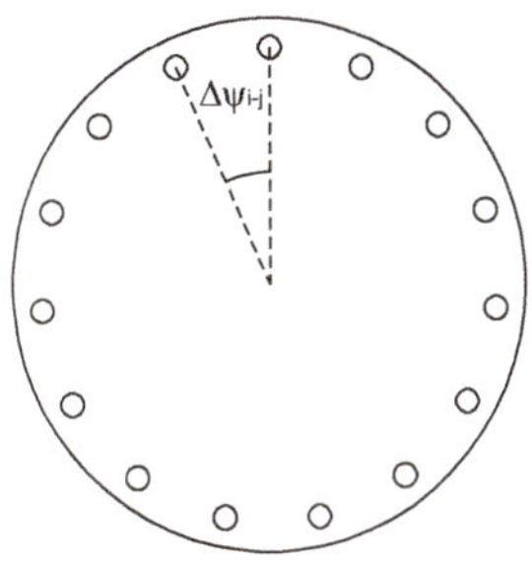

Figure 7. Encoder disk.

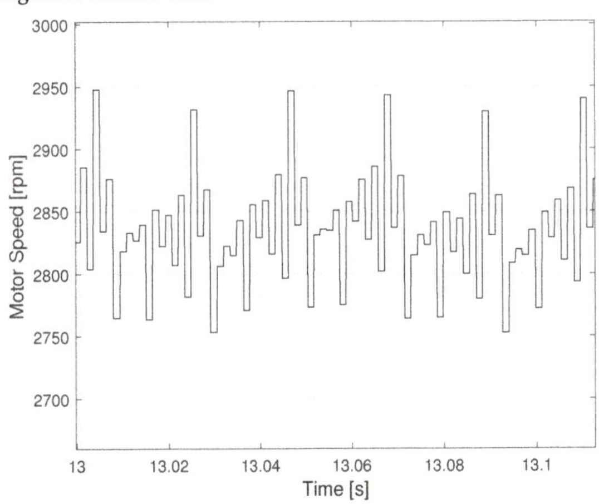

Figure 8. Magnification of motor speed time history.

5. Results and Discussion

In this section, the results obtained with the different identification techniques are reported. Both the identification and validation analyses are carried out in both time and frequency domains.

In Table 4 the steady-state operating points recorded during the staircase experiment are reported. The time domain approach (1) uses all the five operating points to determine four out of the total of six unknown parameters. To determine the remaining two parameters L_a and I_p the transient response from operating point C to D is used.

The other identification approaches focus on operating point C. The reason to choose this point is that it corresponds to around 75% of the maximum supply voltage, which is a reasonable value thinking about the design of a full-scale propulsion plant.

Table 4. Evaluated operating points.

	Operating Points					
	A	B	C	D	E	Unit
$U_{a,0}$	3.91	4.91	5.89	6.88	7.87	[V]
$\omega_{em,0}$	117	163	215	255	295	[rad/s]
$i_{a,0}$	1.00	1.13	1.29	1.44	1.62	[A]

First, the resulting parameter sets of the different approaches are reported in Table 5 to appreciate the difference in terms of numerical value. The parameters derived from the spectral approach (5) are not reported as will be explained later. The table shows that the parameters obtained with the methods were of the same order of magnitude, but differences up to ≈100% were present. The effect of the different sets of parameters on the simulated system behavior is shown in the validation graphs.

Table 5. Identified parameters.

METHOD	I	IV	Unit
L_a	4.87×10^{-4}	6.03×10^{-4}	[H]
K_e	1.37×10^{-2}	1.83×10^{-2}	[Nm/A]
R_a	2.31	1.51	[Ω]
I_p	1.72×10^{-5}	3.18×10^{-5}	[kg m^2]
M_f	1.23×10^{-2}	1.70×10^{-2}	[Nm]
$\frac{K_{Q,J=0}}{\eta_R}$	1.02×10^{-1}	1.36×10^{-1}	[-]

5.1. Results Time Domain Analysis (1)

The time domain identification method was used to derive the parameters from the staircase experiment. The supply voltage U_a during this experiment is shown in Figure 9, while the measured motor speed and current are shown in Figures 10 and 11.

Following the procedure outlined earlier, the five steady-state operating points during the staircase experiment were determined, and the parameters M_f, K_e, R_a, and $K_{Q,J=0}/\eta_R$ were derived by the least-squares method. Subsequently, the transient response of motor speed and current, following the voltage step from C to D, was used to determine the parameters I_p and L_a.

The resulting set of parameters was implemented in the non-linear simulation model, and by using the staircase voltage signal as input, the model and its parameters are verified. The result is shown as the dashed red line in Figures 10 and 11. The motor speed matched the experimental data well: the stationary value errors were within 3% at all voltage levels. Close inspection of the transient responses showed that these were also captured well. The simulated current signal had to be compared with very noisy experimental data, as explained earlier. Nevertheless, the static values seemed to be predicted well. Close inspection of the transient response shows that the simulation model could catch the timing and the initial steep slope of the current, but it was not able to represent the peak values in the current. It is concluded that this is either due to the limitations of the mathematical model, which might be too simple to describe the real physical phenomena, or due to the quality of the measured current signal.

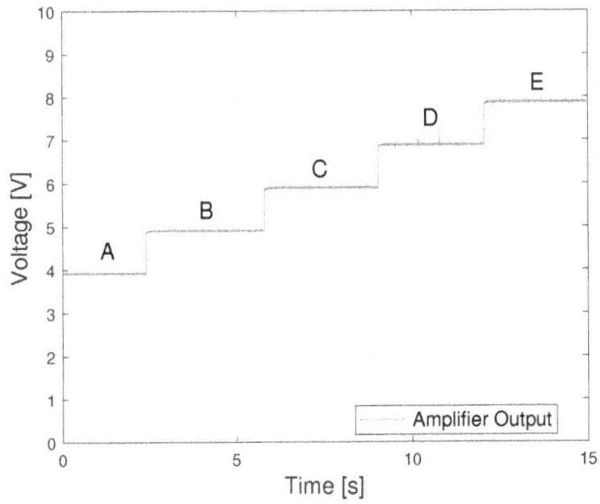

Figure 9. Voltage time history of the staircase test.

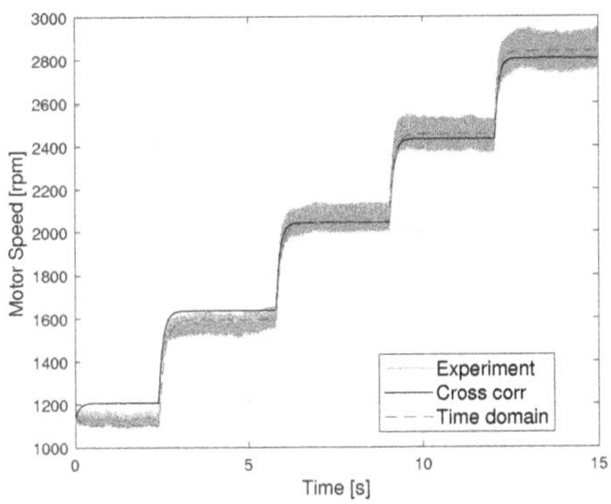

Figure 10. Motor speed time history of the staircase test.

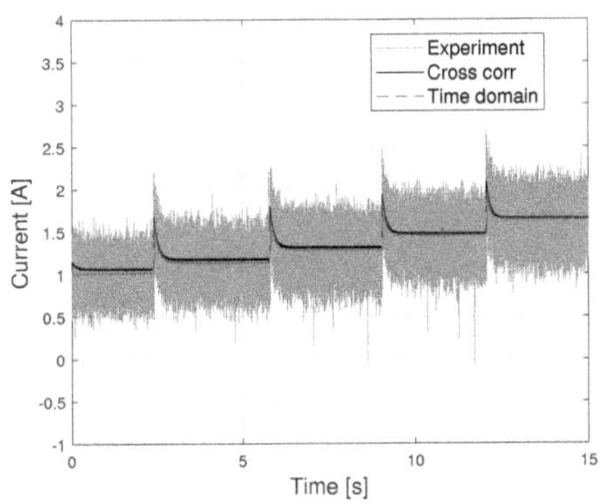

Figure 11. Current time history of the staircase test.

Figures 10 and 11 also show the results of the correlation approach, in continuous black line. However, since the staircase experiment was not used to determine the parameters using the correlation approach, this can be seen as validation of that method.

Figure 10 shows that the correlation approach predicted the motor speed behavior nearby the linearization point well, although the error between simulated and experimental data increased moving further away from the nominal operating point that was used in the sine experiments. Figure 11 shows that, compared to the time domain approach, the correlation approach was better able to predict the transient, although this method was also not able to catch the maximum current value.

To have an independent validation for the time-domain method, an experiment based on a mix of different input voltages, as shown in Figure 12, was used. Figure 13 shows that the parameter sets found by both methods led to similar dynamic behavior as the experiment, although a constant bias of around 50 rpm between simulated and sampled time histories was present. Figure 14 shows that the two parameter sets, in general, gave good correlation with the experiment, but both were unable to capture the maximum amplitudes of the current, which was particularly evident in the "sine wave" part from t = 10–13 s.

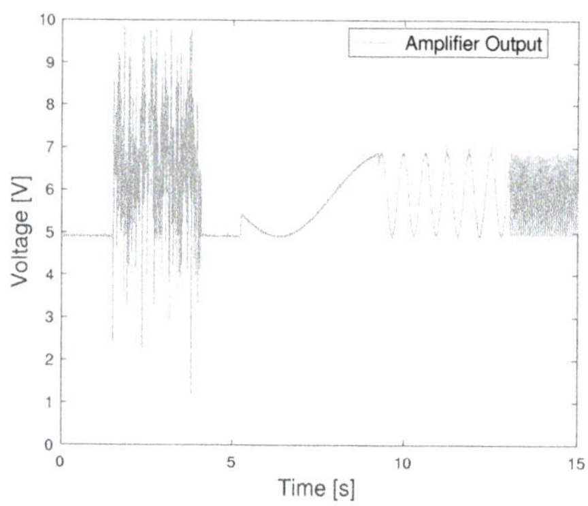

Figure 12. Voltage time history.

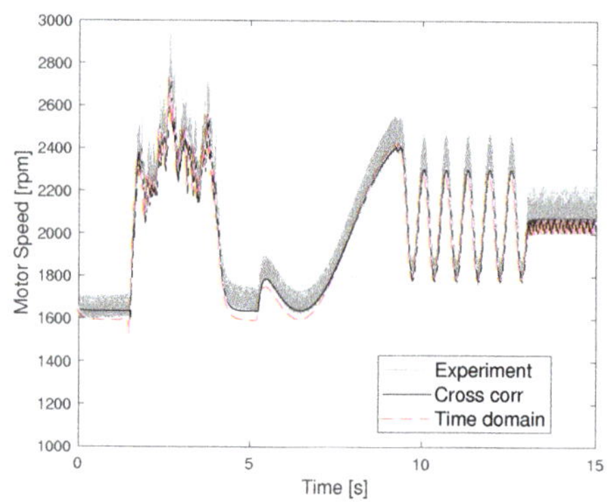

Figure 13. Motor speed time history.

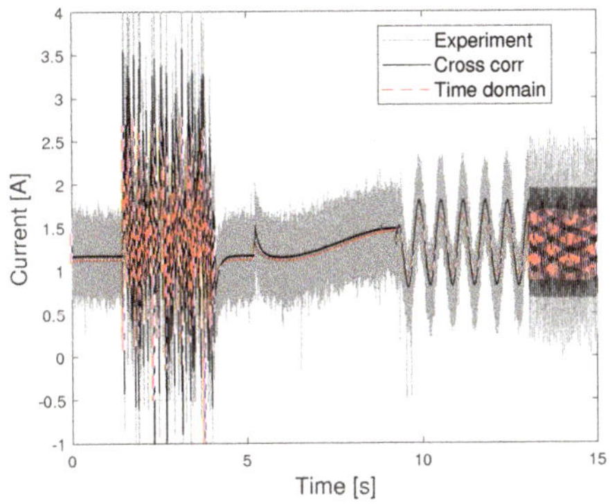

Figure 14. Current time history.

5.2. Results of Frequency Domain Analysis (4 and 5)

The single frequency testing method was applied in the nominal operating point C that is defined in Table 4. The results of the nine experiments are plotted as asterisks data points in the Bode plots shown in Figures 15 and 16. The data points at 1000 and 5000 rad/s were discarded, as closer inspection of the time signals showed that the signal-to-noise ratio was too low to lead to meaningful results.

Based on the data points, the procedure as outlined above was followed, leading to the estimated parameters as listed in Table 5. The following values were iteratively determined from the experimental data points: $s_1 = -2500$ rad/s, $s_2 = -9$ rad/s, $z_1 = -2$ rad/s, $\frac{\delta\omega^*}{\delta U_a^*}(s \to 0) = 1.24$, $\frac{\delta i_a^*}{\delta U_a^*}(s \to 0) = 0.71$. Note that the locations of the poles and zero were read from the dB versions of the Bode plots.

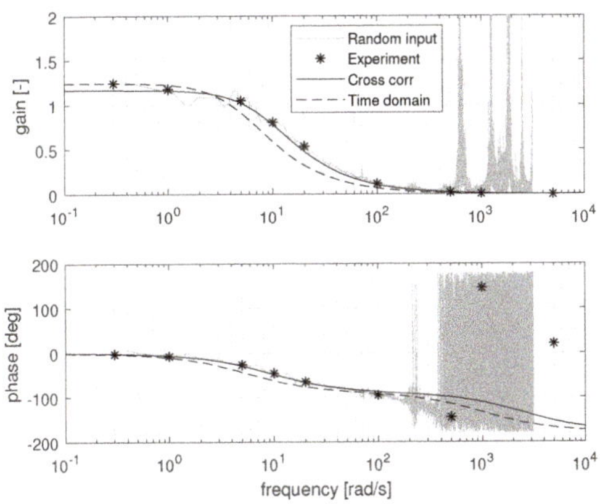

Figure 15. Bode plot of $\frac{\delta\omega^*}{\delta U_a^*}$.

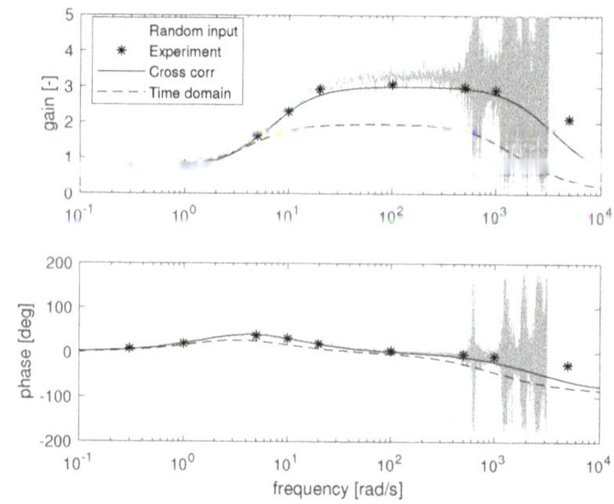

Figure 16. Bode plot of $\frac{\delta i^*}{\delta U_a^*}$.

To verify whether the estimation procedure was followed correctly, the found parameters were implemented in the transfer functions (12) and (13), which are plotted as solid lines in Figures 15 and 16. The agreement in trend and absolute numbers indicates that the procedure was followed correctly and that the linearized model can capture the reality well in the operating point under consideration. Validation in the time domain of the parameter set obtained with this method is reported in the previous section.

The shape of the transfer function for motor speed shows that up to 2 rad/s the response remained flat but then quickly dropped off due to the inertia of the drive train. The transfer function for current showed a flat response up to 1 rad/s and then started to rise due to the zero in the transfer function. Around 20 rad/s, it flattened out due to the inertia of the drive train. Somewhere after 1000 rad/s it dropped off, due to the electric pole, indicating that the current cannot follow the voltage variations anymore.

Figure 15 also shows the transfer function based on the parameter set derived with the time-domain approach. The response to low frequencies was good, but the drop in gain started slightly too early, which aligns with the low estimate of I_p in Table 5. In Figure 16 a substantial deviation from the data points is visible at frequencies higher than 10 rad/s, although the shape is clearly recognizable.

Finally, Figures 15 and 16 also show the results from approach 5. Between 0.4 and 400 rad/s the method resulted in a non-parametric frequency response that aligned well with the asterisk data points. In hindsight, the duration of the experiment should have been extended up to 1–2 min or even longer, instead of 30 s. A more extended trial would allow the estimation of the transfer function up to lower frequencies and would allow for further averaging over multiple data blocks to smooth out irregularities in the results. At frequencies above 400 rad/s, the signal-to-noise ratio dropped leading to bumps in the estimated frequency response.

Although the parameter estimation based on noise injection could be used to assess the unknown parameters from the frequency response, this is not performed here but is left for a further study on the potential of spectral methods for ship drive train identification.

6. Future Outlook

Application of the identification procedures on-board a real ship is expected to make the analytic derivations more complex because the system will, in that case, include other/additional components such as, for instance, a diesel engine and engine speed governor. This introduces at least one extra state equation due to the integral term in the PI(D) governor. An additional state, due the longitudinal equation of motion, and additional parameters would be added if the approach would be extended to free sailing instead of bollard pull conditions. The effect of such additions on the ability to determine parameters needs to be investigated in the future. On the positive side, it has to be noted that in reality it is not likely that all ship drive-train parameters are unknown, which helps to determine estimates of other unknown or more uncertain parameters. More work is required to investigate what parameter estimation procedure would be required for a real ship and drive train.

In the authors' opinion, in the future the presented algorithms could potentially be part of a condition-based maintenance system. By monitoring parameter variations of a propulsion drive train in real time, it could be possible to detect the degradation (or malfunction) of the machinery, and perhaps even to identify the root cause. For instance, an increase in friction coefficient M_f could mean wear in the bearings, an increase in K_Q could mean that the propeller needs to be cleaned, etc. Another possible use of the presented techniques is to assess the correspondence between the design values with the real one, in fact, during the shipbuilding progress some change, or unexpected modification, can modify the original design values.

7. Conclusions and Recommendations

In this paper different parameter identification techniques were discussed and applied to experimentally determine the unknown parameters of a model scale ship drive train in bollard pull conditions.

A set of dedicated experiments was conducted using different DC voltage signals. In all tests the current was affected by a strong noise due to motor cogging. It is therefore recommended to use an electric motor with less strong cogging effect for future experiments. Moreover, the 15 holes encoder was found to give a low-quality motor speed measurement and should be improved.

Three different approaches to determine the unknown DC-electric propulsion plant parameters are discussed including their merits and weaknesses. For now, all three approaches remain candidates to be part of a (real-time) full-scale parameter identification system, which is one of the primary goals.

Two obtained parameter sets have been implemented in a simulation model, and the results were validated against independent measurements, both in the frequency and in the time domains. The time domain results obtained by implementing both parameter sets in the model compared well against the measurements, although differences were present.

In order to move towards firm conclusions about the value of the applied parameter estimation methods for condition monitoring, it is recommended to consider the sensitivity and uncertainty related to the approaches. This recommendation is supported by the relatively large differences between the parameter sets as determined in this paper, and the relatively small differences in time domain response.

Author Contributions: Conceptualization, A.V. and M.M.; methodology, A.V. and M.M.; software, A.V. and M.M.; writing—original draft preparation, A.V. and M.M.; writing—review and editing, A.V. and M.M.; experimental testing A.V. and M.M. All authors have read and agreed to the published version of the manuscript.

Funding: This research was partly supported by Maritime by Holland (in Dutch: NML) and the Ministry of Economic Affairs of the Netherlands. Furthermore, this research was partly funded by University of Genoa within the program "Incentive for research periods abroad".

Acknowledgments: The authors want to acknowledge Eng. Vittorio Garofano of Delft University of Technology for his essential support during the experimental campaign.

Conflicts of Interest: The authors declare no conflict of interest.

Nomenclature

B_{oa}	ship breadth	[m]
C	constant	
c	constant	
D	propeller diameter	[m]

e	general exponent	
F_{BP}	bollar pull force	[N]
I_p	moment of inertia	[kgm^2]
i_a	motor current	[A]
i_{gb}	gearbox ratio	[-]
J	advance ratio	[-]
K	constant	
K_Q	torque coefficient	[-]
K_T	thrust coefficient	[-]
K_e	motor speed constant	[V/rad/s]
K_e	motor back EMF constant	[Nm/A]
k_p	propeller number	[-]
L_{oa}	ship length	[m]
L_a	motor inductance	[H]
$M_{b,em}$	motor torque	[Nm]
M_f	friction torque	[Nm]
M_p	delivered torque	[Nm]
Q	open water torque	[Nm]
R_a	motor resistance	[Ω]
s_1	first pole	[-]
s_2	second pole	[-]
T	propeller thrust	[N]
t	time	[s]
t	thrust deduction factor	[-]
U_a	voltage supply	[V]
Δ	ship displacement	[kg]
ζ	ratio of time constants	[-]
η_R	relative rotative efficiency	[-]
η_{trm}	shaftline efficiency	[-]
ρ	water density	[kg/m^3]
τ_{em}	electric time constant	[s]
τ_ω	mechanical time constant	[s]
$\tau_{\omega,e}$	effective mechanical time constant	[s]
ω_{em}	motor speed	[rad/s]
ω_p	propeller speed	[rad/s]
ω	frequency	[rad/s]

Subscripts and Superscripts

0	nominal
$*$	normalized
δ	small increment

Appendix A. Normalisation and Linearisation

Assume a variable that is the product of powers of other variables:

$$Z = c\, Y^e X \tag{A1}$$

where c is a constant multiplier and e is a constant exponent. In an equilibrium point the variable Z equals

$$Z_0 = c\, Y_0^e X_0 \tag{A2}$$

Normalisation of Equation (A1) by Equation (A2) results in

$$\frac{Z}{Z_0} = \left(\frac{Y}{Y_0}\right)^e \frac{X}{X_0} \tag{A3}$$

If, by definition,

$$X^* = \frac{X}{X_0}, Y^* = \frac{Y}{Y_0}, Z^* = \frac{Z}{Z_0} \tag{A4}$$

then

$$Z^* = Y^{*e} X^* \tag{A5}$$

Now that the constant value c has been removed by the normalization, the next step is to remove the non-linearity from Equation (A5). Differentiation of Equation (A3) by using the chain rule gives

$$\frac{dZ}{Z_0} = \left(\frac{Y}{Y_0}\right)^e \frac{dX}{X_0} + e\left(\frac{Y}{Y_0}\right)^{e-1} \frac{X}{X_0}\frac{dY}{Y_0} \tag{A6}$$

Near equilibrium dX, dY and dZ become small increments δX, δY and δZ. Division of $X = X_0 + \delta X$ by X_0 delivers $\frac{X}{X_0} = 1 + \frac{\delta X}{X_0}$ and likewise $\frac{Y}{Y_0} = 1 + \frac{\delta Y}{Y_0}$. Substitution of this in Equation (A6) gives

$$\begin{aligned}\frac{\delta Z}{Z_0} &= \left(1 + \frac{\delta Y}{Y_0}\right)^e \frac{\delta X}{X_0} \\ &+ e\left(1 + \frac{\delta Y}{Y_0}\right)^{e-1} \left(1 + \frac{\delta X}{X_0}\right)\frac{\delta Y}{Y_0}\end{aligned} \tag{A7}$$

Taylor series expansion of Equation (A7) and neglecting the second and higher order terms leaves

$$\frac{\delta Z}{Z_0} = \frac{\delta X}{X_0} + e\frac{\delta Y}{Y_0} \tag{A8}$$

which by introduction of the shorthand notation for differential increment:

$$\delta Z^* = \frac{\delta Z}{Z_0} = \frac{Z}{Z_0} - 1 \tag{A9}$$

this can be written as

$$\delta Z^* = \delta X^* + e\,\delta Y^* \tag{A10}$$

The latter equation relates the relative change in output Z to the relative change in inputs X and Y, where the constant e, which was present as an exponent in the original Equation (A2), has changed to a constant multiplication factor. Secondly the multiplication of X and Y has turned into a summation. For further background on the linearization process, reference is made to Dorf and Bishop [33] and Franklin et al. [34].

The demonstrated concepts of normalization and linearization are the basis for the following section where they will be applied in the linearization of the system model.

Appendix B. Derivation of Linearized System Model

The electrical circuit of the DC motor is modeled by

$$L_a \frac{di_a}{dt} = U_a - K_e \omega_{em} - R_a i_a \tag{A11}$$

All three right hand side terms vary around equilibrium:

$$U_a = U_{a,0} + \delta U_a, \quad \omega_{em} = \omega_{em,0} + \delta\omega_{em}, \quad i_a = i_{a,0} + \delta i_a$$

In static conditions the right hand side of Equation (A11) equals zero:

$$0 = U_{a,0} - K_e\omega_{em,0} - R_a i_{a,0} \tag{A12}$$

This means that only the small increments are of importance:

$$L_a \frac{di_a}{dt} = \delta U_a^* - K_e\delta\omega_{em}^* - R_a\delta i_a^* \tag{A13}$$

Division of all terms of Equation (A13) by nominal supply voltage minus the nominal emf $(U_{a,0} - K_e\omega_{em,0})$ or alternatively by its equivalent $R_a i_{a,0}$ gives

$$\frac{L_a}{R_a i_{a,0}}\frac{di_a}{dt} = \frac{1}{R_a i_{a,0}}\frac{U_{a,0}}{U_{a,0}}\delta U_a - \frac{K_e}{R_a i_{a,0}}\frac{\omega_{em,0}}{\omega_{em,0}}\delta\omega_{em} - \frac{R_a}{R_a i_{a,0}}\delta i_a \tag{A14}$$

This can be shortened to

$$\tau_{em}\frac{di_a^*}{dt} = \frac{U_{a,0}}{R_a i_{a,0}}\delta U_a^* - \frac{K_e\omega_{em,0}}{R_a i_{a,0}}\delta\omega^* - \delta i_a^* \tag{A15}$$

in which the subscript $_{em}$ is intentionally dropped from $\delta\omega_{em}^*$ because $\delta\omega_{em}^* = \delta\omega_p^*$ and where

$$\tau_{em} = \frac{L_a}{R_a} \tag{A16}$$

The shaft dynamics including constant friction term are described by

$$I_p\frac{d\omega_{em}}{dt} = M_{b,em} - M_f - \frac{M_p}{i_{gb,13}} \tag{A17}$$

in which shaft inertia is assumed constant implying that change of mass of water, entrained by the propeller, is neglected. The brake motor torque is related to current by

$$M_{b,em} = K_e\, i_a \tag{A18}$$

The non-constant torque terms of Equation (A17) vary around equilibrium:

$$M_{b,em} = M_{b,em,0} + \delta M_{b,em} = K_e(i_{a,0} + \delta i_a)$$

and

$$M_p = M_{p,0} + \delta M_p$$

such that:

$$I_p\frac{d\omega_{em}}{dt} = K_e(i_{a,0} + \delta i_a) - M_f - \frac{M_{p,0}}{i_{gb,13}} - \frac{\delta M_p}{i_{gb,13}} \tag{A19}$$

In steady nominal condition the driving torque and the load-torque are equal:

$$0 = K_e i_{a,0} - M_f - \frac{M_{p,0}}{i_{gb,13}} \tag{A20}$$

Subtracting Equation (A20) from Equation (A19) shows that only the small increments are of importance:

$$I_p\frac{d\omega_{em}}{dt} = K_e\delta i_a - \frac{\delta M_p}{i_{gb,13}}$$

Normalizing all terms with nominal motor torque gives

$$\frac{I_p\omega_{em,0}}{K_e i_{a,0}}\frac{d\omega^*}{dt} = \delta i_a^* - \frac{M_{p,0}}{M_{b,em,0}i_{gb,13}}\delta M_p^* \tag{A21}$$

in which the subscript $_{em}$ is intentionally dropped. The integration constant is defined as

$$\tau_\omega = \frac{I_p \omega_{em,0}}{M_{b,em,0}} = \frac{I_p \omega_{em,0}}{K_e\, i_{a,0}} \tag{A22}$$

After noting that the multiplier in the second term of the right hand side of Equation (A21) can be written as

$$\frac{M_{p,0}}{i_{gb,13} M_{b,em,0}} = \eta_{trm,0} \tag{A23}$$

and implementing

$$\delta M_p^* = 2\delta\omega^*$$

the normalised linearised differential equation for shaft rotation is given by

$$\tau_\omega \frac{d\omega^*}{dt} = \delta i_a^* - 2\eta_{trm,0}\delta\omega^* \tag{A24}$$

Introduction of the Laplace operator into Equation (A24) and re-arranging gives

$$\left(\frac{\tau_\omega}{2\eta_{trm,0}} s + 1\right)\delta\omega^* = \frac{1}{2\eta_{trm,0}}\delta i_a^* \tag{A25}$$

which can be shortened by introduction of the effective time-constant:

$$\tau_{\omega,e} = \frac{\tau_\omega}{2\eta_{trm,0}} \tag{A26}$$

such that

$$(\tau_{\omega,e} s + 1)\delta\omega^* = \frac{1}{2\eta_{trm,0}}\delta i_a^* \tag{A27}$$

Similarly, introduction of the Laplace operator in the differential equation for current Equation (A15) and reordering gives

$$\delta i_a^* = \frac{\left(\frac{U_{a,0}}{R_a i_{a,0}}\right)}{(\tau_{em} s + 1)}\delta U_a^* - \frac{\left(\frac{K_e \omega_{em,0}}{R_a i_{a,0}}\right)}{(\tau_{em} s + 1)}\delta\omega^* \tag{A28}$$

Substitution of Equation (A28) into Equation (A27) and reordering gives the transfer function from supply voltage to rotation speed:

$$\frac{\delta\omega^*}{\delta U_a^*} = \frac{\frac{1}{2\eta_{trm,0}} \frac{U_{a,0}}{R_a i_{a,0}}}{\tau_{em}\tau_{\omega,e} s^2 + (\tau_{em} + \tau_{\omega,e})s + 1 + \frac{1}{2\eta_{trm,0}} \frac{K_e \omega_{em,0}}{R_a i_{a,0}}} \tag{A29}$$

In a similar way substitution of Equation (A27) into Equation (A28) and reordering gives the transfer function from supply voltage to current:

$$\frac{\delta i_a^*}{\delta U_a^*} = \frac{(\tau_{\omega,e} s + 1)\frac{U_{a,0}}{R_a i_{a,0}}}{\tau_{em}\tau_{\omega,e} s^2 + (\tau_{em} + \tau_{\omega,e})s + 1 + \frac{1}{2\eta_{trm,0}} \frac{K_e \omega_{em,0}}{R_a i_{a,0}}} \tag{A30}$$

The characteristic equation of the two transfer functions Equations (A29) and (A30) is given by

$$\tau_{em}\tau_{\omega,e} s^2 + (\tau_{em} + \tau_{\omega,e})s + 1 + \frac{1}{2\eta_{trm,0}} \frac{K_e \omega_{em,0}}{R_a i_{a,0}} \tag{A31}$$

If we define

$$C = 1 + \frac{1}{2\eta_{trm,0}} \frac{K_e \omega_{em,0}}{R_a i_{a,0}} \tag{A32}$$

and

$$\zeta = \frac{\tau_{em}}{\tau_{\omega,e}}$$

then the characteristic equation can be written as

$$\zeta \tau_{\omega,e} s^2 + (1 + \zeta) s + \frac{C}{\tau_{\omega,e}}$$

The two exact roots of Equation (A31) can now be determined by the ABC formula:

$$s_{12} = \frac{-(1+\zeta) \pm \sqrt{(1+\zeta)^2 - 4C\zeta}}{2\zeta\tau_{\omega,e}}$$

which can be written as

$$s_{12} = \frac{-(1+\zeta) \pm (1+\zeta)\sqrt{1 - \frac{4C\zeta}{(1+\zeta)^2}}}{2\zeta\tau_{\omega,e}}$$

The electrical time constant is much smaller than the effective time constant for the shaft; therefore, $\zeta \ll 1$. Application of Taylor expansion for the square root operation and leaving out second order terms gives

$$s_{12} \approx \frac{-(1+\zeta) \pm (1+\zeta)\left(1 - 2C\zeta\frac{1}{(1+\zeta)^2}\dots\right)}{2\zeta\tau_{\omega,e}}$$

Another Taylor expansion for the inverse square operation gives

$$s_{12} \approx \frac{-(1+\zeta) \pm (1+\zeta)(1 - 2C\zeta(1 - 2\zeta\dots)\dots)}{2\zeta\tau_{\omega,e}}$$

Further simplification gives the two approximate poles as

$$s_1 \approx \frac{-1}{\zeta\tau_{\omega,e}} = \frac{-1}{\tau_{em}} \tag{A33}$$

and

$$s_2 \approx \frac{-C}{\tau_{\omega,e}} \tag{A34}$$

Besides the two system poles, transfer function (A30) has a single zero which lies at

$$z_1 = \frac{-1}{\tau_{\omega,e}} \tag{A35}$$

The DC-gain of transfer function (A29) is given by

$$\frac{\delta\omega^*}{\delta U_a^*}(s \to 0) = \frac{U_{a,0}}{2\eta_{trm,0}R_a i_{a,0} + K_e\omega_{em,0}} \tag{A36}$$

The DC-gain of transfer function (A30) is given by

$$\frac{\delta i_a^*}{\delta U_a^*}(s \to 0) = \frac{2\eta_{trm,0}U_{a,0}}{2\eta_{trm,0}R_a i_{a,0} + K_e\omega_{em,0}} \tag{A37}$$

Appendix C. Step Response of Motor Speed and Current

The exact response of motor speed to a unit step in voltage is given by

$$\delta\omega^*(t) = K\left(1 + \frac{s_1}{s_2 - s_1}e^{s_2 t} - \frac{s_2}{s_2 - s_1}e^{s_1 t}\right) \quad \text{(A38)}$$

in which $K = \frac{U_{a,0}}{K_e\omega_0 + 2\eta_{trm,0}R_a i_{a,0}}$. However, because $|s_1| >> |s_2|$, the step-response can be approximated by a first order system response:

$$\delta\omega^*(t) \approx K\left(1 - e^{s_2 t}\right) \quad \text{(A39)}$$

The derivation of the step response of current starts with Equation (A30), which can be written as the summation of an overdamped second order lowpass (LP) system and a second order bandpass (BP) system:

$$G(s) = G_{LP}(s) + G_{BP}(s) \quad \text{(A40)}$$

The step response of the lowpass system is given by

$$\delta i^*_{a,LP}(t) = K_{LP}\left(1 + \frac{s_1}{s_2 - s_1}e^{s_2 t} - \frac{s_2}{s_2 - s_1}e^{s_1 t}\right) \quad \text{(A41)}$$

with $K_{LP} = \frac{U_{a,0}2\eta_{trm}}{K_e\omega_0 + 2\eta_{trm,0}R_a i_{a,0}}$. Again, because $|s_1| >> |s_2|$, this can be approximated by a first order system response:

$$\delta i^*_{a,LP}(t) \approx K_{LP}\left(1 - e^{s_2 t}\right) \quad \text{(A42)}$$

The step response of the bandpass part of the system is given by

$$\delta i^*_{a,BP}(t) = K_{BP}\left(\frac{1}{s_2 - s_1}e^{s_2 t} - \frac{1}{s_2 - s_1}e^{s_1 t}\right) \quad \text{(A43)}$$

where $K_{BP} = \frac{U_{a,0}}{L_a i_{a,0}}$. The total response of current to a unit step in voltage is the sum of Equations (A41) and (A43):

$$\delta i^*_a(t) = K_{LP}\left(1 + \frac{s_1}{s_2 - s_1}e^{s_2 t} - \frac{s_2}{s_2 - s_1}e^{s_1 t}\right) + K_{BP}\left(\frac{1}{s_2 - s_1}e^{s_2 t} - \frac{1}{s_2 - s_1}e^{s_1 t}\right) \quad \text{(A44)}$$

or including the simplification:

$$\delta i^*_a(t) \approx K_{LP}\left(1 - e^{s_2 t}\right) + K_{BP}\left(\frac{1}{s_2 - s_1}e^{s_2 t} - \frac{1}{s_2 - s_1}e^{s_1 t}\right) \quad \text{(A45)}$$

References

1. Martelli, M.; Figari, M. Real-Time model-based design for CODLAG propulsion control strategies. *Ocean Eng.* **2017**, *141*, 265–276. [CrossRef]
2. Vrijdag, A.; Stapersma, D.; van Terwisga, T. Control of propeller cavitation in operational conditions. *J. Mar. Eng. Technol.* **2010**, *9*, 15–26. [CrossRef]
3. Cipollini, F.; Oneto, L.; Coraddu, A.; Murphy, A.J.; Anguita, D. Condition-Based Maintenance of Naval Propulsion Systems with supervised Data Analysis. *Ocean Eng.* **2018**, *149*, 268–278. [CrossRef]
4. Mizythras, P.; Boulougouris, E.; Theotokatos, G. Numerical study of propulsion system performance during ship acceleration. *Ocean Eng.* **2018**, *149*, 383–396. [CrossRef]
5. Martelli, M. *Marine Propulsion Simulation;* De Gruyter Open: Berlin, Germany, 2015.
6. Geertsma, R.; Negenborn, R.; Visser, K.; Loonstijn, M.; Hopman, J. Pitch control for ships with diesel mechanical and hybrid propulsion: Modelling, validation and performance quantification. *Appl. Energy* **2017**, *206*, 1609–1631. [CrossRef]
7. Vrijdag, A.; Stapersma, D.; van Terwisga, T. Systematic modelling, verification, calibration and validation of a ship propulsion simulation model. *J. Mar. Eng. Technol.* **2009**, *8*, 3–20. [CrossRef]

8. Ljung, L. *System Identification: Theory for the User*, 2nd ed.; Prentice Hall: Upper Saddle River, NJ, USA, 1999.
9. Weerasooriya, S.; El-Sharkawi, M.A. Identification and control of a DC motor using back-propagation neural networks. *IEEE Trans. Energy Convers.* **1991**, *6*, 663–669. [CrossRef]
10. Liu, X.Q.; Zhang, H.Y.; Liu, J.; Yang, J. Fault detection and diagnosis of permanent-magnet DC motor based on parameter estimation and neural network. *IEEE Trans. Ind. Electron.* **2000**, *47*, 1021–1030. [CrossRef]
11. Lu, W.; Keyhani, A.; Fardoun, A. Neural network-based modeling and parameter identification of switched reluctance motors. *IEEE Trans. Energy Convers.* **2003**, *18*, 284–290. [CrossRef]
12. Cirrincione, M.; Pucci, M.; Cirrincione, G.; Capolino, G. A new experimental application of least-squares techniques for the estimation of the induction motor parameters. *IEEE Trans. Ind. Appl.* **2003**, *39*, 1247–1256. [CrossRef]
13. Wang, K.; Chiasson, J.; Bodson, M.; Tolbert, L.M. A nonlinear least-squares approach for identification of the induction motor parameters. *IEEE Trans. Autom. Control* **2005**, *50*, 1622–1628. [CrossRef]
14. Yoerger, D.R.; Cooke, J.G.; Slotine, J.J.E. The influence of thruster dynamics on underwater vehicle behavior and their incorporation into control system design. *IEEE J. Ocean. Eng.* **1990**, *15*, 167–178. [CrossRef]
15. Healey, A.J.; Rock, S.M.; Cody, S.; Miles, D.; Brown, J.P. Toward an improved understanding of thruster dynamics for underwater vehicles. *IEEE J. Ocean. Eng.* **1995**, *20*, 354–361. [CrossRef]
16. Whitcomb, L.L.; Yoerger, D.R. Comparative experiments in the dynamics and model-based control of marine thrusters. In Proceedings of the OCEANS '95 MTS/IEEE 'Challenges of Our Changing Global Environment', San Diego, CA, USA, 9–12 October 1995; Volume 2, pp. 1019–1028. [CrossRef]
17. Whitcomb, L.L.; Yoerger, D.R. Preliminary experiments in the model-based dynamic control of marine thrusters. In Proceedings of the IEEE International Conference Robotics and Automation, Minneapolis, MN, USA, 22–28 April 1996; Volume 3, pp. 2166–2173. [CrossRef]
18. Whitcomb, L.L.; Yoerger, D.R. Preliminary experiments in model-based thruster control for underwater vehicle positioning. *IEEE J. Ocean. Eng.* **1999**, *24*, 495–506. [CrossRef]
19. Bachmayer, R.; Whitcomb, L. Unsteady dynamics and control of marine thrusters. *J. Acoust. Soc. Am.* **1999**, *105*, 1300. [CrossRef]
20. Bachmayer, R.; Whitcomb, L.; Grosenbaugh, M. An accurate four-quadrant nonlinear dynamical model for marine thrusters: theory and experimental validation. *IEEE J. Ocean. Eng.* **2000**, *25*, 146–159. [CrossRef]
21. Blanke, M.; Lindegaard, K.P.; Fossen, T.I. Dynamic Model for Thrust Generation of Marine Propellers. *IFAC Proc. Vol.* **2000**, *33*, 353–358. [CrossRef]
22. Fossen, T.I.; Blanke, M. Nonlinear output feedback control of underwater vehicle propellers using feedback form estimated axial flow velocity. *IEEE J. Ocean. Eng.* **2000**, *25*, 241–255. [CrossRef]
23. Smogeli, O.N. Control of Marine Propellers from Normal to Extreme Conditions. Ph.D. Thesis, Norwegian University of Science and Technology, Trondheim, Norway, 2006.
24. Pivano, L.; Fossen, T.I.; Johansen, T.A. Nonlinear model identification of a marine propeller over four-quadrant operations. *IFAC Proc. Vol.* **2006**, *39*, 315–320. [CrossRef]
25. Pivano, L.; Fossen, T.I.; Smogeli, Y.N.; Johansen, T.A. Experimental Validation of a Marine Propeller Thrust Estimation Scheme. *Model. Identif. Control* **2007**, *28*, 105–112. [CrossRef]
26. Coraddu, A.; Oneto, L.; Ghio, A.; Savio, S.; Anguita, D.; Figari, M. Machine learning approaches for improving condition-based maintenance of naval propulsion plants. *Proc. Inst. Mech. Eng. Part M J. Eng. Marit. Environ.* **2016**, *230*, 136–153. [CrossRef]
27. Altosole, M.; Campora, U.; Martelli, M.; Figari, M. Performance Decay Analysis of a Marine Gas Turbine Propulsion System. *J. Ship Res.* **2014**, *58*, 117–129. [CrossRef]
28. Carlton, J.S. *Marine Propellers and Propulsion*; Buttersworth-Heineman Ltd.: Oxford, UK, 2007.
29. Stapersma, D.; Vrijdag, A. Linearisation of a ship propulsion system model. *Ocean Eng.* **2017**, *142*, 441–457. [CrossRef]
30. Vrijdag, A.; Stapersma, D. Extension and application of a linearised ship propulsion system model. *Ocean Eng.* **2017**, *143*, 50–65. [CrossRef]
31. Welch, P. The use of fast Fourier transform for the estimation of power spectra: A method based on time averaging over short, modified periodograms. *IEEE Trans. Audio Electroacoust.* **1967**, *15*, 70–73. [CrossRef]
32. Balmer, L. *Signals and Systems: An Introduction*, 2nd ed.; Prentice Hall: Upper Saddle River, NJ, USA, 1997.
33. Dorf, R.C.; Bishop, R.H. *Modern Control Systems*, 9th ed.; Prentice Hall: Upper Saddle River, NJ, USA, 2001.
34. Franklin, G.F.; Powell, J.D.; Emami-Naeini, A. *Feedback Control of Dynamic Systems*; Addison-Wesley: Reading, MA, USA, 1986; Volume 5.

MDPI
St. Alban-Anlage 66
4052 Basel
Switzerland
www.mdpi.com

Journal of Marine Science and Engineering Editorial Office
E-mail: jmse@mdpi.com
www.mdpi.com/journal/jmse

www.ingramcontent.com/pod-product-compliance
Lightning Source LLC
Chambersburg PA
CBHW061505210926
51850CB00008B/1826
* 9 7 8 3 0 3 6 5 9 3 5 6 2 *